Advancements in Soil and Sustainable Agriculture

Advancements in Soil and Sustainable Agriculture

Editors

Antonella Lavini
Mohamed Houssemeddine Sellami

MDPI

Basel • Beijing • Wuhan • Barcelona • Belgrade • Novi Sad • Cluj • Manchester

Editors
Antonella Lavini
Institute for Agricultural and Forestry Systems in the Mediterranean (ISAFOM)
National Research Council of Italy (CNR)
Portici, Italy

Mohamed Houssemeddine Sellami
Institute for Agricultural and Forestry Systems in the Mediterranean (ISAFOM)
National Research Council of Italy (CNR)
Portici, Italy

Editorial Office
MDPI
St. Alban-Anlage 66
4052 Basel, Switzerland

This is a reprint of articles from the Special Issue published online in the open access journal *Soil Systems* (ISSN 2571-8789) (available at: https://www.mdpi.com/journal/soilsystems/special_issues/Advancements_Soil_SustainableAgriculture).

For citation purposes, cite each article independently as indicated on the article page online and as indicated below:

Lastname, A.A.; Lastname, B.B. Article Title. *Journal Name* **Year**, *Volume Number*, Page Range.

ISBN 978-3-0365-9706-5 (Hbk)
ISBN 978-3-0365-9707-2 (PDF)
doi.org/10.3390/books978-3-0365-9707-2

Contents

About the Editors

Antonella Lavini

Antonella Lavini is a distinguished researcher at the Institute for Agricultural and Forestry Systems in the Mediterranean (ISAFOM), part of the Italian National Council of Research. Since 1997, she has contributed expertise to agronomy, holding a Ph.D. in Agronomy from the University of Naples Federico II. Her research focuses on herbaceous crops, particularly quinoa and amaranth, exploring soil–water–plant interactions, agronomic practices, crop tolerance to stresses, and soil quality. Dr. Lavini actively engages in regional, national, and international projects like C4P, TRUSTFARM, SWUP-MED, Protein2Food, and Quinoa Felix, showcasing her global impact. Beyond the lab, she leads technology transfer projects, bridging scientific knowledge with practical applications to advance sustainable agriculture.

Mohamed Houssemeddine Sellami

Mohamed Houssemeddine Sellami, an agronomist and crop scientist with over two decades of experience, is a researcher at the Italian National Council of Research (CNR). Beginning at the National Agronomic Institute of Tunisia, he earned degrees in Plant Protection and Integrated Pest Management. His commitment to sustainable agriculture led to a Ph.D. in Mediterranean Agronomy from the University of Bari. Since joining CNR in 2012, Sellami has excelled in plant breeding, water resources management, proximal sensing, environmental crop eco-physiology, and agronomy. His work addresses food security challenges, optimizing irrigation practices, and developing stress-tolerant crop varieties. Actively participating in projects like UNICAL, CISIA, ACLIMAS, Protein2Food, and BENCHMARKS, Sellami's collaborative efforts contribute significantly to advancing agronomic research on a national and international scale.

Preface

In the face of mounting challenges to global food security and environmental stewardship, sustainable agriculture has emerged as a beacon of hope, illuminating a path towards a more harmonious relationship between humanity and the natural world. At the heart of this transformative shift lies soil health, the very foundation upon which crop productivity and ecological resilience are built. The Special Issue "Advancements in Soil and Sustainable Agriculture" delves into this crucial realm, presenting a collection of groundbreaking research dedicated to safeguarding soil health and charting a course towards a more environmentally conscious agricultural future.

This reprint encapsulates a compendium of cutting-edge research, providing a panoramic view of new and innovative approaches aimed at rendering modern agriculture more attuned to the needs of our planet. At the epicenter of these advancements lies a profound focus on soil health—the linchpin connecting crop productivity, ecological resilience, and the long-term viability of our agricultural practices.

The diverse collection of articles within this Special Issue delves into strategies for fostering soil-friendly agriculture, ranging from novel nutrient management techniques to sophisticated soil productivity enhancement strategies. The authors featured herein are trailblazers in their respective fields, offering insights that transcend disciplinary boundaries and contribute to the collective knowledge base essential for shaping a sustainable agricultural future.

This comprehensive collection serves as a guiding light for researchers, policy makers, and practitioners alike, illuminating the dynamic landscape of sustainable agriculture and paving the way for a more resilient and environmentally conscious future. Its findings underscore the urgent need to transition away from conventional agricultural practices that often prioritize short-term gains at the expense of long-term soil health.

As we navigate the complexities of our changing world, the significance of sustainable agriculture cannot be overstated. "Advancements in Soil and Sustainable Agriculture" seeks to propel this dialogue forward, serving as a catalyst for transformative thinking and action. The collective wisdom contained in these pages is not merely a testament to the current state of research but, more importantly, a compass guiding us towards a future where agriculture and environmental stewardship coalesce harmoniously.

May this reprint inspire curiosity, spark collaboration, and instigate a shared commitment to a more resilient, soil-friendly, and sustainable agricultural paradigm.

Mohamed Houssemeddine Sellami
Editor

soil systems

Editorial

Advancements in Soil and Sustainable Agriculture

Mohamed Houssemeddine Sellami * and Antonella Lavini

National Research Council of Italy (CNR), Institute for Agricultural and Forestry Systems in the Mediterranean (ISAFOM), 80055 Portici, Italy; antonella.lavini@cnr.it
* Correspondence: mohamed.sellami@isafom.cnr.it

1. Introduction

The growing interest in soil health and sustainable agriculture has emerged as a paramount element in addressing the multifaceted challenges facing modern agriculture [1]. In an era marked by mounting pressures for increased food production, heightened environmental conservation, and urgent climate change mitigation, innovative strategies have surfaced to accord soil health and productivity a central role [2]. This shift in perspective acknowledges the indispensable significance of soil as a fundamental substrate for crop growth as well as its crucial role in carbon sequestration and the maintenance of ecological equilibrium. Notably, this paradigmatic transformation has fostered the adoption of a spectrum of innovative approaches [3]. Precision agriculture, enabled by technological advancements, has revolutionized resource management by optimizing resource allocation, fine-tuning nutrient application, and curtailing detrimental environmental impacts, all while facilitating the sustainable stewardship of soil [4,5]. In tandem, organic farming practices have gained prominence, underlining the imperative of eschewing synthetic chemicals, nurturing natural soil ecosystems, and augmenting the organic matter content in soils [6]. Complementing these strategies is the integration of biofertilizers, which harness the inherent power of beneficial microorganisms to stimulate nutrient cycling and enhance soil structure [7,8]. These dynamic developments collectively epitomize a burgeoning era of soil-centered agriculture, one that wholeheartedly aligns with the overarching objective of advancing agricultural productivity while concurrently upholding the principles of long-term environmental sustainability [9,10]. Within this context, this Special Issue titled "Advancements in Soil and Sustainable Agriculture" directs its attention toward novel and inventive methods for rendering contemporary agriculture more soil-friendly as well as strategies for enhancing soil productivity. This Special Issue comprises eleven papers, consisting of eight research articles and three review articles, authored by a diverse group of scientists.

Citation: Sellami, M.H.; Lavini, A. Advancements in Soil and Sustainable Agriculture. *Soil Syst.* **2023**, *7*, 98. https://doi.org/10.3390/soilsystems7040098

Received: 25 October 2023
Accepted: 27 October 2023
Published: 31 October 2023

2. A Comprehensive Review of Published Articles

As per the scholarly articles released in this Special Issue from 2022 to 2023, this Editorial categorizes the research papers into four specific themes:

Nutrient Management and Soil Health: This set of articles is dedicated to the relationship between nutrient management and its influence on various soil health indicators. The research papers authored by Gagnon and Ziadi (Contribution 1), Arrobas et al. (Contribution 2), de São José et al. (Contribution 3), and Denton-Thompson and Sayer (Contribution 4) are grouped within this thematic, exploring how nutrient management practices, including the use of soil amendments, biofertilizers, micronutrients, and residue management, impact soil characteristics, metal availability, and plant health. In their study, Gagnon and Ziadi (Contribution 1) examine the residual effects of applying paper mill biosolids and forest-derived liming materials on soil properties and metal availability. Their findings underscore that the application of these materials enhances soil properties without causing an increase in metal availability, showcasing the potential of sustainable nutrient management practices. Arrobas et al. (Contribution 2) assess the effects of humic and fulvic acids on soil and olive plant properties, revealing a minimal impact on tissue elemental composition and dry matter yield. However, their research underscores the need for a more comprehensive understanding of how to optimize the advantages of these soil amendments for crop production and sustainability. de São José et al. (Contribution 3) investigate the impact of managing eucalyptus harvest residues on soil organic carbon content and stock. Their study evaluates the use of the carbon management index (CMI) to determine the relationship between residue management and soil health. Their research emphasizes the importance of maintaining certain components (bark and branches) in order to increase soil carbon retention. In addition, Denton-Thompson and Sayer (Contribution 4) conduct a review of microbial nutrient mobilization in natural ecosystems and its potential to improve crop nutrition. Their work emphasizes the benefits of co-applying biofertilizers with conventional fertilizers to enhance nutrient uptake. This research underscores the critical importance of comprehending microbial nutrient mobilization for enhancing both crop nutrition and soil sustainability.

Soil management and Agricultural sustainability: This theme encompasses an examination of diverse agricultural practices, including tillage systems, crop rotations, and biochar utilization, and their influence on critical aspects of soil, such as carbon sequestration, microbial activity, crop yields, and overall soil sustainability. Notably, the research papers authored by Gualberto et al. (Contribution 5), Sellami and Terribile (Contribution 6), and Bogale et al. (Contribution 7) are grouped within this thematic area, delving into the repercussions of these agricultural practices on soil carbon sequestration, microbiological activity, yield, and overall soil sustainability. In the work by Gualberto et al. (Contribution 5), the authors investigate the impact of alterations in land use on soil organic carbon fractions in Northeastern Brazil. Their study showcases how distinct land management approaches can uphold soil health indicators, indicating that specific crop rotations and practices play a pivotal role in preserving topsoil quality and retaining essential active carbon components within the soil. In the study conducted by Sellami and Terribile (Contribution 6), the authors provide a bibliometric analysis of the evolution of soil health research. Their analysis spotlights the mounting emphasis on comprehending the consequences of soil management for biochemical processes, microbiological activities, and greenhouse gas emissions. This work underscores the significance of holistic, integrated strategies for sustainable soil management. Furthermore, Bogale et al. (Contribution 7) review the independent and symbiotic effects of soil tillage systems and biochar on soil carbon sequestration and crop production. Their research indicates that the combined influence of soil tillage techniques, such as conservation tillage, and the incorporation of biochar can foster soil carbon sequestration while enhancing crop physiology, yield, nutrient uptake, and soil health indicators. This approach offers a promising solution for advancing agricultural sustainability.

Soil Contamination and Phytoremediation: This thematic area centers on soil contamination and the application of phytoremediation as a strategy for improving soil health and mitigating contamination. Specifically, the research articles authored by Joya-Barrero et al. (Contribution 8), Ahmad et al. (Contribution 9), and Faria et al. (Contribution 10) fall under this thematic category, with a primary focus on soil contamination and the application of phytoremediation to improve soil health and mitigate contamination. In their study, Joya-Barrero et al. (Contribution 8) investigate the sources of cadmium in cacao crop soils and suggest strategies for reducing its presence. This research emphasizes the importance of understanding both natural and anthropogenic sources of soil contaminants and proposes practical solutions like selecting low-accumulator crops and amending soils with microorganisms to reduce metal content. The work of Ahmad et al. (Contribution 9) concentrates on the phytoextraction capacities of different crops, such as Brassica species, with respect to remediating soil contaminated with toxic metals. The outcomes highlight the potential of these crops as efficient agents for remediating soil polluted by toxic metals, thereby underscoring their pivotal role in promoting soil sustainability. Furthermore, Faria et al. (Contribution 10) investigate the impact of arbuscular mycorrhiza fungal communities on the uptake of essential and toxic metals by wheat plants. This research underscores the critical role of these fungal communities in enhancing crop resilience to metal toxicity, thereby contributing to sustainable agricultural practices and the overall health of soil ecosystems.

Soil Nitrogen Management and Environmental Impact: Under this thematic area, a sole article authored by De Marco (Contribution 11) is included. It delves into the nitrogen budget and employs statistical entropy analysis pertaining to the Tiber River catchment. This study amalgamates diverse methodologies to assess the implications of altering nitrogen usage scenarios, underscoring the significance of comprehending the factors behind varying nitrogen budgets at both the catchment and sub-catchment scales. This research makes a valuable contribution to endeavors aiming to diminish nitrogen accumulation and mitigate environmental hazards in heavily impacted ecosystems.

3. Conclusions

The amalgamation of these diverse themes epitomizes the intricate tapestry of contemporary soil and sustainable agriculture research. Within the articles published in this Special Issue "Advancements in Soil and Sustainable Agriculture" lies a shared commitment to confront the multifaceted challenges of modern agriculture. One of the overarching goals of these diverse studies is to promote the sustainability of the Earth's most fundamental resource: soil. Soil sustainability is, without a doubt, at the heart of these endeavors. It is the very foundation upon which agriculture, the backbone of human civilization, is built. Therefore, as the world grapples with ever-mounting pressures, ranging from the increasing demand for food production to the urgent need for environmental conservation and climate change mitigation, the importance of nurturing our soil's health and longevity cannot be overstated. Through the collaborative efforts of scientists, researchers, and agricultural practitioners, these themes underscore the significance of innovative solutions that are essential to securing a more sustainable and environmentally conscious agricultural future. These investigations, collectively, offer profound insights into the intricate dynamics of soil health and its intimate connection with the ecological well-being of our planet. They emphasize the vital role of holistic, integrated strategies in promoting not only agricultural productivity but also the long-term sustainability of our planet's delicate ecosystems. As we navigate the complexities of the 21st century, this collective pursuit embodies our shared commitment to nurturing and preserving the very earth beneath our feet, ensuring that it remains fertile, vibrant, and sustainable for generations to come. In the ever-evolving narrative of agriculture, it is clear that the chapters dedicated to soil sustainability are not only essential but also pivotal to the success and well-being of humanity and our planet.

Author Contributions: Conceptualization, M.H.S.; methodology, M.H.S.; investigation, M.H.S.; writing—original draft preparation, M.H.S. and A.L.; writing—review and editing, M.H.S.; visualization, M.H.S.; supervision, M.H.S. All authors have read and agreed to the published version of the manuscript.

Funding: This research received no external funding.

Conflicts of Interest: The authors declare no conflict of interest.

List of Contributions

1. Gagnon, B.; Ziadi, N. Paper Mill Biosolids and Forest-Derived Liming Materials Applied on Cropland: Residual Effects on Soil Properties and Metal Availability. *Soil Syst.* **2023**, *7*, 40. https://doi.org/10.3390/soilsystems7020040.
2. Arrobas, M.; de Almeida, S.F.; Raimundo, S.; da Silva Domingues, L.; Rodrigues, M.Â. Leonardites Rich in Humic and Fulvic Acids Had Little Effect on Tissue Elemental Composition and Dry Matter Yield in Pot-Grown Olive Cuttings. *Soil Syst.* **2022**, *6*, 7. https://doi.org/10.3390/soilsystems6010007.
3. de São José, J.F.B.; Vargas, L.K.; Lisboa, B.B.; Vieira, F.C.B.; Zanatta, J.A.; Araujo, E.F.; Bayer, C. Soil Carbon Stock and Indices in Sandy Soil Affected by Eucalyptus

Harvest Residue Management in the South of Brazil. *Soil Syst.* **2023**, *7*, 93. https://doi.org/10.3390/soilsystems7040093.
4. Denton-Thompson, S.M.; Sayer, E.J. Micronutrients in Food Production: What Can We Learn from Natural Ecosystems? *Soil Syst.* **2022**, *6*, 8. https://doi.org/10.3390/soilsystems6010008.
5. Gualberto, A.V.S.; de Souza, H.A.; Sagrilo, E.; Araujo, A.S.F.; Mendes, L.W.; de Medeiros, E.V.; Pereira, A.P.d.A.; da Costa, D.P.; Vogado, R.F.; da Cunha, J.R.; et al. Organic C Fractions in Topsoil under Different Management Systems in Northeastern Brazil. *Soil Syst.* **2023**, *7*, 11. https://doi.org/10.3390/soilsystems7010011.
6. Sellami, M.H.; Terribile, F. Research Evolution on the Impact of Agronomic Practices on Soil Health from 1996 to 2021: A Bibliometric Analysis. *Soil Syst.* **2023**, *7*, 78. https://doi.org/10.3390/soilsystems7030078.
7. Bogale, A.A.; Melash, A.A.; Percze, A. Symbiotic and Asymmetric Causality of the Soil Tillage System and Biochar Application on Soil Carbon Sequestration and Crop Production. *Soil Syst.* **2023**, *7*, 48. https://doi.org/10.3390/soilsystems7020048.
8. Joya-Barrero, V.; Huguet, C.; Pearse, J. Natural and Anthropogenic Sources of Cadmium in Cacao Crop Soils of Santander, Colombia. *Soil Syst.* **2023**, *7*, 12. https://doi.org/10.3390/soilsystems7010012.
9. Ahmad, I.; Malik, S.A.; Saeed, S.; Rehman, A.-u.; Munir, T.M. Phytoremediating a Wastewater-Irrigated Soil Contaminated with Toxic Metals: Comparing the Efficacies of Different Crops. *Soil Syst.* **2022**, *6*, 77. https://doi.org/10.3390/soilsystems6040077.
10. Faria, J.M.S.; Teixeira, D.M.; Ferreira, D.; Barrulas, P.; Brito, I.; Pinto, A.P.; Carvalho, M. Manganese Uptake to Wheat Shoot Meristems Is Differentially Influenced by Arbuscular Mycorrhiza Fungal Communities Adapted to Acidic Soil. *Soil Syst.* **2022**, *6*, 50. https://doi.org/10.3390/soilsystems6020050.
11. De Marco, A.; Fornasier, M.F.; Screpanti, A.; Lombardi, D.; Vitale, M. Nitrogen Budget and Statistical Entropy Analysis of the Tiber River Catchment, a Highly Anthropized Environment. *Soil Syst.* **2022**, *6*, 17. https://doi.org/10.3390/soilsystems6010017.

References

1. Davis, A.G.; Huggins, D.R.; Reganold, J.P. Linking Soil Health and Ecological Resilience to Achieve Agricultural Sustainability. *Front. Ecol. Environ.* **2023**, *21*, 131–139. [CrossRef]
2. Handayani, I.P.; Hale, C. Healthy Soils for Productivity and Sustainable Development in Agriculture. *IOP Conf. Ser.* **2022**, *1018*, 012038. [CrossRef]
3. FAO. *Healthy Soils Are the Basis for Healthy Food Production*; FAO: Rome, Italy, 2015; Available online: https://www.fao.org/soils-2015/news/news-detail/en/c/277682/ (accessed on 24 October 2023).
4. Rayhan Shaheb, M.d.; Sarker, A.; Shearer, S.A. Precision Agriculture for Sustainable Soil and Crop Management. In *Soil Science—Emerging Technologies, Global Perspectives and Applications*; IntechOpen: London, UK, 2022; Volume 49. [CrossRef]
5. Nath, S. A vision of precision agriculture: Balance between agricultural sustainability and environmental stewardship. *Agron. J.* **2023**, *Early View*, 1–18. [CrossRef]
6. Gamage, A.; Gangahagedara, R.; Gamage, J.; Jayasinghe, N.; Kodikara, N.; Suraweera, P.; Merah, O. Role of Organic Farming for Achieving Sustainability in Agriculture. *Farming Syst.* **2023**, *1*, 100005. [CrossRef]
7. Bhardwaj, D.; Ansari, M.W.; Sahoo, R.K.; Tuteja, N. Biofertilizers function as key player in sustainable agriculture by improving soil fertility, plant tolerance and crop productivity. *Microb. Cell. Fact.* **2014**, *13*, 66. [CrossRef] [PubMed]
8. Mącik, M.; Gryta, A.; Frąc, M. Biofertilizers in Agriculture: An Overview on Concepts, Strategies and Effects on Soil Microorganisms. *Adv. Agron.* **2020**, *162*, 31–87. [CrossRef]
9. Trap, J.; Éric, B. Intensifying the Soil Ecological Functions for Sustainable Agriculture: Acting with Stakeholders. *Curr. Res. Environ. Sustain.* **2023**, *5*, 100225. [CrossRef]
10. Mehmet Tuğrul, K. *Soil Management in Sustainable Agriculture. Sustainable Crop Production*; IntechOpen: London, UK, 2020. [CrossRef]

soil systems

Article

Leonardites Rich in Humic and Fulvic Acids Had Little Effect on Tissue Elemental Composition and Dry Matter Yield in Pot-Grown Olive Cuttings

Margarida Arrobas [1], Surian Fernanda de Almeida [2], Soraia Raimundo [1], Lucas da Silva Domingues [2] and Manuel Ângelo Rodrigues [1,*]

[1] Centro de Investigação de Montanha (CIMO), Instituto Politécnico de Bragança, Campus de Sta Apolónia, 5300-253 Bragança, Portugal; marrobas@ipb.pt (M.A.); soraia-raimundo@live.com.pt (S.R.)
[2] Campus Dois Vizinhos, Universidade Tecnológica Federal do Paraná, Dois Vizinhos 85660-000, PR, Brazil; suryan_nanda@hotmail.com (S.F.d.A.); lucasdomingues@utfpr.edu.br (L.d.S.D.)
* Correspondence: angelor@ipb.pt; Tel.: +351-273303260

Abstract: The use of humic substances in agriculture has increased in recent years, and leonardite has been an important raw material in the manufacture of commercial products rich in humic and fulvic acids. Leonardite-based products have been used to improve soil properties and to help plants cope with abiotic and biotic stresses. In this study, the effects of two commercial leonardites and an organic compost, in addition to a control treatment, were assessed for pot-grown olive plants over a period of fourteen months on soil properties, tissue elemental composition and dry matter yield (DMY). Three organic amendments were applied at single and double rates of that set by the manufacturer. The study was arranged in two experiments: one containing the seven treatments mentioned above and the other containing the same treatments supplemented with mineral nitrogen (N), phosphorus (P) and potassium (K) fertilization. Overall, organic compost increased soil organic carbon by ~8% over the control. In the experiment without NPK supplementation, N concentrations in shoots and P in roots were the highest for the compost application (leaf N 12% and root P 32% higher than in the control), while in the experiment with NPK supplementation, no significant differences were observed between treatments. Total DMY was ~10% higher in the set of treatments with NPK in comparison to treatments without NPK. Leonardites did not affect significantly any measured variables in comparison to the control. In this study, a good management of the majority of environmental variables affecting plant growth may have reduced the possibility of obtaining a positive effect on plant nutritional status and growth from the use of commercial leonardites. The leonardites seemed to have caused a slight effect on biological N immobilization. This is not necessarily an advantage or a drawback; it is rather a feature that must be understood to help farmers make better use of these products.

Keywords: plant biostimulants; humic substances; organic compost; *Olea europaea* L.

Citation: Arrobas, M.; de Almeida, S.F.; Raimundo, S.; da Silva Domingues, L.; Rodrigues, M.Â. Leonardites Rich in Humic and Fulvic Acids Had Little Effect on Tissue Elemental Composition and Dry Matter Yield in Pot-Grown Olive Cuttings. *Soil Syst.* **2022**, *6*, 7. https://doi.org/10.3390/soilsystems6010007

Academic Editors: Antonella Lavini and Mohamed Houssemeddine Sellami

Received: 15 December 2021
Accepted: 6 January 2022
Published: 11 January 2022

1. Introduction

Soil organic matter is the most widely used soil fertility index and a measure of its suitability for plant growth. Organic matter improves physical, chemical and biological soil properties, resulting in increased water holding capacity and aeration, pH stabilization, trace element immobilization and enhanced microbial activity and nutrient bioavailability [1]. For centuries, the use of farmyard manure and other organic materials in agricultural fields was the main method of improving soil fertility and maintaining crop productivity. Currently, the addition of organic amendments continues to be a recommended practice to restore fertility and to increase crop growth and yield, particularly in degraded soils [2,3].

In recent years, agriculture has undergone increasing specialization, and the rearing of livestock has tended to separate itself from crop production. There are regions of the world where landless livestock production systems are dominant, which create complications in waste management, sometimes resulting in serious problems of environmental contamination [4,5]. In other regions, animals and their associated farmyard manures are practically nonexistent. Although organic materials resulting from agro-industrial activities are sometimes available, which can be composted and used in agriculture [6,7], their availability is always limited given the extent of cultivated areas that need them. This is one of the primary causes of the low levels of organic matter in soils, especially when associated with other inadequate soil management practices, such as excessive tillage [8,9].

In recent years, the use of plant biostimulants in agriculture has increased greatly. Plant biostimulants are materials of a diverse nature, the most salient of which are protein hydrolysates and other N-containing compounds, seaweed extracts, inorganic compounds, chitosan, humic and fulvic acids and beneficial microorganisms [10,11]. Although commercial uses and specifications may vary depending on the nature of the plant biostimulant, they are not characterized by providing nutrients directly to the plants due to the low rates in which they are used but rather by favouring processes in the soil or in the plants that result in increased nutrient use efficiency and/or tolerance to several abiotic and biotic stresses [12,13].

Products rich in humic substances (HS) are an important group within the category of plant biostimulants [10,11]. HS includes humic acid, fulvic acid and humin fractions. Due to the insoluble nature of humin fractions, the commercial biostimulant activity of HS is focused on humic and fulvic acids [14]. They are normally applied to the soil as fertigation and less directly to the crop as a foliar spray. The reduced availability of conventional organic amendments has increased interest in such products, since it is expected that they have an effect on soil properties that can make up for the lack of manure or compost [15,16].

Commercial HSs are complex mixtures of distinct types of biomolecules which render their mode of action difficult to understand. The mechanisms by which HS elicit biostimulatory activity on plants remain elusive despite great efforts made over recent years [14,17]. Although positive results from the use of humic substances are not guaranteed, since they seem to be dependent on several agro-ecological variables that are difficult to optimize, a great number of positive results in soils and/or plants has been recorded [18,19]. Enhancement of plant growth by the application of humic substances has often been attributed to increased nutrient use efficiency and nutrient cycling due to the stimulus of microbial activity [13,20]. HSs have also been related to crop protection against environmental stresses, since they can enhance the activity of key enzymes in the metabolism of phenylpropanoids, which play a central role in the production of phenolic compounds involved in secondary metabolism, providing a strong argument for stress response modulation [17,21]. The negative action of HS has been associated with inordinate interactions with essential proteins or with the presence of entrapped small soil phenols exhibiting strong phytotoxicity in plants [22].

Commercial HSs are obtained from different raw materials, such as composted by-products, vermicomposts, peat and leonardite. Due to natural abundance and cost-effectiveness and because it contains high amounts of humic acids, leonardite has become an important raw material for the manufacture of commercial products [16,23]. Some of the positive effects of HS contained in leonardite may be ascribed to a general improvement in soil fertility, resulting in higher nutrient availability for plants. However, more consistent results from the use of HS tend to be recorded when plants are subjected to some form of environmental stress [20,24].

Nursery-grown and potted plants are very demanding in their requirements, giving rise to many difficulties in managing the salinity of the medium and the availability of water and nutrients for plants [25,26]. Due to ease and low cost with which HS can be added to a growing medium, it was hypothesized for this study that the use of two HS-rich leonardites in potted olive cuttings could benefit plant growth and make the cultivation process easier to implement. Thus, the main goal of the study was to compare two commercial leonardites rich in humic and fulvic acids with a composted organic amendment and a control treatment. An experiment was carried out using pre-rooted olive cuttings and growing them until they reached commercial size. Two commercial leonardites and the organic amendment were applied at single and double rates of that set by the manufacturer. The experimental apparatus was arranged as two independent experiments: In one, organic materials were used alone; in the other, they were used as a supplement to mineral NPK fertilization. This experimental design allowed the integration of a considerable number of treatments, 14 in total, without excessively complicating the interpretation of results.

2. Materials and Methods

2.1. Experimental Setup

This study reports the results of two experiments that took place in Bragança, northeastern Portugal. According to the Köppen–Geiger climate classification, the region benefits from a warm summer Mediterranean climate (Csb). The mean annual temperature is 12.3 °C, and annual precipitation is 758.3 mm [27].

Pot experiments were carried out indoors in a greenhouse covered by a double-wall polycarbonate panel. Lateral and zenith openings ensured aeration of the greenhouse and thermal conditions suitable for cultivation. The greenhouse was also equipped by a reflective screen to assist in the regulation of internal temperature, which automatically slid across when the temperature reached 28 °C.

The pots were filled with 3 kg of dried and sieved (2 mm mesh) soil and mixed with 50 g of perlite. The soil used in these experiments was collected from the surface layer (0.0–0.20 m) in a plot left fallow during the previous year. It is classified as a Regosol of colluvial origin, and the texture is sandy-clay loam (24.2% clay, 21.7% silt and 54.2% sand). Soil organic carbon is low (11.7 g kg^{-1}), and pH(H_2O) is close to neutrality (6.8). The levels of extractable P and K (Egnér–Riehm) are both classified as medium, 85.7 mg P_2O_5 kg^{-1} and 94.0 mg K_2O kg^{-1}, respectively. Cation exchange capacity is 17.9 $cmol_c$ kg^{-1} (11.63, 4.24, 0.45 and 1.58 $cmol_c$ kg^{-1} of Ca, Mg, K and Na, respectively).

In these experiments, three organic amendments (Nutrimais®, Humitec® and Humic gold®) and a compound NPK (10% N, 10% P_2O_5, and 10% K_2O) fertilizer were used. Nutrimais is a dehydrated and pelletized product, resulting from composting forestry, agro-industrial and domestic waste. It is recommended for agricultural use at rates of 3500 to 7000 l ha^{-1}. Humitec and Humic gold are two commercial formulations of leonardite, obtained from fossilized forest wood from about 40 to 60 million years ago, and recommended as soil conditioners. Humitec is recommended for direct application to the soil at rates of 75 to 275 kg ha^{-1}. Humic gold is recommended for application in fertigation at rates of 2 to 6 kg ha^{-1}. The most relevant properties of these fertilizing materials are presented in Table 1.

Table 1. Properties of the organic amendments used in these experiments (data provided by the manufacturers).

	Humitec	Humic Gold	Nutrimais
Moisture (%)	-	-	10.5
Organic matter (%)	55	-	52.5
Total organic carbon (%)	23.2	-	29.2
Total humic extract (%)	40	72	-
Humic acids (%)	30	56	-
Fulvic acids (%)	10	16	-
Total N (%)	2	-	2.4
Sulfur (SO_3) (%)	5	-	-
Silicon (SiO_2) (%)	24	-	-
Potassium (K_2O) (%)	-	8	1.8
Phosphorus (P_2O_5) (%)	-	-	1.5
Calcium (CaO) (%)			15.2
Magnesium (MgO)			0.7

The experiments were conducted on a completely randomized design, with seven treatments each and four replicates. Nutrimais® (Nu), Humitec® (Ht) and Humic gold® (Hg) were applied at two rates (a single and a double rate of those set by the vendor). Thus, the rates used were 5 (Nu1) and 10 (Nu2) t ha^{-1}, 250 (Ht1) and 500 (Ht2) kg ha^{-1} and 5 (Hg1) and 10 (Hg2) kg ha^{-1}. The conversion of the rates of organic amendments recommended for use in the field to those used in pots took into account the individual area occupied by each pot. The pots were placed in a square with a 0.267 m side, each pot occupying the equivalent of 0.071 m^2. Thus, each pot received organic amendments in an amount equivalent to a fraction of 1/140,000 of a hectare. Thus, each pot received 35.7 (Nu1), 71.4 (Nu2), 1.79 (Ht1), 3.57 (Ht2), 0.036 (Hg1) and 0.071 (Hg2) g of commercial product. A non-fertilized control (C) was also included in the experimental design.

The second experiment was arranged in the same manner. The seven treatments described for the first experiment were supplemented with mineral NPK fertilization. Thus, the treatments of this experiment were named as Nu1+, Nu2+, Ht1+, Ht2+, Hg1+, Hg2+ and C+. These pots received the same rates of organic amendments reported for the first experiments in addition to 5 g pot^{-1} of NPK fertilizer, which corresponds to a field fertilization of ~70 kg ha^{-1} of N, P_2O_5 and K_2O.

Rooted cuttings of homogeneous size (~10 cm in height) of the cultivar Cobrançosa were used in this study. The trials were set up on 4 August 2019. Subsequently, the pots were kept free of weeds and watered regularly with 200 mL of water. Irrigation frequency was dependent on phenological stages and environmental conditions. The experiment was completed on 7 October 2020.

2.2. Sample Collection and Analysis

Soil samples were taken by carefully recovering all soil from the pots and by trying to avoid breaking the roots. Soil samples were then properly homogenized and 200 g subsamples sent to the laboratory. Afterwards, the roots were washed with a low-pressure water jet in a metallic box with a 1 mm mesh in an attempt to recover as many roots as possible. The plant was divided into roots, stems and leaves, and the three parts were dried and weighed separately. Plant tissues were oven-dried at 65 °C to a constant weight and then ground in a 1-millimetre mesh mill.

Soil samples were oven-dried at 40 °C and analyzed for pH (H_2O and KCl) (soil: solution, 1:2.5), cation-exchange capacity (ammonium acetate, pH 7.0), organic carbon (wet digestion, Walkley–Black method) and extractable P and K (Egnér–Riehm method). Soil boron (B) was extracted by hot water, and the extracts were analysed by the azomethine-H method. For more details on these analytical procedures, the reader is referred to van Reeuwijk [28]. The availability of other soil micronutrients (copper (Cu), iron (Fe), zinc (Zn) and manganese (Mn)) was determined by atomic absorption spectrometry after extraction with ammonium acetate and EDTA, according to the method previously described by Lakanen and Erviö [29].

Elemental tissue analyses were performed by Kjeldahl (N), colorimetry (B and P), flame emission spectrometry (K) and atomic absorption spectrophotometry (Ca, Mg, Cu, Fe, Zn and Mn) methods after nitric digestion of the samples [30].

2.3. Data Analysis

Data were analysed for normality and homogeneity of variances using the Shapiro–Wilk and Bartlett's test, respectively. Thereafter, one-way ANOVA was performed. When significant differences were found among treatments, the means were separated by Tukey HSD ($\alpha = 0.05$) test.

3. Results

In the experiment without NPK supplementation, significant differences ($p < 0.05$) between treatments were found (Figure 1). The application of Nutrimais increased DMY in comparison to the control and other treatments such as Hu1 and Hg1. The differences were mainly due to the contribution of leaves to total DMY. In the experiment in which the application of organic amendments was supplemented with NPK, significant differences between treatments ($p < 0.05$) were not found; however, overall the DMY of these treatments was ~10% higher than that of treatments without NPK supplementation. In both experiments, the contribution of roots to DMY was approximately one-third of the contribution of stems and leaves.

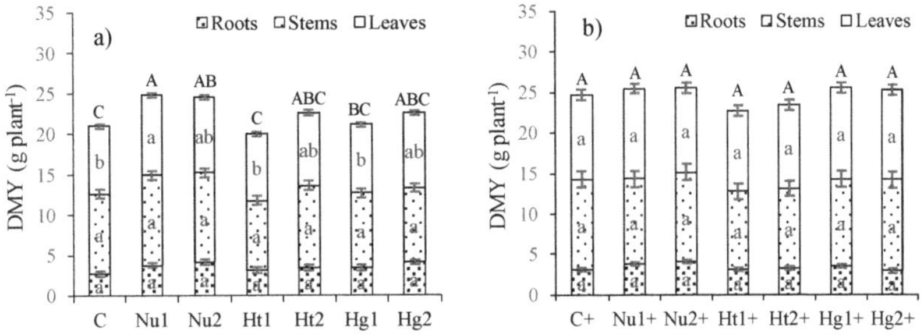

Figure 1. Dry matter yield (DMY) in roots, stems, leaves and total in response to the application of organic amendments (C, Control; Nu, Nutrimais; Ht, Humitec; and Hg, Humic gold, applied at two rates (1,2)) in experiments (**a**) without and (**b**) with NPK(+) mineral fertilization. Lowercase letters show the results of analysis of variance and means separation (Tukey HSD, $\alpha = 0.05$) for each plant part and uppercase letters for total DMY.

In the experiment without NPK addition, Nutrimais applied at the rate 2 (Nu2) produced the highest average leaf N concentration (23.7 g kg^{-1}), a value significantly higher than that of the control (20.9 g kg^{-1}) and other treatments such as Ht2 (17.5 g kg^{-1}) and Hg2 (18.6 g kg^{-1}) (Table 2). Comparing the two rates of Humitec and also the two rates of Humic gold, it seems that the higher rate reduced leaf N concentration, although without significant differences. In the experiment without NPK supplementation, no significant differences between treatments were found for any of the other nutrients. Leaf P and K, the other nutrients applied in the experimental design, varied from 1.5 to 2.0 g kg^{-1} and 8.2 to 9.2 g kg^{-1}, respectively. In the experiment with NPK supplementation, no significant differences were recorded for any of the elements determined. The effect of organic amendments on leaf nutrient concentration was less than experimental variability. The range of variation of the macronutrients N, P and K was 23.4–24.5 g kg^{-1}, 1.6–2.3 g kg^{-1} and 8.7–9.7 g kg^{-1}, respectively.

Stem N concentrations, although lower than those of the leaves, followed a somewhat similar pattern when comparing treatments (Table 3). The highest values were recorded for Nu2 (9.8 g kg^{-1}), which were significantly higher than those of the Ht2 treatment (7.9 g kg^{-1}). As in the leaves, there seems to be a tendency for higher rates of Humitec (Hu2) and Humic gold (Hg2) to produce stem N concentrations lower than those recorded in Hu1 and Hg1 treatments. For the other nutrients and for all nutrients in the experiment with NPK supplementation, significant differences between treatments were not observed. As mentioned for leaf nutrient concentration, the effect of organic amendments was not enough to stand out from experimental variability.

Table 2. Leaf nutrient concentration in response to the application of organic amendments (C, Control; Nu, Nutrimais; Ht, Humitec; and Hg, Humic gold, applied at two rates (1,2)) in the experiments without and with NPK(+) mineral fertilization. For each independent experiment, means followed by the same letter are not significantly different by Tukey HSD test (α = 0.05).

	N	P	K	Ca	Mg	B	Fe	Mn	Zn	Cu
Treatments			g kg^{-1}					mg kg^{-1}		
C	20.9 b	1.5 a	8.2 a	3.9 a	0.9 a	20.6 a	186.8 a	39.1 a	12.4 a	9.0 a
Nu1	21.3 ab	2.0 a	9.2 a	4.5 a	1.0 a	18.8 a	158.2 a	29.4 a	17.0 a	8.5 a
Nu2	23.7 a	2.0 a	9.0 a	4.8 a	1.0 a	21.8 a	161.8 a	41.0 a	20.3 a	8.3 a
Ht1	20.2 bc	1.9 a	8.9 a	4.7 a	1.0 a	21.7 a	182.3 a	34.4 a	16.7 a	8.1 a
Ht2	17.5 c	2.0 a	9.2 a	4.5 a	1.1 a	19.9 a	180.8 a	30.7 a	18.6 a	7.4 a
Hg1	21.0 ab	1.9 a	9.2 a	4.6 a	1.1 a	20.7 a	149.1 a	32.1 a	15.2 a	8.1 a
Hg2	18.6 bc	1.7 a	8.2 a	4.3 a	1.3 a	18.7 a	151.0 a	31.2 a	15.0 a	8.1 a
Prob. < P	<0.0001	0.6929	0.4674	0.8649	0.7053	0.1134	0.0866	0.1066	0.1220	0.9852
Std error	0.58	0.24	0.45	0.46	0.17	0.85	10.28	3.00	1.77	1.23
NPK addition										
C+	24.5 a	1.6 a	8.8 a	4.5 a	1.1 a	18.9 a	160.3 a	30.1 a	20.5 a	7.4 a
Nu1+	23.9 a	2.3 a	9.3 a	4.9 a	1.2 a	22.7 a	135.8 a	44.0 a	18.7 a	8.0 a
Nu2+	24.3 a	2.3 a	9.6 a	4.8 a	1.2 a	19.3 a	132.9 a	37.5 a	16.0 a	7.9 a
Ht1+	24.0 a	1.9 a	8.7 a	5.0 a	1.2 a	21.1 a	191.0 a	42.6 a	16.5 a	8.6 a
Ht2+	24.4 a	1.9 a	9.5 a	4.6 a	1.2 a	20.6 a	168.1 a	51.0 a	17.2 a	7.8 a
Hg1+	23.4 a	1.6 a	8.7 a	4.5 a	1.0 a	21.8 a	151.0 a	40.9 a	13.3 a	7.9 a
Hg2+	23.9 a	1.8 a	9.7 a	4.6 a	1.1 a	22.3 a	192.0 a	47.5 a	13.2 a	11.2 a
Prob. < P	0.3391	0.2545	0.4550	0.6578	0.4789	0.0727	0.1397	0.1171	0.2014	0.1519
Std error	0.35	0.24	0.44	0.24	0.07	0.93	17.06	4.69	2.06	0.93

Table 3. Stem nutrient concentration in response to the application of organic amendments (C, Control; Nu, Nutrimais; Ht, Humitec; and Hg, Humic gold, applied at two rates (1,2)) in experiments without and with NPK(+) mineral fertilization. For each independent experiment, means followed by the same letter are not significantly different by Tukey HSD test (α = 0.05).

	N	P	K	Ca	Mg	B	Fe	Mn	Zn	Cu
Treatments			g kg^{-1}					mg kg^{-1}		
C	9.3 ab	1.3 a	5.3 a	2.5 a	0.7 a	16.9 a	292.9 a	14.5 a	11.8 a	9.6 a
Nu1	9.3 ab	1.3 a	7.3 a	2.7 a	0.7 a	14.3 a	285.1 a	15.4 a	12.1 a	8.9 a
Nu2	9.8 a	1.6 a	7.4 a	3.0 a	0.8 a	15.8 a	222.6 a	14.5 a	15.8 a	8.2 a
Ht1	8.1 ab	1.4 a	6.9 a	2.7 a	0.7 a	16.2 a	262.1 a	14.9 a	14.7 a	6.9 a
Ht2	7.9 b	1.5 a	6.7 a	2.6 a	0.7 a	16.6 a	261.3 a	13.8 a	16.0 a	8.2 a
Hg1	9.2 ab	1.4 a	7.0 a	2.8 a	0.7 a	15.9 a	209.5 a	12.6 a	14.1 a	8.4 a
Hg2	8.6 ab	1.3 a	7.0 a	2.4 a	0.6 a	17.1 a	206.9 a	9.5 a	11.8 a	8.2 a
Prob. < P	0.0225	0.4461	0.6957	0.7196	0.8532	0.2151	0.1490	0.2690	0.0967	0.3281
Std error	0.38	0.12	0.86	0.24	0.10	0.75	25.73	1.67	1.23	0.73
NPK addition										
C+	9.8 a	1.3 a	6.6 a	2.7 a	0.7 a	14.3 a	194.9 a	14.3 a	11.1 a	9.6 a
Nu1+	10.6 a	1.5 a	7.4 a	2.8 a	0.8 a	16.2 a	217.3 a	13.8 a	10.2 a	10.9 a
Nu2+	10.2 a	1.4 a	6.9 a	2.6 a	0.7 a	16.2 a	187.5 a	10.7 a	9.8 a	11.4 a
Ht1+	10.0 a	1.5 a	7.3 a	2.7 a	0.7 a	16.2 a	279.3 a	14.6 a	11.6 a	12.8 a
Ht2+	10.0 a	1.5 a	6.9 a	2.7 a	0.7 a	16.3 a	219.0 a	16.4 a	12.2 a	9.9 a
Hg1+	10.2 a	1.6 a	6.9 a	3.0 a	0.7 a	14.8 a	200.1 a	13.7 a	8.7 a	11.0 a
Hg2+	9.7 a	1.5 a	7.2 a	2.4 a	0.6 a	15.9 a	208.2 a	9.7 a	10.6 a	12.0 a
Prob. < P	0.7240	0.1978	0.8315	0.5016	0.8007	0.1754	0.1422	0.4308	0.3428	0.2483
Std error	0.39	0.07	0.37	0.17	0.06	0.61	21.86	2.25	1.05	0.91

As in aboveground tissue, average root N concentration was the highest in the Nu2 treatment (16.3 g kg^{-1}), a value significantly higher than that of the Ht2 treatment (13.3 g kg^{-1}) (Table 4). Moreover, as in the aboveground tissue, the average values of treatments Ht2 (13.3 g kg^{-1}) and Hg2 (13.7 g kg^{-1}) were lower than those of Ht1 (14.1 g kg^{-1}) and Hg1 (15.6 g kg^{-1}). Root P concentrations varied significantly between treatments, Nut2 displaying the highest value (2.5 g kg^{-1}) and C, Ht1 and Ht2 displaying the lower one (1.7 g kg^{-1}) in the experiment without NPK addition. Root Mg concentration was significantly higher in Nu2 (4.3 g kg^{-1}) in comparison to the control (3.1 g kg^{-1}). For the other nutrients and for all nutrients in the experiment with NPK supplementation, significant differences between treatments were not found. Root N, P and K concentrations in the experiment with NPK addition were in the ranges 15.4–18.4 g kg^{-1}, 1.8–2.4 g kg^{-1} and 12.0–14.1 g kg^{-1}, respectively.

Table 4. Root nutrient concentration in response to the application of organic amendments (C, Control; Nu, Nutrimais; Ht, Humitec; and Hg, Humic gold, applied at two rates (1,2)), in experiments without and with NPK(+) mineral fertilization. For each independent experiment, means followed by the same letter are not significantly different by Tukey HSD test (α = 0.05).

	N	P	K	Ca	Mg	B	Fe	Mn	Zn	Cu
Treatments			g kg^{-1}					mg kg^{-1}		
C	15.0 ab	1.7 b	11.3 a	4.8 a	3.1 b	19.8 a	5152.6 a	132.5 a	31.8 a	90.9 a
Nu1	16.2 ab	1.9 ab	11.3 a	5.6 a	3.6 ab	19.1 a	5902.8 a	145.9 a	35.5 a	92.4 a
Nu2	16.3 a	2.5 a	11.2 a	5.7 a	4.3 a	20.5 a	6920.9 a	167.6 a	43.1 a	95.7 a
Ht1	14.1 ab	1.7 b	11.4 a	4.5 a	3.2 ab	18.0 a	5906.6 a	145.6 a	36.6 a	79.8 a
Ht2	13.3 b	1.7 b	11.0 a	4.8 a	3.3 ab	18.0 a	6759.6 a	156.6 a	41.7 a	97.4 a
Hg1	15.6 ab	1.8 ab	12.6 a	5.1 a	3.3 ab	17.7 a	6463.0 a	142.6 a	34.4 a	95.9 a
Hg2	13.7 ab	1.9 ab	11.4 a	4.7 a	3.6 ab	20.7 a	6026.7 a	166.7 a	43.1 a	101.1 a
Prob. < P	0.0471	0.0486	0.9233	0.0615	0.0462	0.1199	0.4635	0.4365	0.5449	0.5921
Std error	0.71	0.18	0.93	0.28	0.24	0.87	604.96	12.75	4.92	7.66
NPK addition										
C+	15.4 a	1.8 a	13.3 a	5.8 a	3.7 a	17.6 a	5636.2 a	147.8 a	33.4 a	69.8 a
Nu1+	17.1 a	2.0 a	13.9 a	5.5 a	3.8 a	20.7 a	4973.9 a	142.8 a	35.5 a	82.2 a
Nu2+	18.4 a	2.4 a	12.6 a	5.9 a	3.9 a	20.6 a	5545.7 a	160.3 a	41.5 a	91.6 a
Ht1+	17.4 a	2.0 a	13.3 a	5.2 a	3.4 a	19.0 a	4042.7 a	152.0 a	33.6 a	80.9 a
Ht2+	17.7 a	2.2 a	14.1 a	5.1 a	3.3 a	17.9 a	5119.3 a	149.3 a	40.8 a	95.6 a
Hg1+	17.4 a	1.9 a	12.7 a	5.7 a	3.4 a	20.1 a	4531.8 a	151.4 a	33.2 a	87.5 a
Hg2+	16.9 a	2.0 a	12.0 a	5.1 a	3.4 a	17.4 a	4767.0 a	108.6 a	37.7 a	99.1 a
Prob. < P	0.1968	0.3780	0.8096	0.0682	0.1098	0.0558	0.5068	0.1070	0.1771	0.2304
Std error	0.72	0.14	1.09	0.19	0.15	0.86	583.17	11.32	2.64	7.99

Soil organic carbon varied significantly between treatments in the experiments without and with NPK addition (Table 5). Higher values were found in Nu2 (12.3 g kg^{-1}) and Nu2+ (12.6 g kg^{-1}) treatments in each of the experiments. Extractable P varied also between treatments, and Nu2 (167.6 mg P_2O_5 kg^{-1}) and Nu2+ (226.0 mg P_2O_5 kg^{-1}) showed higher values of each experiment. Overall, the values of extractable P in the experiment with NPK addition were higher than those of the experiment where only organic amendments were applied. Extractable K followed a trend similar to extractable P. The higher values of the experiments were found in treatments Nu2 (163.3 mg K_2O kg^{-1}) and Nu2+ (264.0 mg K_2O kg^{-1}). No significant differences were found between treatments for exchangeable Ca^{++} and Mg^{++} and CEC, as well as for extractable micronutrients. The effect of the treatments on these variables was poor and less than experimental variability.

Table 5. Selected soil properties in response to the application of organic amendments (C, Control; Nu, Nutrimais; Ht, Humitec; and Hg, Humic gold, applied at two rates (1,2)), in experiments without and with NPK(+) mineral fertilization. For each independent experiment, means followed by the same letter are not significantly different by Tukey HSD test ($\alpha = 0.05$).

			Extractable		Exchangeable			Extractable				
	Organic Carbon	pH	P (P_2O_5)	K (K_2O)	Ca^{++}	Mg^{++}	CEC	B	Fe	Zn	Cu	Mn
Treatments	g kg^{-1}	H_2O	mg kg^{-1}			$cmol_+$ kg^{-1}				mg kg^{-1}		
C	11.0 bc	6.7 a	78.9 c	95.3 bc	8.8 a	5.8 a	15.3 a	0.9 a	109.0 a	3.8 a	42.8 a	168.8 a
Nu1	10.8 bc	6.8 a	123.8 b	122.0 b	9.3 a	6.2 a	16.4 a	0.8 a	109.1 a	4.0 a	45.7 a	175.8 a
Nu2	12.3 a	6.9 a	167.6 a	163.3 a	8.5 a	5.3 a	14.9 a	1.0 a	110.3 a	4.2 a	29.4 a	167.7 a
Ht1	10.4 c	6.9 a	57.6 c	90.3 c	8.4 a	5.1 a	14.1 a	0.8 a	101.3 a	3.2 a	32.0 a	167.4 a
Ht2	10.4 c	6.9 a	58.9 c	86.3 c	7.8 a	4.2 a	12.6 a	0.8 a	99.6 a	3.4 a	33.7 a	161.3 a
Hg1	11.1 b	6.9 a	71.8 c	103.0 bc	8.3 a	5.4 a	14.5 a	0.9 a	110.2 a	4.3 a	44.6 a	170.2 a
Hg2	10.8 bc	6.9 a	59.5 c	89.7 c	9.0 a	7.0 a	16.6 a	0.7 a	98.7 a	3.3 a	29.9 a	157.8 a
Prob. < P	<0.0001	0.0774	<0.0001	<0.0001	0.7194	0.1133	0.2865	0.3813	0.0902	0.1177	0.0997	0.1473
Std error	0.17	0.05	7.98	6.41	0.62	0.60	1.16	0.08	3.48	0.32	4.82	4.26
NPK addition												
C+	10.7 c	6.7 a	132.3 b	163.3 a	9.4 a	6.3 a	16.5 a	0.8 a	109.2 a	3.8 a	30.9 a	156.9 a
Nu1+	11.5 bc	6.8 a	158.4 b	218.7 a	9.3 a	6.6 a	17.1 a	1.0 a	104.5 a	3.9 a	32.1 a	151.2 a
Nu2+	12.6 a	6.7 a	226.0 a	264.0 a	8.5 a	6.0 a	15.9 a	0.9 a	102.7 a	4.1 a	28.6 a	158.8 a
Ht1+	11.1 bc	6.7 a	139.8 b	190.7 a	7.9 a	5.6 a	14.5 a	0.8 a	100.1 a	4.4 a	41.0 a	160.2 a
Ht2+	11.6 b	6.7 a	181.6 ab	214.3 a	8.6 a	5.6 a	15.3 a	0.9 a	129.3 a	4.6 a	48.2 a	186.0 a
Hg1+	11.5 bc	6.9 a	136.8 b	153.3 a	9.0 a	5.8 a	15.8 a	0.9 a	110.5 a	4.6 a	42.4 a	167.6 a
Hg2+	11.3 bc	6.8 a	142.1 b	190.0 a	7.8 a	5.5 a	14.2 a	0.8 a	94.1 a	3.9 a	33.4 a	158.1 a
Prob. < P	<0.0001	0.0542	0.0004	0.0827	0.6668	0.8893	0.7648	0.0805	0.1807	0.1151	0.1341	0.4111
Std error	0.17	0.04	11.24	24.07	0.74	0.69	1.40	0.06	8.47	0.24	5.11	10.85

4. Discussion

In the experiment without the addition of NPK, the application of Nutrimais increased DMY in comparison to the control. In the experiment with NPK addition, differences between treatments were not found, and the average values were ~10% higher than those of the treatments without NPK addition. The use of organic amendments can positively influence several soil properties, which may result in the enhancement of plant growth [2,7]. In these experiments, under controlled growing conditions, the positive effect of Nutrimais in DMY was probably due to the supply of nutrients, a thesis supported by the absence of a positive effect of Nutrimais on DMY when the pots were supplemented with the NPK fertilizer.

The analysis of the elemental composition of plant tissues showed that Nutrimais consistently increased tissue N levels (leaves, stems and roots). In the experiment with NPK supplementation, tissue N levels were globally higher than in the experiment without NPK but with no differences between treatments. As no differences in shoots were observed in the concentration of other nutrients, the result suggests that the positive effect of Nutrimais on DMY was due to the supply of N. Nitrogen is one of the main limiting factors for plant growth in most agricultural ecosystems [31–34]. Soils do not accumulate N in readily available forms, with plants being dependent on regular applications of N as a fertilizer [1]. The availability of N to plants from organic substrates depends on their N content and C/N ratio. C/N ratios above or below 25 are usually associated to net N immobilization or mineralization, respectively [1]. Nutrimais presented an N concentration of 24.1 g kg^{-1} and C/N ratio of 11.94 (Table 1), suggesting net mineralization in the short term, which was most likely the reason for the increase in N concentration in plant tissues and DMY.

Leaf P concentrations did not vary with the application of Nutrimais. Although P is a limiting factor in plant growth in many regions of the world, where crop yield is dependent on the regular application of P as a fertilizer [35–37]. In northeastern Portugal, in soils with similar properties to those used in this experiment, it has been difficult to observe a response in crops relative to P applications [38,39]. This suggests that these soils provide non-limiting amounts of P for plant growth. The roots from Nutrimais pots showed higher P concentrations than those of the other treatments. It has also been shown from previous studies that olive plants can accumulate P in roots when the nutrient is highly available in the soil, while the concentration of P in aboveground plant tissues is maintained at adequate levels [25,40]. This may have reduced the importance of P on plant growth in this experiment. Nevertheless, it is clear that Nutrimais provided the system with P.

The application of Nutrimais increased K levels in the soil, but this was not enough to increase the concentration of K in plant tissues. These soils normally supply K in adequate amounts to plants. In previous studies, there has also been a reduced response of plants observed relative to the application of K [41,42], as mentioned above for P.

Nutrimais increased the content of soil organic carbon in the pots with and without NPK supplementation. The increase in soil organic matter is an important objective in the use of organic amendments and stems from the direct supply of carbon and a slow process of mineralization in the soil [1]. Leonardites did not have such an effect because, at the rates they were used, the carbon added was not enough to be detected by soil analysis.

The leonardites Hu and Hg tended to reduce average tissue N concentrations at the higher rate they were used. This seems to be evidence of biological net N immobilization. Humic and fulvic acids, by stimulating soil microbiology, may have had this short-term effect, contributing to the immobilization of N. This is not necessarily negative. It is a property of organic amendments with little available N [1], but it is important to take this property into account to avoid N shortage during the growing season. In potted cultivation, especially in nurseries, it is common to use slow or controlled release fertilizers as a method of reducing leaching or other forms of nutrient loss [24]. These kinds of leonardites may have a similar role in regulating the N cycle.

Leonardites are also known for their ability to immobilize trace metals, as they contain oxidized functional groups that can bind different metal ions or heavy metals [15,43]. This effect, however, was not observed in this study, since the concentration of metals in plant tissues did not vary significantly with the experimental treatments. The reduced rate at which these commercial products are recommended and used in this particular study was probably the reason for the absence of any detectable effect.

Humic substances are recommended for increasing crop P uptake since they are capable of competing with P to be bound to soil adsorption complexes [44]. In agreement with this, Kaya et al. [45] observed that leonardite enhanced leaf P and yield in maize under P and water stress in calcareous soils. In this study, however, no benefits were found for P nutrition from the use of leonardites, perhaps because the plants were not grown under water stress conditions and these soils may have provided enough P for plant growth, even without P fertilization as above mentioned.

Although some studies have reported different beneficial effects for plants resulting from the use of HS [15,20,46], in other studies, it was shown that there is a possibility of HS exerting a negative impact on plant growth and development [22]. Atiyeh et al. [47] reported that the inhibitory effect of HS on plant growth is dosage dependent and that the negative action of HS was associated with inordinate interactions with essential proteins. It has also been hypothesized that another structural feature in terms of HS toxicity is the presence of entrapped small soil phenols, which exhibit a strong phytotoxicity by affecting glycolysis and pentose phosphate pathways in plants [22,48]. In this study, however, no relevant signs of positive or negative effects on plants or soil were detected when leonardites were used at single and double recommended rates.

5. Conclusions

Nutrimais produced typical results for an organic amendment, with an increase in organic carbon and nutrients in the soil, which, in turn, increased the concentrations of some nutrients in plant tissues and dry matter yield in the experiment without NPK addition. The positive effect on dry matter yield was probably due to the increase in nitrogen availability, the most limiting nutrient in these experiments.

In these pot experiments, a good control of the majority of environmental variables affecting plant growth may have been the reason for not having found any beneficial effects of the use of these commercial leonardites. Leonardites are rich in humic substances, and their effect on plants tends to become more evident when plants are grown under some kind of environmental stress.

Leonardites are not used to provide nutrients due to their low nutrient concentration and the rates at which they are usually applied. Used at such low rates in our experiment, they also did not cause any measurable effect on increasing soil organic carbon. Indeed, the results seem to show a slight effect of biological nitrogen immobilization from the use of leonardites, which may be due to some stimulus in soil biological activity. This is not necessarily an advantage or a drawback; it is simply a feature that must be understood. Despite there being promising indications about these products found in some studies, in others they have not been so clear; thus, more studies are needed to set safe guidelines for their use by farmers.

Author Contributions: Conceptualization, M.Â.R.; methodology, M.A.; investigation, S.F.d.A. and S.R.; resources, M.Â.R. and M.A.; data curation, M.Â.R. and L.d.S.D.; writing—original draft preparation, M.A. and S.F.d.A.; writing—review and editing, M.Â.R., M.A. and L.d.S.D.; supervision, M.A. and M.Â.R.; project administration, M.A.; funding acquisition, M.Â.R. and M.A. All authors have read and agreed to the published version of the manuscript.

Funding: This research was funded by the Foundation for Science and Technology (FCT, Portugal) and FEDER under Programme PT2020 for financial support to CIMO (UIDB/00690/2020). The research was integrated in the activities of the operational group "Novas práticas em olivais de sequeiro: estratégias de mitigação e adaptação às alterações climáticas", funded by PT2020 and EAFRD (European Agricultural Fund for Rural Development).

Data Availability Statement: Data sharing not applicable. No new data were created or analyzed in this study. Data sharing is not applicable to this article.

Conflicts of Interest: The authors declare no conflict of interest.

References

1. Weil, R.R.; Brady, N.C. *Nature and Properties of Soils*, 15th ed.; Pearson: London, UK, 2017.
2. Obriot, F.; Stauffer, M.; Goubard, Y.; Cheviron, N.; Peres, G.; Eden, M.; Revallier, A.; Vieublé-Gonod, L.; Houot, S. Multi-criteria indices to evaluate the effects of repeated organic amendment applications on soil and crop quality. *Agric. Ecosyst. Environ.* **2016**, *232*, 165–178. [CrossRef]
3. Montiel-Rozas, M.M.; Domínguez, M.T.; Madejón, E.; Madejón, P.; Pastorelli, R.; Renella, G. Long-term effects of organic amendments on bacterial and fungal communities in a degraded Mediterranean soil. *Geoderma* **2018**, *332*, 20–28. [CrossRef]
4. Francis, D.D.; Vigil, M.F.; Mosier, A.R. Gaseous losses of nitrogen other than through denitrification. In *Nitrogen in Agricultural Systems*; Schepers, J.S., Raun, W.R., Eds.; ASA-CSSA-SSSA: Madison, WI, USA, 2008; pp. 255–279.
5. Cao, X.; Reichel, R.; Wissel, H.; Kummer, S.; Brüggemann, N. High carbon amendments increase nitrogen retention in soil after slurry application—an incubation study with silty loam soil. *J. Soil Sci. Plant Nutr.* **2021**. [CrossRef]
6. Afonso, S.; Pereira, E.; Arrobas, M.; Rodrigues, M.A. Recycling nutrient-rich hop leaves by composting with wheat straw and farmyard manure in suitable mixtures. *J. Environ. Manag.* **2021**, *284*, 112105. [CrossRef] [PubMed]
7. Arrobas, M.; Carvalho, J.T.N.; Raimundo, S.; Poddere, G.; Rodrigues, M.A. The safe use of compost derived from municipal solid waste depends on its composition and conditions of application. *Soil Use Manag.* **2021**. [CrossRef]
8. Almagro, M.; de Vente, J.; Boix-Fayos, C.; García-Franco, N.; Aguilar, J.M.; González, D.; Solé-Benet, A.; Martínez-Mena, M. Sustainable land management practices as providers of several ecosystem services under rainfed Mediterranean agroecosystems. *Mitig Adapt. Strat. Glob. Chang.* **2016**, *21*, 1029–1043. [CrossRef]
9. Rodrigues, M.A.; Arrobas, M. Cover cropping for increasing fruit production and farming sustainability. In *Fruit Crops: Diagnosis and Management of Nutrient Constraints*; Srivastava, A.K., Hu, C., Eds.; Elsevier: Amsterdam, The Netherlands, 2020; pp. 279–295.
10. du Jardin, P.; Xu, L.; Geelen, D. Agricultural functions and action mechanisms of plant biostimulants (PBs): An Introduction. In *The Chemical Biology of Plant Biostimulants*; Geelen, D., Xu, L., Eds.; Wiley: Hoboken, NJ, USA, 2020; pp. 3–29.
11. Rouphael, Y.; Colla, G. Biostimulants in agriculture. *Front. Plant Sci.* **2020**, *11*, 1–7. [CrossRef] [PubMed]
12. Patel, J.S.; Selvaraj, V.; Gunupuru, L.R.; Rathor, P.K.; Prithiviraj, B. Combined application of *Ascophyllum nodosum* extract and chitosan synergistically activates host-defense of peas against powdery mildew. *BMC Plant Biol.* **2020**, *20*, 113. [CrossRef]
13. Sun, Q.; Liu, J.; Huo, L.; Li, Y.C.; Li, X.; Xia, L.; Zhou, Z.; Zhang, M.; Li, B. Humic acids derived from Leonardite to improve enzymatic activities and bioavailability of nutrients in a calcareous soil. *Int. J. Agric. Biol. Eng.* **2020**, *13*, 200–205. [CrossRef]

14. Lamar, R.T. Possible role for electron shuttling capacity in elicitation of PB activity of humic substances on plant growth enhancement. In *The Chemical Biology of Plant Biostimulants*; Geelen, D., Xu, L., Eds.; Wiley: Hoboken, NJ, USA, 2020; pp. 97–121.
15. Sugier, D.; Kołodziej, B.; Bielińska, E. The effect of leonardite application on *Arnica montana* L. yielding and chosen chemical properties and enzymatic activity of the soil. *J. Geochem. Explor.* **2013**, *129*, 76–82. [CrossRef]
16. Canellas, L.P.; Olivares, F.L.; Aguiar, N.O.; Jones, D.L.; Nebbioso, A.; Mazzei, P.; Piccolo, A. Humic and fulvic acids as biostimulants in horticulture. *Sci. Hortic.* **2015**, *196*, 15–27. [CrossRef]
17. du Jardin, P. Plant biostimulants: Definition, concept, main categories and regulation. *Sci. Hortic.* **2015**, *196*, 3–14. [CrossRef]
18. Denre, M.; Ghanti, G.; Sarkar, K. Effect of humic acids application on accumulation of mineral nutrition and pungency in garlic (*Allium sativum* L.). *Int. J. Biotech. Mol. Biol. Res.* **2014**, *5*, 7–12.
19. Sandepogu, M.; Shukla, P.S.; Asiedu, S.; Yurgel, S.; Prithiviraj, B. Combination of *Ascophyllum nodosum* extract and humic acid improves early growth and reduces post-harvest loss of lettuce and spinach. *Agriculture* **2019**, *9*, 240. [CrossRef]
20. Litardo, R.C.M.; Bendezú, S.J.G.; Zenteno, M.D.C.; Pérez-Almeida, I.B.; Parismoreno, L.L.; García, E.D.L. Effect of mineral and organic amendments on rice growth and yield in saline soils. *J. Saudi Soc. Agric. Sci.* **2021**. [CrossRef]
21. Schiavon, M.; Pizzeghello, D.; Muscolo, A.; Vaccaro, S.; Francioso, O.; Nardi, S. High molecular size humic substances enhance phenylpropanoid metabolism in maize (*Zea mays* L.). *J. Chem. Ecol.* **2010**, *36*, 662–669. [CrossRef] [PubMed]
22. Lee, J.G.; Yoon, H.Y.; Cha, J.-Y.; Kim, E.-K.; Kim, P.J.; Jeon, J.-R. Artificial humification of lignin architecture: Top-down and bottom-up approaches. *Biotechnol. Adv.* **2019**, *37*, 107416. [CrossRef] [PubMed]
23. Özyazici, G. Yield and quality of black cumin (Nigella sativa L.) according to leonardite and nitrogen doses. *Appl. Ecol. Environ. Res.* **2020**, *18*, 7057–7075. [CrossRef]
24. Wang, J.; Wu, J.; Lu, J.; Yuan, G. Effects of leonardite on the coastal saline soil improvement. *Chem. Ecol.* **2020**, *36*, 750–765. [CrossRef]
25. Lopes, J.I.; Correia, C.M.; Gonçalves, A.; Silva, E.; Martins, S.; Arrobas, M.; Rodrigues, M.A. Arbuscular mycorrhizal fungi inoculation reduced the growth of pre-rooted olive cuttings in a greenhouse. *Soil Syst.* **2021**, *5*, 30. [CrossRef]
26. Rodrigues, M.A.; Piroli, L.B.; Forcelini, D.; Raimundo, S.; Domingues, L.C.; Cassol, L.C.; Correia, C.M.; Arrobas, M. Use of commercial mycorrhizal fungi in stress-free growing conditions of potted olive cuttings. *Sci. Hortic.* **2021**, *275*, 109712. [CrossRef]
27. IPMA (Instituto Português do Mar e da Atmosfera). *Normais Climatológicas*; Instituto Português do Mar e da Atmosfera: Lisabona, Portugal, 2021. Available online: https://www.ipma.pt/pt/oclima/normais.clima/ (accessed on 2 December 2021).
28. van Reeuwijk, L.P. *Procedures for Soil Analysis*; Technical Paper 9; ISRIC FAO: Wageningen, The Netherlands, 2002.
29. Lakanen, E.; Erviö, R. A Comparison of Eight Extractants for the Determination of Plant Available Micronutrients in Soils. *Acta Agral. Fenn.* **1971**, *123*, 223–232.
30. Temminghoff, E.E.J.M.; Houba, V.G. *Plant Analysis Procedures*; Kluwer Academic Publishers: Dordrecht, The Netherlands, 2004.
31. Rodrigues, M.A.; Afonso, S.; Ferreira, I.Q.; Arrobas, M. Response of stevia to nitrogen fertilization and harvesting regime in Northeastern Portugal. *Arch. Agron. Soil Sci.* **2017**, *63*, 626–637. [CrossRef]
32. Arrobas, M.; Afonso, S.; Rodrigues, M.A. Diagnosing the nutritional condition of chestnut groves by soil and leaf analyses. *Sci. Hortic.* **2018**, *228*, 113–121. [CrossRef]
33. Arrobas, M.; Santos, D.; Ribeiro, A.; Pereira, E.; Rodrigues, M.A. Soil and foliar nitrogen and boron fertilization of almond trees grown under rainfed conditions. *Eur. J. Agron.* **2019**, *106*, 39–48. [CrossRef]
34. Ferreira, I.Q.; Arrobas, M.; Moutinho-Pereira, J.M.; Correia, C.M.; Rodrigues, M.A. The effect of nitrogen applications on the growth of young olive trees and nitrogen use efficiency. *Turk. J. Agric.* **2020**, *44*, 278–289. [CrossRef]
35. Vistoso, E.; Iraira, S.; Sandaña, P. Phosphorus use efficiency in permanent pastures in Andisols. *J. Soil Sci. Plant Nutr.* **2021**, *21*, 2587–2599. [CrossRef]
36. Li, G.; Huang, G.; Li, H.; van Ittersum, M.K.; Leffelaar, P.A.; Zhang, F. Identifying potential strategies in the key sectors of China's food chain to implement sustainable phosphorus management: A review. *Nutr. Cycl. Agroecosys.* **2016**, *104*, 341–359. [CrossRef]
37. Tian, D.; Li, Z.; O'Connor, D.; Shen, Z. The need to prioritize sustainable phosphate-based fertilizers. *Soil Use Manag.* **2020**, *36*, 351–354. [CrossRef]
38. Arrobas, M.; Afonso, S.; Ferreira, I.Q.; Moutinho-Pereira, J.M.; Correia, C.M.; Rodrigues, M.A. Liming and application of nitrogen, phosphorus, potassium and boron on a young plantation of chestnut. *Turk. J. Agric.* **2017**, *41*, 441–451. [CrossRef]
39. Rodrigues, M.A.; Ferreira, I.Q.; Afonso, S.; Arrobas, M. Sufficiency ranges and nutrient removals in lemon balm based on crop response to applied nitrogen, phosphorus, potassium and boron. *J. Plant Nutr.* **2018**, *41*, 996–1008. [CrossRef]
40. Ferreira, I.Q.; Rodrigues, M.A.; Moutinho-Pereira, J.M.; Correia, C.; Arrobas, M. Olive tree response to applied phosphorus in field and pot experiments. *Sci. Hortic.* **2018**, *234*, 236–244. [CrossRef]
41. Arrobas, M.; Ferreira, I.Q.; Afonso, S.; Rodrigues, M.A. Sufficiency ranges and crop nutrient removals for peppermint (*Mentha* x *piperita* L.) established from field and pot fertilizer experiments. *Commun Soil Sci. Plant Anal.* **2018**, *49*, 1719–1730. [CrossRef]
42. Ferreira, I.Q.; Arrobas, M.; Moutinho-Pereira, J.M.; Correia, C.; Rodrigues, M.A. Olive response to potassium applications under different water regimes and cultivars. *Nutr. Cycl. Agroecosys.* **2018**, *112*, 387–401. [CrossRef]
43. Cieschi, M.T.; Lucena, J.L. Leonardite iron humate and synthetic iron chelate mixtures in *Glycine max* nutrition. *J. Sci. Food Agric.* **2021**, *101*, 4207–4219. [CrossRef] [PubMed]
44. Purwanto, B.H.; Wulandari, P.; Sulistyaningsih, E.; Utami, S.N.H.; Handayani, S. Improved corn yields when humic acid extracted from composted manure is applied to acid soils with phosphorus fertilizer. *Appl. Environ. Soil Sci.* **2021**. [CrossRef]

45. Kaya, C.; Şenbayram, M.; Akram, N.A.; Ashraf, M.; Alyemeni, M.N.; Ahmad, P. Sulfur-enriched leonardite and humic acid soil amendments enhance tolerance to drought and phosphorus deficiency stress in maize (*Zea mays* L.). *Sci. Rep.* **2020**, *10*, 6432. [CrossRef] [PubMed]
46. Salin, A.; Karadogan, T.; Tonguc, M. Effects of leonardite applications on yield and some quality parameters of potatoes (*Solanum tuberosum* L.). *Turk. J. Field Crop.* **2013**, *18*, 20–26.
47. Atiyeh, R.M.; Lee, S.; Edwards, C.A.; Arancon, N.Q.; Metzger, J.D. The influence of humic acids derived from earthworm-processed organic wastes on plant growth. *Bioresour. Technol.* **2002**, *84*, 7–14. [CrossRef]
48. Muscolo, A.; Panuccio, M.R.; Sidari, M. The effect of phenols on respiratory enzymes in seed germination. Respiratory enzyme activities during germination of Pinus laricio seeds treated with phenols extracted from different forest soils. *Plant Growth Regul.* **2001**, *35*, 31–35. [CrossRef]

soil systems

Review

Micronutrients in Food Production: What Can We Learn from Natural Ecosystems?

Sarah M. Denton-Thompson * and Emma J. Sayer *

Lancaster Environment Centre, Lancaster University, Lancaster LA1 4YQ, UK
* Correspondence: s.denton-thompson@lancaster.ac.uk (S.M.D.-T.); e.sayer@lancaster.ac.uk (E.J.S.);
Tel.: +44-(0)1524-510140 (E.J.S.)

Abstract: Soil micronutrients limit crop productivity in many regions worldwide, and micronutrient deficiencies affect over two billion people globally. Microbial biofertilizers could combat these issues by inoculating arable soils with microorganisms that mobilize micronutrients, increasing their availability to crop plants in an environmentally sustainable and cost-effective manner. However, the widespread application of biofertilizers is limited by complex micronutrient–microbe–plant interactions, which reduce their effectiveness under field conditions. Here, we review the current state of seven micronutrients in food production. We examine the mechanisms underpinning microbial micronutrient mobilization in natural ecosystems and synthesize the state-of-knowledge to improve our overall understanding of biofertilizers in food crop production. We demonstrate that, although soil micronutrient concentrations are strongly influenced by soil conditions, land management practices can also substantially affect micronutrient availability and uptake by plants. The effectiveness of biofertilizers varies, but several lines of evidence indicate substantial benefits in co-applying biofertilizers with conventional inorganic or organic fertilizers. Studies of micronutrient cycling in natural ecosystems provide examples of microbial taxa capable of mobilizing multiple micronutrients whilst withstanding harsh environmental conditions. Research into the mechanisms of microbial nutrient mobilization in natural ecosystems could, therefore, yield effective biofertilizers to improve crop nutrition under global changes.

Keywords: soil micronutrient availability; food crop production; microbial biofertilizer; microbial nutrient mobilization; plant micronutrient uptake; biofortification

Citation: Denton-Thompson, S.M.; Sayer, E.J. Micronutrients in Food Production: What Can We Learn from Natural Ecosystems? *Soil Syst.* **2022**, *6*, 8. https://doi.org/10.3390/soilsystems6010008

Academic Editors: Antonella Lavini and Mohamed Houssemeddine Sellami

Received: 29 November 2021
Accepted: 7 January 2022
Published: 12 January 2022

1. Introduction

Human population growth and increasing levels of consumption are placing ever greater demands on farmers to increase crop production [1–3]. However, land availability for food production is limited, resulting in intensive agricultural practices that affect soil health and the wider environment [2,4]. It is therefore vital that global food production is increased sustainably to maintain soil health [4,5]. Nutrient availability in soils is one of the key factors underpinning food production [3,6]. As well as the more widely researched macronutrients, nitrogen (N), potassium (K) and phosphorus (P), six trace elements are consistently described as important micronutrients for crop growth and yield: iron (Fe), zinc (Zn), manganese (Mn), molybdenum (Mo), boron (B), and copper (Cu) [7–10]. In addition, cobalt (Co) is beneficial to plants and plays a crucial role in nitrogen fixation in leguminous crops [9]. For simplicity, we refer henceforth to these seven trace elements as 'micronutrients'. Micronutrients in soils are largely derived from their parent material and required in minimal quantities by crops (i.e., <1 part per million). However, soil properties and conventional agricultural management practices, such as the use of higher yielding crop cultivars and the application of NPK inorganic fertilizers, can result in an insufficient micronutrient supply to crops [7–9]. A deficiency in even one of these nutrients can substantially reduce crop growth and yield, although an excessive supply, resulting in toxicity can be equally detrimental [7,9,10]. In general, most plants have a narrow optimum micronutrient range, making it difficult to maintain the balance between deficiency and toxicity [7,9,10]. Thus, achieving a balanced supply of micronutrients in soils is essential for combatting human and animal dietary deficiencies in many parts of the world.

1.1. Micronutrients and Hidden Hunger

It is estimated that over two billion people worldwide suffer from 'hidden hunger', a deficiency in one or more micronutrients [11]. Deficiencies in Fe, Zn and Co (in the form of Vitamin B12) are amongst the most prevalent, especially in populations where diets mainly consist of cereals and legumes [12–14]. Similar micronutrient deficiencies were also noted in livestock and are most often seen in free-ranging herds with reduced supplementary feeding [15]. Hidden hunger in livestock and human populations that are heavily dependent upon crop-based foodstuffs demonstrates a clear need to address disparities in the micronutrient uptake and content of crops used in food production. Soil micronutrient content can be enhanced by fertilizer application, but this requires costly repeated applications with variable results and questionable environmental implications [16]. The excessive application of micronutrients would pose an equal, if not greater, environmental threat than over-fertilization with macronutrients, since toxicity thresholds for micronutrients are reached at considerably lower doses [7,9,10]. Therefore, to address micronutrient deficiency in crop production in a safe and sustainable manner, alternatives to inorganic fertilizer application are needed.

1.2. Microbial Biofortification

Biofortification is a promising avenue to address micronutrient deficiencies in agricultural crops. Crop biofortification encompasses a range of strategies that ultimately aim to improve the micronutrient content of crop plants, either by increasing micronutrient availability in arable soils, enhancing plant nutrient uptake, or both [2,16,17]. Plant breeding and genetic modification can create new cultivars with enhanced nutrient absorption, but these approaches are often costly, time-consuming and restricted by law [2,3]. By contrast, microbial biofortification is a recent strategy involving the enhancement of native soil microbial populations or the inoculation of arable soils with 'microbial biofertilizers' (hereafter biofertilizers) composed of plant growth-promoting microorganisms (PGPMs) [2,6,18,19]. Microbial biofortification focuses on enhancing the natural actions of microbes, which improve micronutrient availability and uptake, making it considerably more cost-effective, sustainable, and less environmentally damaging [2,16,20,21]. Microbial methods are also particularly effective for addressing micronutrient imbalances because they can be used to tackle both toxicity and deficiency [22–24].

Although microbial biofortification has evident theoretical benefits, its practical application proves difficult because the approach is underpinned by interactions among micronutrients, microbial communities, and crop species, which are still poorly understood [3,19,20]. Finding suitable microbial species for biofertilizers whilst ensuring that arable soil conditions support and enhance their action is particularly challenging. Thus, to ensure the widespread application of biofertilizers, we need to address several key knowledge gaps, such as effectiveness of biofertilizers under field conditions, the impacts of microbial or crop diversity, and the largely unknown effects of introduced microbial species on the native soil microbiota [6,19,20]. One way forward is to assess how plant–microbe interactions influence micronutrient availability in semi-natural ecosystems where environmental conditions, microbial communities and plant diversity are highly variable. We can then use this information to identify suitable combinations of microbiota and soil conditions for developing biofertilizers. Therefore, the primary aims of this review are to (i) assess the current state of seven micronutrients in food production, (ii) examine the interactions underpinning microbial biofortification in non-agroecosystems, and (iii) synthesize the state-of-knowledge on micronutrients to improve our overall understanding of microbial biofortification in food crop production. Although it is clear that many of the micronutrients reviewed here can also occur at toxic levels in some soils, our review focuses on micronutrient deficiency in arable soils, hence the exclusion of micronutrients such as chlorine and nickel, for which deficiency rarely occurs and is not as widely researched [25–27]. We focused on synthesizing the past and current literature to provide an integrative review of micronutrient biofertilizers (following [28]). Our literature search was carried out using Web of Science and Google Scholar, comprising (1) a general search for papers containing the terms 'micronutrient', 'agriculture', 'arable', 'biofertilizer', 'inoculant', 'soil' AND 'microb*' OR 'microorganism' followed by (2) a more specific search for each micronutrient by ecosystem (e.g., soil AND microb* AND iron AND grassland).

2. Iron (Fe)

Iron (Fe) is the most abundant micronutrient and the fourth most abundant element overall in the Earth's crust [29]. Aside from parent material weathering, atmospheric deposition and the degradation of organic matter also provide soil Fe inputs to a lesser degree [25,29,30]. Iron is an essential component of human nutrition because it is involved in oxygen transport; in plants, Fe is required for enzyme production, photosynthesis, and N metabolism [9]. The total concentration of Fe in soils is generally sufficient to meet plant needs, but various soil properties can render the majority of Fe immobile, resulting in low Fe availability in c. 30% of soils worldwide [9,31]. Ensuring that crops receive sufficient Fe for optimum growth and yield is achieved by enhancing Fe availability, the efficiency of Fe absorption by crops, or a combination of both.

The availability of Fe in soils is governed by its chemical form, which in turn is dictated by numerous abiotic and biotic factors [9,32,33]. Iron primarily exists in ferrous (Fe II) and ferric (Fe III) forms, which can bind to clay fractions abiotically via cation exchange, or precipitate to form oxides, hydroxides and oxyhydroxides, which can render Fe unavailable under neutral or alkaline conditions [25,29]. Iron is widely used in biotic microbial redox reactions to yield energy for organic carbon degradation [25,34]. The reduction of ferric Fe(III) to ferrous Fe(II), which is more soluble and, therefore, readily available, primarily occurs under anaerobic conditions in water-saturated soils. However, under aerobic conditions, the oxidation of soluble Fe(II) to the more insoluble Fe(III) prevails, reducing Fe availability [25,30,34]. Consequently, agricultural management practices that influence soil pH, redox potential, saturation, or aeration can affect the availability of Fe in soils.

2.1. *Fe Availability and Acquisition in Arable Soils*

The greater availability of Fe in acidic soils is one of the key issues in combatting Fe deficiencies in crops because low soil pH conditions are often not beneficial for crop plants and can reduce the availability of other nutrients [9]. The use of liming to counteract soil acidification is a common agricultural practice, but for every unit of increase in pH (between pH 4–9) Fe solubility can decrease by up to a thousandfold [8]. Other agricultural practices such as tillage and irrigation can affect soil pH by altering aeration and water saturation, which can affect Fe mobilization or immobilization [8,35]. Consequently, in cultivated soils that are naturally calcareous, or where management practices alter soil pH, soluble Fe concentrations may be suboptimal for meeting crop requirements [32,33,36]. The interaction between soil pH and soil organic matter (hereafter SOM) has a lesser but equally significant impact on Fe availability. Generally, soil Fe retention and availability increases with SOM content, which is problematic in arable soils where SOM is often heavily depleted [8,35,37]. Soil organic matter is both a source of Fe and of reducing agents, and hence the microbial mineralisation of SOM increases Fe concentrations and provides the necessary conditions for maximum Fe solubility [8]. Organic acids derived from SOM can increase Fe availability by reducing soil pH and by forming soluble Fe complexes [8,33]. However, other compounds in SOM can bind Fe to form increasingly insoluble organo–mineral complexes as soil pH increases [8,9,33]. Nonetheless, arable soils with low SOM content are more likely to be susceptible to Fe loss or fixation to stable compounds [8,35,37].

As well as the indirect impacts of alterations to soil conditions, arable management practices also directly influence Fe availability. In many countries, the application of inorganic fertilizers to correct Fe deficiency in crops has become increasingly common [9]. However, increasing Fe application may have a limited effect on heavily limed or naturally calcareous soils because the high soil pH favours Fe oxidation to insoluble forms. The application of organic fertilizers (e.g., farmyard manure) declined in recent decades with the widescale movement away from mixed farming [38], but preliminary evidence supports their use to combat Fe deficiency, which warrants further investigation [39]. Hence, soil management to increase Fe availability is possible, but when the necessary changes to soil conditions cannot be accomplished, deficiencies can instead be addressed by altering the nutrient acquisition strategies of plants and microbes.

2.1.1. Fe Acquisition by Crops: Strategies and Efficiency

Two factors can interfere with acquisition and use of Fe by crops, even when Fe availability in soils is not limiting. Firstly, plant breeding to increase crop yield and improve resistance to pathogens or pests can select for traits that increase the plant's micronutrient requirements, creating Fe deficiencies, even though soil Fe availability was considered sufficient for past cultivars [9]. Secondly, the application of herbicides such as glyphosate, diclofop-methyl and chlorsulfuron can interfere with root growth and Fe translocation from the roots to shoots and grain, resulting in plant Fe deficiency [33,40,41]. Plants employ two strategies to improve Fe acquisition under deficient conditions (Figure 1). Strategy I, used by nongraminaceous monocots and dicots (e.g., legumes), involves the release of protons into the rhizosphere, which increases soil acidity and mobilizes ferric Fe(III). Enzymes (chelate reductases) at the root–soil interface then reduce Fe(III) to ferrous Fe(II), which can be absorbed by the plant via ferrous transporters [29,42,43]. Strategy II plants are graminaceous monocots (e.g., grasses and cereal crops); the plants exude Fe-chelating organic substances (phytosiderophores), which form complexes with Fe(III) that plants can then absorb via plasma membrane transport systems without reduction [29,42,43].

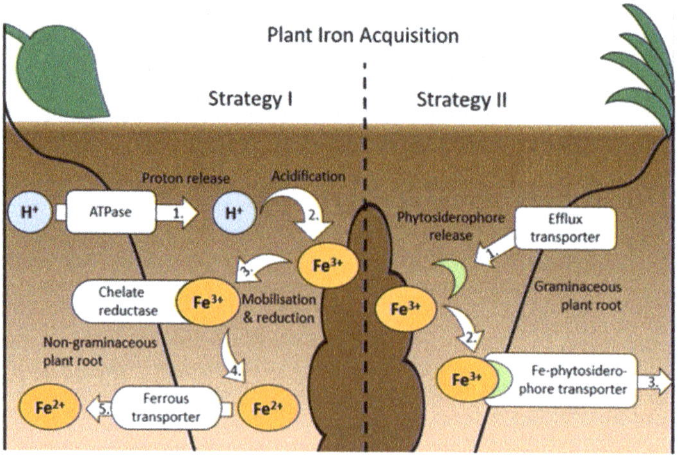

Figure 1. Diagram of the two iron (Fe) acquisition strategies employed by plants: orange ellipses show the chemical form of Fe, blue circles represent hydrogen (H^+) ions, white boxes with arrows indicate molecule transport points into/out of the root, and numbers indicate individual steps of the process (based on [29,42,43]).

Although strategy I plant species mobilize Fe by adjusting unfavourable soil pH, this method is less effective in well-aerated calcareous soils and is considered less efficient overall than the method utilised by strategy II plants [42,44]. Phytosiderophores can be exuded in large quantities by strategy II plants and are capable of chelating Fe regardless of pH. However, phytosiderophores are only released diurnally, with production peaking a few hours after dawn, and can be rapidly decomposed by certain microbial taxa [42,44,45]. Furthermore, when Fe availability is low, plants and soil microbes can compete for available Fe, which has prompted increased research into microbial strategies for increasing Fe availability [29,43,45].

2.1.2. Fe Mobilization by Microbes

Iron is widely used in microbial redox reactions to yield energy for organic carbon degradation [33,45,46]. The dissimilatory reduction of Fe(III) to Fe(II) occurs under anaerobic conditions, and microbes can also perform assimilatory Fe reduction under aerobic conditions both indirectly via acidification of the rhizosphere and directly by producing siderophores [33,45,46]. Microbial siderophores function in the same way as phytosiderophores but are considerably more efficient, possessing a greater stability and a higher affinity for forming Fe(III) complexes [33,46,47]. To date, over 500 types of microbial siderophores have been identified, with many microbial taxa capable of producing or utilising multiple types [33,43,47]. The strong affinity of microbial siderophores for Fe allows them to compete with plants by 'stealing' Fe from low-affinity phytosiderophores [33,45,47]. Despite this capability, numerous studies demonstrated a beneficial relationship between the presence of PGPMs and Fe accumulation by plants [43,48–50].

Microbial biofortification of Fe in plants can occur via several mechanisms: the presence of PGPMs can induce increased root hair proliferation and branching, trigger plant biochemical responses to Fe limitation, and prevent Fe acquisition by phytopathogenic microbes [43,48–50]. In addition, there is also considerable evidence that plants can utilise microbial siderophores, which appear to be dictated by the plants' Fe acquisition strategies. Chelate reductases in the roots of strategy I plants are capable of accepting Fe(III) complexes from microbial siderophores for reduction, producing freely available Fe(II), which can provide a significant fraction of plant Fe requirements [33,42,48]. The utilisation of microbial siderophores by strategy II plants is more variable due to the competition between microbial siderophores and phytosiderophores. However, some microbial siderophores, such as rhizoferrin, have an Fe affinity equal to phytosiderophores; therefore, they can act as an additional source of Fe to strategy II plants, either by direct uptake of Fe from microbial siderophores by the plant or (to a greater extent) Fe exchange from microbial siderophores to phytosiderophores [33,42,50,51]. Therefore, microbial biofortification could significantly improve crop Fe acquisition if the selected PGPMs induce plant responses to Fe limitation or produce siderophores that promote plant Fe uptake.

2.2. Fe in Natural Ecosystems

Studies of biotic interactions with Fe in (semi-)natural ecosystems provide valuable insights into microbial taxa suitable for biofertilizers because the soils, microbial communities, and vegetation have been much less affected by human activities than in agroecosystems [35,37,52]. Forest soils received considerable attention for bioprospecting, most likely because they possess high microbial diversity [35,37,52], which increases the likelihood of finding suitable PGPMs. Numerous dissimilatory Fe-reducing bacterial taxa isolated from tropical forest soils are already acclimated to periodic waterlogging [53–55], providing an insight into how microbial Fe reduction may be influenced by soil aeration or water saturation. Fluctuations in Fe redox reactions in tropical forest soils closely follow rainfall patterns, with long dry periods leading to increased aeration and decreased Fe reduction [56]. Thus, biofertilizers based on taxa from tropical soils may be best applied immediately after rainfall or irrigation to maximise Fe reduction, and thus crop Fe acquisition. However, since many tropical forest soils are rich in Fe, microbial taxa isolated from tropical soils might not be suitable for soils with limited Fe availability. Instead, taxa capable of siderophore production under Fe-limited conditions may prove more effective. Siderophore-producing microbes isolated from the soils of montane forest ecosystems are not only capable of increasing Fe availability and acquisition but can do so even when exposed to extreme environmental stress [57,58]. Although research into the utilisation of these microbial species as biofertilizers only began in the last decade, preliminary results indicate significant potential for their application in combatting Fe deficiency in arable crops [57,58].

Other ecosystems exposed to extreme or fluctuating temperatures are potential sources of microbial taxa for novel biofertilizers, especially for use in agroecosystems in colder climates. Bacterial strains capable of producing siderophores at both low (e.g., 4 °C) and relatively high (e.g., 30 °C) temperatures were isolated from cold deserts and glaciers [59,60]. Surprisingly, grassland ecosystems received less attention as potential sources of biofertilizers, although siderophore-producing fungal species may be sourced from grassland soils [61]. Many fungal species from Ascomycota and Basidiomycota can produce the siderophore ferricrocin, which plants can utilise to obtain Fe; however, some of these fungal species are pathogenic, and so a careful separation from PGPM species is required [61].

Finally, research into the degradation of SOM in natural ecosystems could be informative because microbial SOM mineralisation under water-saturated conditions releases Fe from organic complexes and increases the availability of soluble organic carbon [54,56]. As the amount of SOM in agroecosystems is often limited [35,37,42], the effects of organic matter applications on Fe availability in arable soils are still unclear. Nonetheless, studies in forest and grassland soils demonstrate that Fe reduction increases with the availability of labile carbon due to its stimulatory effects on microbial activity [53,54,62]. Even at higher soil pH, Fe reduction can be further enhanced by regular additions of organic carbon, especially under anaerobic conditions [53,54]. As SOM can also act as a source of soluble Fe, the addition of organic fertilizers to arable soils could significantly and sustainably increase Fe availability to crops, with or without biofertilizers.

3. Zinc (Zn)

Zinc (Zn) deficiency is now considered a global-scale crisis, and correcting its low availability in arable soils has become a widely researched topic [9,63–65]. In plants, Zn is essential for growth [9,65], heat stress tolerance [63], pathogen resistance, and reproduction [64,66]. In humans, Zn is a vital component of DNA, RNA and over 300 bodily enzymes, with evidence also suggesting a role in gene expression [63,67]. However, many countries reported extensive Zn deficiency in arable soils [63], which not only leads to yield reductions of up to 80%, but to widespread Zn deficiency in approximately one third of the global human population, especially in populations dependent on cereal-based diets [64,68].

Zinc levels in soils are maintained by atmospheric inputs, which generally exceed outputs or losses via leaching and biological uptake [25,69]. However, although the total Zn concentrations in most soils appear sufficiently high to meet plant needs, approximately 90% of Zn in soils worldwide is thought to exist in a form unavailable to plants [25,64,66], and c. 30–50% of soils possess insufficient available Zn to meet plant requirements [25,66,70]. Unlike some other micronutrients, Zn can persist in numerous free ionic forms and complexes with other metals, as well as in crystalline forms bound to clay fractions and as a component of insoluble or soluble (organic) complexes. As such, soil type and soil properties also have a strong impact on Zn levels, with lower Zn concentrations in sandy soils and higher concentrations in organic and calcareous soils [25,69]. Zinc availability declines with increasing pH due to increased fixation in soil minerals and the formation of insoluble Zn compounds (e.g., hydroxides, calcium zincate) in soils with pH > 7.5. Therefore, in highly calcareous or heavily limed soils, Zn availability and uptake by crops is greatly limited [9,25,71]. Aside from liming, several other arable management practices can also reduce Zn availability [23,63]: excessive P fertilization can interfere with plant Zn acquisition and applying organic material with a high ligand content can reduce Zn availability via the formation of insoluble organo–mineral complexes [9,66,70]. Furthermore, flooding (e.g., in rice paddies) and the selection of crop cultivars with high Zn-demand can also result in insufficient Zn availability [9,63,66].

Although plants are capable of chelating Zn in the same way as Fe through rhizospheric acidification and the production of phytosiderophores, these plant strategies are often insufficient for addressing Zn deficiencies [19,63,64]. Consequently, the application of inorganic Zn fertilizers is common practice in many countries, with one application theoretically providing enough Zn to last for up to six crop rotations [63]. However, due to the rapid fixation and immobilization of Zn in most soils, only 1–20% of the applied Zn is absorbed by crops, rendering this approach financially costly and environmentally unsustainable [63,71,72]. Therefore, research efforts turned to developing biofertilizers that can mobilize insoluble Zn pools already present in soils [3,63,66].

3.1. Zn Mobilization by Microbes

The microbial biofortification of crops with Zn received extensive research and is hailed as the most promising solution to Zn deficiency in arable agriculture [3]. The crop uptake of Zn via microbial siderophores is far less difficult to implement than Fe uptake [3,19,66]; numerous bacterial strains, as well as arbuscular mycorrhizal fungi (AMF) and fungal species from the genus Trichoderma, can acidify the rhizosphere through organic acid or proton release and produce siderophores that enable them to solubilize Zn [3,19,21,64]. Furthermore, endophytic fungal species such as AMF produce hyphae capable of physically accessing Zn pools that plant roots would otherwise be unable to reach; the chelated Zn is absorbed into the hyphal network and transported directly back to the plant's roots [3,19,64]. Due to the efficiency and effectiveness of microbial Zn solubilization, the application of biofertilizers successfully alleviated Zn deficiency in many crops [3,64,66].

In soils where Zn deficiency is caused by low concentrations of total Zn rather than low availability, the application of Zn fertilizers in combination with biofertilizers prevents the applied Zn from being instantly immobilized [73–75]. Using this dual application, farmers can biofortify their crops whilst applying less Zn, which is more financially and environmentally sustainable [75,76]. However, the highly variable sensitivity of microbes to Zn toxicity and variable microbial responses among taxa can limit the co-application of biofertilizers and conventional Zn fertilizers [77,78]. Zinc nano-fertilizers were recently suggested as a more sustainable alternative, with successful crop Zn deficiency remediation at low application rates, and either neutral or positive impacts on native soil microbiota [79–82]. In summary, the co-application of Zn nano-fertilizers with biofertilizers could present a way to alleviate crop Zn deficiencies, while maintaining healthy soil microbial communities. However, although numerous Zn-solubilizing microbes were identified, some strains may be more effective than others under given environmental conditions. Therefore, natural ecosystems remain an important potential source for new and alternative Zn-biofortifying microbes.

3.2. Lessons for Zn Biofortification

As Zn is one of the most common heavy metals in municipal waste, sewage, and composted residues [83,84], research investigating soil Zn concentrations in semi-natural ecosystems mostly focuses on issues of contamination, and studies of Zn-solubilizing microbes are few and far between. However, two studies offer some valuable insights into the efficacy of biofertilizers. A study of the effects of historical sewage sludge application on microbial communities in grassland and arable soils showed that altered microbial community structure was associated with elevated soil Zn concentrations, even 11 years after the final sewage application [85]. Hence, substantial past Zn fertilizer additions to arable soils could affect the effectiveness of biofertilizers, a potential issue that warrants further investigation. A study in desert soils identified two bacterial strains in the roots of chickpea (*Cicer arietinum* L.) plants, which were capable of significantly increasing plant Zn uptake [86]. Bacteria sourced from such a harsh environment could be used to create biofertilizers with greater resistance to extreme temperatures and drought. Hence, the bioprospecting of semi-natural ecosystems may yield further Zn-solubilizing species to create biofertilizers for arable food production under challenging environmental conditions [86].

4. Boron (B) and Molybdenum (Mo)

The co-limitation by boron (B) and molybdenum (Mo) is common in leguminous crops [87–89]. Food crop deficiencies in B and Mo occur worldwide, but leguminous (e.g., beans, peas, lentils) and other dicotyledon crops (e.g., sunflower, oilseed rape) are typically more prone to B and Mo deficiency than cereal crops, whereas fruiting tree and vine species are susceptible to B deficiency alone [9,89,90]. Boron deficiency can disrupt crop growth, reduce crop yield, elongate roots, and reduce root nodulation in legumes [9,89,91]. However, the biological function of B in humans is yet to be established, and B deficiency is rarely (if ever) recorded [92]. Molybdenum deficiency in crops also disrupts growth and reduces yield, affecting N fixation and the production of multiple enzymes, including those that facilitate P uptake [9,93,94]. The deficiency of Mo in humans can occur and is linked to oral, throat and gastric cancers; however, its prevalence is extremely low [95,96]. Therefore, in the context of food production, B and Mo deficiencies are a more significant issue for crop growth and yield than for human nutrition.

Soil pH is the primary factor controlling the availability of B and Mo; Mo availability increases with pH, whereas B availability tends to increase up to pH 7, and then declines [30,97]. Under acidic conditions, B is highly mobile and prone to leaching, but under alkaline conditions it forms insoluble complexes with SOM and (oxy-)hydroxides or becomes fixed onto clay fractions [9,25,30]. Conversely, Mo is readily available in a pH range of 7.5–9 but becomes fixed to SOM and Fe hydroxides under acidic conditions [9,25,98]. Therefore, B deficiency is typically observed in calcareous or heavily limed soils and sandy soils prone to leaching, whereas Mo deficiency is common in acidic and heavily leached soils. These differences in the availability of B and Mo create issues for alleviating the deficiencies of one nutrient in arable soils, whilst controlling for the toxicity of the other. However, low B and Mo availability in arable soils is a more common and widespread issue than toxicity [9,89,98]. Since B and Mo availability is primarily controlled by soil pH and leaching, arable management practices affecting these two soil conditions, such as liming, irrigation and tillage, can affect their availability [9,25,98–100]. Low SOM content in conventional arable systems reduces the availability of both nutrients, and thus the addition of organic matter can increase both B and Mo availability via the formation of soluble complexes [25,99,101].

4.1. Microbial B and Mo Requirements

Soil microbes facilitate the release of B and Mo through organic matter degradation and mobilization from insoluble complexes [102]. Bacteria and fungi require B and Mo to produce numerous enzymes, and these micronutrients also play very specific roles in the symbiotic relationship between leguminous plant species and rhizobial bacteria [89,102], in which the rhizobia fix and convert atmospheric N to ammonia [91,103,104]. Rhizobia infect the roots of legumes by suppressing the plant's pathogenic defences using polysaccharides attached to their cell surface. In B-deficient rhizobia, polysaccharide synthesis is reduced by approximately 65–80%, which hinders infection and limits the successful establishment of the symbiotic relationship [91,105]. Furthermore, as Mo is an essential component of nitrogenase, the N-fixing enzyme produced by rhizobia, Mo-limited bacteria fail to supply N to the plant [95,98,106]. Hence, microbial deficiency in either of these micronutrients can substantially reduce leguminous crop growth and yield [89,98].

4.2. Management of B and Mo in Arable Soils

The application of B and Mo as inorganic fertilizers was used to correct deficiencies in arable soils and food crops worldwide [9]. Inorganic B can be applied by seed priming, or directly to soils or foliage. Seed priming is cost-effective but less successful for fortification, whereas foliar application corrects deficiency in crops but not in soil microbes [9,100,107]. Soil B application is very effective for increasing crop yield, fortifying grain, and increasing microbial activity and abundance under optimal soil conditions, but the quantities and chemical form of the applied B must be carefully monitored to prevent toxicity [100,102,108].

Additionally, in calcareous or heavily limed soils, most of the applied inorganic B is likely to rapidly become unavailable, rendering this method both costly and ineffective [100,101]. Inorganic Mo fertilizer is often applied either in combination with, or as an accidental contaminant of, other fertilizers, but can also be applied separately as a foliar spray [9,109]. Foliar Mo spray can fortify leguminous crops and enable symbiotic translocation of Mo to the rhizobia colonizing their roots, increasing root nodulation by more than 50% [98]. However, in acidic soils, foliar application is much less effective and does not provide Mo for free-living microbes in the soil, which may be important for non-leguminous crops [98,110].

Low B and Mo availability in soils might be best alleviated by combining inorganic fertilizers with biofertilizers. For example, the co-application of inorganic B fertilizer and biofertilizers produced the greatest increase in broccoli growth, yield, and weight [111]. The joint application of inorganic Mo fertilizer and biofertilizers can boost soil microbial activity, increase yield, and quadruple root nodulation in leguminous crops [88,112,113]. Thus, the co-application of biofertilizers with inorganic fertilizers offers a promising way to combat B and Mo deficiencies in both food crops and soil microbial communities. However, few (if any) microbial taxa were found to be capable of increasing B and Mo availability or uptake in arable soils without the co-addition of inorganic fertilizers.

4.3. B and Mo in Natural Ecosystems

Given the clear requirements of soil microorganisms for both micronutrients, bioprospecting in natural ecosystems could yield microbial taxa suitable for application as biofertilizers [109,114,115]. Studies relating to the microbial usage and mobilization of B in natural ecosystems are limited and have yet to yield B-mobilizing microbial taxa [57,58]. Research into Mo-mobilizing taxa in natural ecosystems is also limited, but it is possible that free-living, N-fixing microbes in forest soils are capable of releasing chelating agents to acquire Mo for nitrogenase production [116], and there is evidence that *Azotobacter vinelandii* produces a Mo-chelating siderophore or 'molybdophore' [117]. Hence, there is still much work to be carried out to identify and characterize the microbial acquisition of B and Mo in natural ecosystems, but such research could yield microbial strains suitable for use as biofertilizers.

5. Manganese (Mn)

Manganese (Mn) is essential for most living organisms and is required by all plants for photosynthesis, chloroplast breakdown and synthesis, and enzyme structure and function [25,118,119]. In humans, Mn is needed for reproduction, carbohydrate and lipid metabolism, and neurological functioning [120]. Crops grown for food production are the primary source of Mn for humans, providing over 50% of our dietary intake [118,121]. Although Mn deficiency has not yet been observed in the human population [120], it occurs in c. 10% of arable soils worldwide; global fruit, cereal, and certain vegetable crops are all prone to Mn deficiency, which causes interveinal leaf chlorosis and necrotic spotting, reduced tillering, the inhibition of root growth, stunted plant growth and suboptimal nutrient assimilation [9,25,118,121]. As such, correcting Mn deficiency is vital for maintaining crop yields, food production and human Mn intake [118,121].

Manganese is the fifth most abundant metal and twelfth most abundant element in soils, and thus the total concentrations are often sufficient to meet plant requirements [120]. However, Mn availability decreases at soil pH > 7.5 as it adsorbs strongly into various (hydr-)oxides, clay fractions, organic compounds, and calcium carbonate [25]. Consequently, Mn deficiency typically occurs in arable crops grown in calcareous or heavily limed soils [9,25,118,121]. However, even under alkaline conditions, Mn-containing organo–mineral and anionic complexes can remain relatively soluble and contribute to Mn availability for plants and microbes [25]. In soils with pH < 5.5, Mn adsorption is greatly reduced and its availability to plants and microbes increases [25,119]. However, soil aeration also strongly influences Mn availability by affecting the microbial oxidation or reduction of

Mn. Under aerobic conditions, the oxidation of Mn(II) to Mn(III), and then to Mn(IV) by bacteria and fungi, reduces its availability; under anaerobic conditions, Mn is reduced, which increases its availability [25,122]. Hence, arable management practices that alter soil pH or aeration (e.g., reduced organic matter application, tillage, and irrigation) can cause substantial losses of Mn by leaching [25]. Finally, the application of herbicides, such as glyphosate, can inhibit crop acquisition of Mn [9,25,118,121]. Hence, arable management can substantially influence Mn availability by altering abiotic and biotic soil factors.

When soil Mn availability becomes limiting, crop plants employ alternative measures to acquire sufficient Mn for their biological functions [123,124]. Plants can acquire Mn in much the same way as Fe, either via acidification of the rhizosphere (strategy I) or the release of phytosiderophores (strategy II) [123,124]. However, Mn mobilization by organic acids released into the rhizosphere is relatively low [123] and phytosiderophore affinity for Mn is weak; therefore, other micronutrients are more likely to be chelated in place of Mn [124]. Consequently, the plant acquisition of Mn is heavily dependent on soil pH, aeration, and microbial Mn reduction, and agricultural intervention is required to fortify crops with Mn in arable soils where these factors are unfavourable [25,71,123].

5.1. *Management of Mn in Arable Soils*

The use of inorganic fertilizers has become common practice for correcting Mn deficiency in arable crops worldwide [9,118,121]. The soil application of inorganic Mn is perhaps most prevalent due its low cost and ease of implementation, but foliar application is far more effective for supporting crop growth, increasing yield, and fortifying grains with Mn, although multiple applications are required [9,121]. Seed coating (with Mn) and osmopriming (seed soaking in Mn solution) were trialled in rice and wheat crops, with coating being better for crop growth and osmopriming boosting yields close to or greater than foliar applications [118,121]. However, the effectiveness of all Mn application methods varies depending on the target crop [118,121], and not all methods are affordable for farmers on lower incomes [9]. Furthermore, aside from foliar application, the uptake of inorganic Mn fertilizers by crops can still be greatly limited when glyphosate is applied [9]. The inclusion of leguminous crops, legume intercropping, or legumes in herbal ley rotations can increase Mn availability for subsequent arable crops by reducing soil pH; however, these practices have fallen out of favour in many countries [1,125,126]. Overall, the widespread application of inorganic Mn fertilizers is often financially unsustainable and frequently ineffective. However, inorganic Mn fertilizers may be unnecessary where total soil Mn concentrations are high but inaccessible to plants, since biofertilizers could substantially increase plant access to Mn.

Biofertilizers capable of fortifying crops with Mn can be split into two groups: Mn-mobilizing fungal taxa and Mn-reducing rhizobacteria. Crop inoculation with AMF or fungal taxa applied as biopesticides is thought to aid Mn availability and acquisition by acidifying the rhizosphere and promoting numerous plant growth traits [127,128], although evidence for its effectiveness is mixed. Inoculation with fungal taxa alone does not significantly increase plant Mn acquisition [127], and the co-inoculation of AMF and Mn-reducing rhizobacteria can even reduce Mn uptake, possibly because AMF suppress Mn-reducing rhizobacterial populations [128,129]. However, the inoculation with Mn-reducing rhizobacteria alone can greatly improve Mn availability and acquisition by crops, and aid in the control of rhizosphere phytopathogens such as take-all fungal species, *Rhizoctonia* fungi and *Fusarium* wilt fungi [6,71,128–131]. It should be noted, however, that Mn-reducing rhizobacterial strains were primarily tested at soil pH < 7.5 [6,128,129,131], and few studies tested their effectiveness under alkaline conditions [132]. Therefore, further testing should focus on assessing the efficacy of rhizobacteria for improving Mn acquisition in alkaline soils, as well as the use of Mn-mobilizing AMF. Nonetheless, biofertilizers comprising Mn-reducing rhizobacteria are highly promising for combatting Mn deficiency in food crops.

5.2. Mn in Natural Ecosystems

The utilization of Mn by fungi and bacteria in forest ecosystems provides some useful insights into the potential issues surrounding the use of microbes for crop biofortification [133–136]. Manganese plays a key role in litter decomposition by fungi, which typically use Mn for two key functions: as a structural component for lignin-degrading enzymes, such as Mn-peroxidase, and as an electron receiver/donator during the decomposition process [133,137]. Initially, fungi acquire either free Mn(II) from easily-degradable fractions of leaf litter or source Mn from the surrounding soil to synthesize enzymes, but if Mn concentrations are insufficient, the fungi compete with plants to acquire Mn [133]. Furthermore, during the later stages of litter degradation, Mn is cycled enzymatically between readily available Mn(II) and stabilized Mn(III) to oxidize, and thus degrade, lignin [134–137]. The effect that this redox cycling has on the Mn availability for plants is unclear, but since readily available Mn(II) is thought to be the predominant form of Mn in fresh leaf litter, its (albeit temporary) conversion to unavailable Mn(III) could limit Mn availability to plants [137]. Finally, once leaf litter degradation is complete, Mn accumulates in the form of Mn(III), which, under aerobic conditions, is oxidized to Mn(IV), further limiting its availability to plants [136,137]. Although research into Mn use by AMF in natural ecosystems is limited, there are likely to be some similarities with Mn use by fungal decomposers [138]. Overall, if AMF Mn cycling in agroecosystems is similar to the fungal cycling of Mn during litter decomposition, it might explain why the application of AMF as biofertilizers can reduce Mn availability to food crops [129].

It was recently discovered that certain strains of bacteria in soils are also capable of oxidizing Mn(II) to yield energy and to protect themselves from harsh environmental conditions via the formation of Mn(III/IV) oxide casings [122,137]. However, most research into Mn-reducing strains of rhizobacteria was conducted in agroecosystems [128,129,132]. Therefore, bioprospecting in the soils of semi-natural ecosystems could yield more Mn-reducing rhizobacteria suitable for application as biofertilizers.

6. Cobalt (Co)

Cobalt (Co) is considered essential for livestock and humans [9,25,139], and although Co is not essential to plants, it is widely regarded as beneficial [140]. In plants, Co is involved in multiple processes, including stem growth, leaf expansion, and bud development [140]. Cobalt is also required for the activation of several enzymes, including those involved in N-fixation in legumes, and is thus particularly beneficial to leguminous crops [9,139,141–143]. Importantly, bacteria and archaea are the only organisms capable of utilizing Co to produce cobalamin, also known as vitamin B12, a vital nutrient for mammals, which causes pernicious anaemia in humans suffering from its deficiency [9,144,145]. The primary reason for poor vitamin B12 production in leguminous crops is the deficiency of Co in the soil [146].

Cobalt availability is not primarily dictated by soil pH [25,139]. Although increasing soil pH can reduce Co availability, other soil factors, such as Fe or Mn hydrous oxide abundance, SOM content, soil texture, Mn concentrations, and microbial activity, can equally affect Co availability [25,139,147,148]. Since Co is not essential for crop growth, and microbial requirements for this micronutrient are low, Co availability in soils is usually sufficient, and Co deficiency is rarely recorded outside the livestock sector [25,139,149]. However, it is important to note that soil contamination from roadside emissions, industrial processes, and sludge application (as an arable management practice) can all result in soil Co concentrations reaching toxic, and eventually biocidal, levels [25,150]. Therefore, the careful management of Co in arable soils is required to ensure microbial and crop plant uptake of these micronutrients is sufficient without becoming excessive, a function that could be fulfilled by biofertilizers.

Microbial Management of Co in Arable Soils

To date, there is no research into biofertilizers for managing Co availability in arable soils, even though there are clear links between Co availability and microbial activity. Microbes are thought to induce Co mobilization indirectly by altering soil pH and reducing Fe/Mn hydrous oxides via redox reactions, which inadvertently released precipitated Co [139,151]. Depending on whether Co availability is limiting or approaching toxic levels, microbes can directly mobilize or immobilize Co to facilitate or reduce its uptake by organisms regardless of pH, but the mechanisms that microbes employ to achieve this are still unclear [147–149,151]. The quality of SOM could also affect the microbial mobilization of Co because readily available carbon sources can increase microbial activity, and thus Co mobilization, whereas more complex organic compounds have the opposite effect [148,151]. Consequently, the application of appropriate organic fertilizers could both enhance Co availability when it becomes limiting and reduce Co mobilization when Co toxicity inhibits microbial activity or crop growth. Finally, there is also some evidence to suggest that crop plants can alter the composition of the rhizosphere community to promote taxa that express genes for Co resistance, which could mitigate Co toxicity [152]. The strong evidence for the microbial control of Co in soils suggests that novel biofertilizers could prove highly effective for managing Co availability, especially if interactions between crop type and biofertilizers are taken into account.

7. Copper (Cu)

Copper holds a unique role in arable agriculture because, unlike other micronutrients, it is applied both to correct deficiency and as a pesticide [25]. The use of Cu as a pesticide is one of the causes of widespread soil contamination with Cu [25,153,154], but here we focus on Cu deficiency and the role of biofertilizers for maintaining favourable soil Cu availability.

Copper is essential for numerous crucial biological processes. Copper deficiency is the leading nutritional deficiency in agricultural animals globally, and recent research links Cu deficiency in humans to ischemic heart disease, osteoporosis, and Alzheimer's disease [155]. In plants, Cu plays a role in reproduction, photosynthesis, and disease resistance, and Cu deficiency causes stunted crop growth, yields reductions of up to 20%, and reduces grain or fruit quality [9,25]. Crop deficiency in Cu can occur in soils with low Cu concentrations or low availability [9,25,156]. The low availability of Cu is often a result of leaching, but also occurs at neutral or alkaline pH levels and in soils with a high SOM content [9,25,154,156]. The biotic complexation of Cu with SOM and the abiotic adsorption of Cu onto SOM, (oxy-)hydroxides, clay fractions, and carbonates have the greatest impact on Cu availability because complexation and adsorption reduce the overall Cu solubility [9,25,154]. Approximately 20–40% of arable soils have a low Cu availability, which can be exacerbated by cultivating crops that are sensitive to Cu deficiency (e.g., cereals, vegetables, citrus trees), liming, the application of macronutrient fertilizers, and the increased availability of other micronutrients (e.g., Mn, Zn, Fe) [9,25,156].

Solutions for Cu Deficiency

A prevalent solution to Cu deficiency in arable soils is the application of Cu fertilizers or pesticides, repeated as and when they are needed [9]. However, since this approach is partly to blame for widespread Cu accumulation and the subsequent toxicity in arable soils, agronomists are exploring more sustainable alternatives. For example, nanoparticle fertilizers and pesticides release Cu into soils at a slow but continuous rate, requiring fewer applications and potentially reducing environmental impacts [157,158]. However, environmental conditions and arable management practices can affect the efficacy of Cu nanoparticles to the point where detrimental impacts on soil microbial communities are observed [158,159]. By contrast, there is little research into biofertilizers to address Cu deficiencies in agricultural crops. Studies addressing the microbial biofortification of multiple micronutrients identified numerous taxa capable of increasing Cu availability and biofortifying crop plants with Cu [20,127,131,160]. Since many soil microbes require Cu as an enzyme cofactor, it is plausible that such species would be efficient Cu mobilizers, although more research into their underlying mechanisms is required [79,161]. Overall, biofertilizers could provide a sustainable method for biofortifying arable food crops with Cu, but their development is currently limited by a paucity of information on microbial Cu mobilization.

8. Conclusions

Numerous environmental factors and intensive agricultural practices can contribute to low micronutrient availability in arable soils, but biofertilizers offer an effective and sustainable way of fortifying crops with micronutrients to improve growth and yield, as well as alleviating hidden hunger in the global human population (Figure 2). Inorganic fertilizers can be costly, unsustainable, and often ineffective, but microorganisms capable of increasing micronutrient acquisition by plants can be used to produce biofertilizers and fortify food crops. By synthesizing the literature, we drew three key conclusions to guide future research and the refinement of biofertilizers (Figure 2):

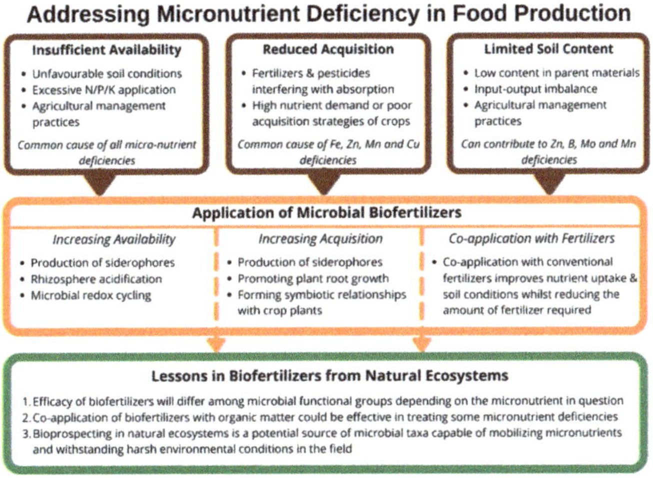

Figure 2. Conceptual diagram showing the main causes of micronutrient limitation, strategies for alleviating micronutrient deficiencies using microbial biofertilizers, and research considerations for future research into biofertilizers, based on observations from natural ecosystems.

(1) Microbial functional groups suitable for use as biofertilizers will differ depending on the micronutrient in question, and biotic interactions could undermine their effectiveness. Research to identify microorganisms for developing new micronutrient biofertilizers should not only characterize the mechanisms microbes employed to mobilize and acquire micronutrients, but also assess potential interactions among different microbial functional groups and ascertain the potential for plant–microbial nutrient competition.

(2) The co-application of biofertilizers with inorganic fertilizers proves effective for treating deficiencies of Zn, B, and Mo, and reduces the amount of inorganic fertilizers needed. Substituting conventional inorganic fertilizers with nano-fertilizers is a growing area of interest, and research into their co-application with biofertilizers could reduce the associated risks of toxicity. Although the co-application of biofertilizers with organic fertilizers is under-researched, work on microbial interactions with easily degradable SOM in natural ecosystems suggests that the co-application of biofertilizers and organic fertilizers could boost microbial activity and micronutrient availability.

(3) Finally, bioprospecting in natural ecosystems is a potential source of novel microbial taxa that are both capable of mobilizing numerous micronutrients and withstanding

harsh environmental conditions. Biofertilizers, including such organisms, could mitigate the impacts of climate changes, as well as the negative impacts of conventional arable management practices on soil conditions, which can reduce microbial diversity and abundance. Therefore, the bioprospecting for microbial taxa suitable for reproduction and application as biofertilizers should focus on selecting species that are both highly effective and tolerant of unfavourable environmental conditions.

Author Contributions: S.M.D.-T., E.J.S. conceived the study and wrote the text; S.M.D.-T. performed the literature review, led the writing, and created the diagrams. Both authors contributed to the interpretations and conclusions. All authors have read and agreed to the published version of the manuscript.

Funding: This research received no external funding.

Conflicts of Interest: The authors declare no conflict of interest.

References

1. Watson, C.A.; Oborn, I.; Edwards, A.C.; Dahlin, A.S.; Eriksson, J.; Lindstrom, B.E.M.; Linse, L.; Owens, K.; Topp, C.F.E.; Walker, R.L. Using Soil and Plant Properties and Farm Management Practices to Improve the Micronutrient Composition of Food and Feed. *J. Geochem. Explor.* **2012**, *121*, 15–24. [CrossRef]
2. Carvalho, S.M.P.; Vasconcelos, M.W. Producing More with Less: Strategies and Novel Technologies for Plant-Based Food Biofortification. *Food Res. Int.* **2013**, *54*, 961–971. [CrossRef]
3. Ku, Y.S.; Rehman, H.M.; Lam, O.M. Possible Roles of Rhizospheric and Endophytic Microbes to Provide a Safe and Affordable Means of Crop Biofortification. *Agronomy* **2019**, *9*, 764. [CrossRef]
4. Gomiero, T. Soil Degradation, Land Scarcity and Food Security: Reviewing a Complex Challenge. *Sustainability* **2016**, *8*, 281. [CrossRef]
5. Power, A.G. Ecosystem Services and Agriculture: Trade-offs and Synergies. *Philos. Trans. R. Soc. B* **2010**, *365*, 2959–2971. [CrossRef]
6. Scagliola, M.; Valentinuzzi, F.; Mimmo, T.; Cesco, S.; Crecchio, C.; Pii, Y. Bioinoculants as Promising Complement of Chemical Fertilizers for a More Sustainable Agricultural Practice. *Front. Sustain. Food Syst.* **2021**, *4*, 305. [CrossRef]
7. Brady, N. *The Nature and Properties of Soils*, 8th ed.; Macmillan Publishing Co.: New York, NY, USA, 1974.
8. Fageria, N.K.; Baligar, V.C.; Clark, R.B. Micronutrients in Crop Production. *Adv. Agron.* **2002**, *77*, 185–268.
9. Alloway, B.J. *Micronutrient Deficiencies in Global Crop Production*; Springer Science & Business Media B.V.: Berlin, Germany, 2008.
10. He, Z.L.; Yang, X.E.; Stoffella, P.J. Trace Elements in Agroecosystems and Impacts on the Environment. *J. Trace Elem. Med. Biol.* **2005**, *19*, 125–140. [CrossRef]
11. Lowe, N.M. The global challenge of hidden hunger: Perspectives from the field. *Proc. Nutr. Soc.* **2021**, *80*, 283–289. [CrossRef]
12. World Health Organisation (WHO) Guidelines on Food Fortification with Micronutrients. 2006. Available online: https://www.who.int/publications/i/item/9241594012 (accessed on 1 July 2021).
13. World Health Organisation (WHO) Vitamin and Mineral Nutrition Information System (VMNIS). 2021. Available online: https://www.who.int/teams/nutrition-and-food-safety/databases/vitamin-and-mineral-nutrition-information-system (accessed on 1 July 2021).
14. Ritchie, H.; Roser, M. Our World in Data: Micronutrient Deficiency. 2017. Available online: https://ourworldindata.org/micronutrient-deficiency (accessed on 1 July 2021).
15. Fisher, G.E.J. Micronutrients and Animal Nutrition and the Link between the Application of Micronutrients to Crops and Animal Health. *Turk. J. Agric. For.* **2008**, *32*, 221–233.
16. White, P.J.; Broadley, M.R. Biofortification of Crops with Seven Mineral Elements Often Lacking in Human Diets—Iron, Zinc, Copper, Calcium, Magnesium, Selenium and Iodine. *New Phytol.* **2009**, *182*, 49–84. [CrossRef]
17. World Health Organisation (WHO) Biofortification of Staple Crops. 2019. Available online: https://www.who.int/elena/titles/biofortification/en/ (accessed on 1 July 2021).
18. Abhilash, P.C.; Dubey, R.K.; Tripathi, V.; Gupta, V.K.; Singh, H.B. Plant Growth-Promoting Microorganisms for Environmental Sustainability. *Trends Biotechnol.* **2016**, *34*, 847–850. [CrossRef]
19. Singh, D.; Prasanna, R. Potential of Microbes in the Biofortification of Zn and Fe in Dietary Food Grains. A Review. *Agron. Sustain. Dev.* **2020**, *40*, 15. [CrossRef]
20. Behera, B.; Das, T.K.; Raj, R.; Ghosh, S.; Raza, B.; Sen, S. Microbial Consortia for Sustaining Productivity of Non-Legume Crops: Prospects and Challenges. *Agric. Res.* **2021**, *10*, 1–14. [CrossRef]
21. Mitter, E.K.; Tosi, M.; Obregon, D.; Dunfield, K.E.; Germida, J.J. Rethinking Crop Nutrition in Times of Modern Microbiology: Innovative Biofertilizer Technologies. *Front. Sustain. Food Syst.* **2021**, *5*, 29. [CrossRef]
22. Pii, Y.; Mimmo, T.; Tomasi, N.; Terzano, R.; Cesco, S.; Crecchio, C. Microbial Interactions in the Rhizosphere: Beneficial Influences of Plant Growth-Promoting Rhizobacteria on Nutrient Acquisition Process. A Review. *Biol. Fertil. Soil.* **2015**, *51*, 403–415. [CrossRef]
23. Verma, S.; Kuila, A. Bioremediation of Heavy Metals by Microbial Process. *Environ. Technol. Innov.* **2019**, *14*, 100369. [CrossRef]
24. Yadav, K.K.; Gupta, N.; Kumar, V.; Singh, J.K. Bioremediation of Heavy Metals from Contaminated Sites Using Potential Species: A Review. *Indian J. Environ. Prot.* **2017**, *37*, 65–84.

25. Kabata-Pendias, A. *Trace Elements in Soils and Plants*; Taylor & Francis Group: Baton Rouge, LA, USA, 2010.
26. Siqueira Freitas, D.; Wurr, R.; Reis dos Rodrigues, A.; De Barros Reis, F.; Soares de Carvalho, T.; Schulze, J.; Carbone Carneiro, M.A.; Guimaraes Guilherme, L.R. Hidden Nickel Deficiency? Nickel Fertilization Via Soil Improves Nitrogen Metabolism and Grain Yield in Soybean Genotypes. *Front. Plant Sci.* **2018**, *9*, 614. [CrossRef]
27. Barman, M.; Datta, S.P.; Rattan, R.K.; Meena, M.C. Critical Limits of Deficiency of Nickel in Intensively Cultivated Alluvial Soils. *J. Soil Sci. Plant Nutr.* **2019**, *20*, 284–292. [CrossRef]
28. Sayer, E.J. The Anatomy of an Excellent Review Paper. *Funct. Ecol.* **2018**, *32*, 2278–2281. [CrossRef]
29. Lemanceau, P.; Bauer, P.; Kraemer, S.; Briat, J.-F. Iron Dynamics in the Rhizosphere as a Case Study for Analysing Interactions between Soils, Plants and Microbes. *Plant Soil* **2009**, *321*, 513–535. [CrossRef]
30. Whitehead, D.C. *Nutrient Elements in Grassland: Soil-Plant-Animal Relationships*; CABI Publishing: Wallingford, UK, 2000.
31. Meena, V.S.; Meena, S.K.; Verma, J.P.; Kumar, A.; Aeron, A.; Mishra, P.K.; Bisht, J.K.; Pattanayak, A.; Naveed, M.; Dotaniya, M.L. Plant Beneficial Rhizospheric Microorganism (PBRM) Strategies to Improve Nutrients Use Efficiency: A Review. *Ecol. Eng.* **2017**, *107*, 8–32. [CrossRef]
32. Buni, A. Effects of Liming Acidic Soils on Improving Soil Properties and Yield of Haricot Bean. *J. Environ. Anal. Toxicol.* **2014**, *4*, 1–4.
33. Mimmo, T.; Del Buono, D.; Terzano, R.; Tomasi, N.; Vigani, G.; Crecchio, C.; Pinton, R.; Zocchi, G.; Cesco, S. Rhizospheric Organic Compounds in the Soil-Microorganism-Plant System: Their Role in Iron Availability. *Eur. J. Soil Sci.* **2014**, *65*, 629–642. [CrossRef]
34. Calabrese, S.; Barcellos, D.; Thompson, A.; Porporato, A. Theoretical Constrains on Fe Reduction Rates in Upland Soils as Function of Hydroclimatic Conditions. *J. Geophys. Res. Biogeosciences* **2020**, *125*, e2020JG005894. [CrossRef]
35. Abbasi, M.K.; Zafar, M.; Sultan, T. Changes in Soil Properties and Microbial Indices across Various Management Sites in the Mountain Environments of Azad Jammu and Kashmir. *Commun. Soil Sci. Plant Anal.* **2010**, *41*, 768–782. [CrossRef]
36. Deiss, L.; Kleina, G.B.; Moraes, A.; Franzluebbers, A.J.; Motta, A.C.V.; Dieckow, J.; Sandini, I.E.; Anghinoni, I.; Carvalho, P.C.F. Soil Chemical Properties Under No-Tillage as Affected by Agricultural Trophic Complexity. *Eur. J. Soil Sci.* **2019**, *71*, 1090–1105. [CrossRef]
37. Bhople, B.S.; Sharma, S. Seasonal Variation of Rhizospheric Soil Properties under Different Land Use Systems at Lower Shivalik Foothills of Punjab, India. *Agrofor. Syst.* **2020**, *94*, 1959–1976.
38. Schiere, J.B.; Ibrahim, M.N.M.; Van Keulen, H. The Role of Livestock for Sustainability in Mixed Farming: Criteria and Scenario Studies under Varying Resource Allocation. *Agric. Ecosyst. Environ.* **2002**, *2*, 139–153. [CrossRef]
39. Sharma, S.; Dhaliwal, S.S. Effects of Sewage Sludge and Rice Straw Compost on Yield, Micronutrient Availability and Soil Quality under Rice-Wheat System. *Commun. Soil Sci. Plant Anal.* **2019**, *50*, 1943–1954. [CrossRef]
40. Eker, S.; Ozturk, L.; Yazici, A.; Erenoglu, B.; Romheld, V.; Cakmak, I. Foliar-Applied Glyphosate Substantially Reduced Uptake and Transport of Iron and Manganese in Sunflower (*Helianthus annus* L.) Plants. *J. Agric. Food Chem.* **2006**, *54*, 10019–10025. [CrossRef] [PubMed]
41. Osborne, L.D.; Robson, A.D.; Bowran, D.G. The Impact of Chlorsulfuron and Diclofop-Methyl on Nutrient Uptake by Wheat. *Aust. J. Agric. Res.* **1993**, *44*, 1757–1766. [CrossRef]
42. Kraemer, S.M.; Crowley, D.E.; Kretzschmar, R. Geochemical Aspects of Phytosiderophore-Promoted Iron Acquisition by Plants. *Adv. Agron.* **2006**, *91*, 1–46.
43. Ferreira, M.J.; Silva, H.; Cunha, A. Siderophore-Producing Rhizobacteria as a Promising Tool for Empowering Plants to Cope with Iron Limitation in Saline Soils: A Review. *Pedosphere* **2019**, *29*, 409–420. [CrossRef]
44. Curie, C.; Briat, J.F. Iron Transport and Signalling in Plants. *Annu. Rev. Plant Biol.* **2003**, *54*, 183–206. [CrossRef]
45. Marschner, P.; Crowley, D.; Rengel, Z. Rhizosphere Interactions between Microorganisms and Plants Governs Iron and Phosphorus Acquisition along the Root Axis—Model and Research Methods. *Soil Biol. Biochem.* **2011**, *43*, 883–894. [CrossRef]
46. Schroder, I.; Johnson, E.; De Vries, S. Microbial Ferric Iron Reductases. *FEMS Microbiol. Rev.* **2003**, *27*, 427–447. [CrossRef]
47. Colombo, C.; Palumbo, G.; He, J.-Z.; Pinton, R.; Cesco, S. Review on Iron Availability in Soil: Interaction of Fe Minerals, Plants, and Microbes. *J. Soils Sedim.* **2014**, *14*, 538–548. [CrossRef]
48. Roriz, M.; Carvalho, S.M.P.; Castro, P.M.L.; Vasconcelos, M.W. Legume Biofortification and the Role of Plant Growth-Promoting Bacteria in a Sustainable Agricultural Era. *Agronomy* **2020**, *10*, 435. [CrossRef]
49. Saha, M.; Sarkar, S.; Sarkar, B.; Sharma, B.K.; Bhattacharjee, S.; Tribedi, P. Microbial Siderophores and Their Potential Applications: A Review. *Environ. Sci. Pollut. Res.* **2016**, *23*, 3984–3999. [CrossRef]
50. Vaid, S.K.; Kumar, A.; Sharma, A.; Srivastava, P.C.; Shukla, A.K. Role of Some Plant Growth Promotory Bacteria in Enhanced Fe Uptake of Wheat. *Commun. Soil Sci. Plant Anal.* **2017**, *48*, 756–768. [CrossRef]
51. Yehuda, Z.; Shenker, M.; Romheld, V.; Marschner, H.; Hadar, Y.; Chen, Y. The Role of Ligand Exchange in the Uptake of Iron from Microbial Siderophores by Gramineous Plants. *Plant Physiol.* **1996**, *112*, 1273–1280. [CrossRef]
52. Ali, S.; Hayat, R.; Begum, F.; Bohannan, B.J.M.; Inebert, L.; Meyer, K. Variation in Soil Physical, Chemical and Microbial Parameters Under Different Land Uses in Bagrot Valley, Gilgit, Pakistan. *J. Chem. Soc. Pak.* **2017**, *39*, 97–107.
53. Bhattacharyya, A.; Campbell, A.N.; Tfaily, M.M.; Lin, Y.; Kukkadapu, R.K.; Silver, W.L.; Nico, P.S.; Pett-Ridge, J. Redox Fluctuations Control the Coupled Cycling of Iron and Carbon in Tropical Forest Soils. *Environ. Sci. Technol.* **2018**, *52*, 14129–14139. [CrossRef]
54. Chacon, N.; Silver, W.L.; Dubinsky, E.A.; Cusack, D.F. Iron Reduction and Soil Phosphorus Solubilization in Humid Tropical Forest Soils: The Roles of Labile Carbon Pools and an Electron Shuttle Compound. *Biogeochemistry* **2006**, *78*, 67–84. [CrossRef]
55. Dubinsky, E.A.; Silver, W.L.; Firestone, M.K. Tropical Forest Soil Microbial Communities Couple Iron and Carbon Biogeochemistry. *Ecology* **2010**, *91*, 2604–2612. [CrossRef]

56. Barcellos, D.; O'Connell, C.S.; Silver, W.; Meile, C.; Thompson, A. Hot Spots and Hot Moments of Soil Moisture explain Fluctuations in Iron and Carbon Cycling in a Humid Tropical Forest Soil. *Soil Syst.* **2018**, *2*, 59. [CrossRef]
57. Adhikari, P.; Pandey, A. Bioprospecting Plant Growth Promoting Endophytic Bacteria Isolated from Himalayan Yew (*Taxus wallichiana* Zucc.). *Microbiol. Res.* **2020**, *239*, 126536. [CrossRef]
58. Dastager, S.G.; Deepa, C.K.; Pandey, A. Plant Growth Promoting Potential of Pontibacter niistensis in Cowpea (*Vigna unguiculata* (L.) Walp.). *Appl. Soil Ecol.* **2011**, *49*, 250–255. [CrossRef]
59. Balcazar, W.; Rondon, J.; Rengifo, M.; Ball, M.M.; Melfo, A.; Gomez, W.; Yarzabal, L.A. Bioprospecting Glacial Ice for Plant growth promoting Bacteria. *Microbiol. Res.* **2015**, *177*, 1–7. [CrossRef]
60. Yadav, A.N.; Sachan, S.G.; Verma, P.; Saxena, A.K. Bioprospecting of Plant Growth Promoting Psychrotrophic Bacilli from the Cold Desert of North Western Indian Himalayas. *Indian J. Exp. Biol.* **2016**, *52*, 142–150.
61. Boiteau, R.M.; Shaw, J.B.; Pasa-Tolic, L.; Koppenaal, D.W.; Jansson, J.K. Micronutrient Metal Speciation is Controlled by Competitive Organic Chelation in Grassland Soils. *Soil Biol. Biochem.* **2018**, *120*, 283–291. [CrossRef]
62. Kusel, K.; Wagner, C.; Trinkwalter, T.; Gossner, A.S.; Baumler, R.; Drake, H.L. Microbial Reduction of Fe(III) and Turnover of Acetate in Hawaiian Soils. *FEMS Microbiol. Ecol.* **2002**, *40*, 73–81. [CrossRef]
63. Singh, B.; Natesan, S.K.A.; Singh, B.K.; Usha, K. Improving Zinc Efficiency of Cereals under Zinc Deficiency. *Curr. Sci.* **2005**, *88*, 36–44.
64. Ullah, A.; Farooq, M.; Rehman, A.; Hussain, M.; Siddique, K.H.M. Zinc Nutrition in Chickpea (*Cicer arietinum*): A Review. *Crop Pasture Sci.* **2020**, *71*, 199–218. [CrossRef]
65. Nepomuceno, R.A.; Brown, C.B.; Gargarino, A.M.P.; Pedro, M.S.; Brown, M.B. Growth Enhancement of Rice (*Oryza sativa* L.) by Zinc-Solubilising Bacteria Isolated from Vesicular-Arbuscular Mycorrhizal Root Inoculant (VAMRI). *Philipp. J. Crop Sci.* **2020**, *45*, 34–40.
66. Imran, M.; Arshad, M.; Khalid, A.; Kanwal, S.; Crowley, D.E. Perspectives of Rhizosphere Microflora for Improving Zn Bioavailability and Acquisition by Higher Plants. *Int. J. Agric. Biol.* **2014**, *16*, 653–662.
67. Bala, R.; Kalia, A.; Dhaliwal, S.S. Evaluation of Efficacy of ZnO Nanoparticles as Remedial Zinc Nanofertilizer for Rice. *J. Soil Sci. Plant Nutr.* **2019**, *19*, 379–389. [CrossRef]
68. Ramesh, A.; Sharma, S.K.; Sharma, M.P.; Yadav, N.; Joshi, O.P. Inoculation of Zinc Solubilising Bacillus araybhattai Strains for Improved Growth, Mobilization and Biofortification of Zinc in Soybean and Wheat Cultivated in Vertisols of Central India. *Appl. Soil Ecol.* **2014**, *73*, 87–96. [CrossRef]
69. Strachel, R.; Zaborowska, M.; Wyszkowska, J. Deliberations on Zinc—A Trace Mineral of a Toxic Element? *J. Elem.* **2016**, *21*, 625–639.
70. Yang, X.-W.; Tian, X.-H.; Lu, X.-C.; Cao, Y.-X.; Chen, Z.-H. Impacts of Phosphorus and Zinc Levels on Phosphorus and Zinc Nutrition and phytic Acid Concentration in Wheat (*Triticum aestivum* L.). *Sci. Food Agric.* **2011**, *91*, 2322–2328. [CrossRef] [PubMed]
71. Rengel, Z. Availability of Mn, Zn and Fe in the Rhizosphere. *J. Soil Sci. Plant Nutr.* **2015**, *15*, 397–409. [CrossRef]
72. Dubey, A.N.; Chattopadhyaya, N.; Mandal, N. Variation in Soil Microbial Population and Soil Enzymatic Activities Under Zincated Nanoclay Polymer Composites (ZNCPCs), Nano-ZnO and Zn Solubilizers in Rice Rhizosphere. *Agric. Res.* **2021**, *10*, 21–31. [CrossRef]
73. Ash, M.; Yadav, J.; Yadav, J.S. Yield Attributes of Rice (*Oryza Sativa*) as Affected by Integrated Use of Zinc Oxide and Zinc Solubilizers. *Indian J. Agric. Sci.* **2020**, *90*, 2180–2184.
74. Subramanian, K.S.; Tenshia, V.; Jayalakshmi, K.; Ramachandran, V. Biochemical Changes and Zinc Fractions in Arbuscular Mycorrhizal Fungi (*Glomus intraradices*) Inoculated and Uninoculated Soils under Differential Zinc Fertilization. *Appl. Soil Ecol.* **2009**, *43*, 32–39. [CrossRef]
75. Suganya, A.; Saravanan, A.; Baskar, M.; Pandiyarajan, P.; Kavimani, R. Agronomic Biofortification of Maize (*Zea mays* L.) with Zinc by Using of Graded Levels of Zinc in Combination with Zinc Solubilising Bacteria and Arbuscular Mycorrhizal Fungi. *J. Plant Nutr.* **2020**, *44*, 988–994. [CrossRef]
76. Vaid, S.K.; Srivastava, P.C.; Sharma, A.; Rawat, D.; Mathpal, B.; Shankhadhar, S.C.; Shukla, A.K. Residual Effect of Zinc Applied to Rice on Zinc Nutrition of Succeeding Wheat Crop Inoculated with Zinc Solubilizing Microbial Consortium. *Isr. J. Plant Sci.* **2019**, *66*, 227–237. [CrossRef]
77. Liu, Y.M.; Cao, W.Q.; Chen, X.X.; Yu, B.G.; Lang, M.; Chen, X.P.; Zou, C.Q. The Responses of Soil Enzyme Activities, Microbial Biomass and Microbial Community Structure to Nine Years of Varied Zinc Application Rates. *Sci. Total Environ.* **2020**, *737*, 140245. [CrossRef]
78. Qin, N.; Sun, M.T.; Zhu, L.X.; Ge, C.M.; Wang, J. Residual Effects of Soil Zn Fertilization on Soil Characteristics, Yield and Quality of *Platycodon grandiflorum*. *Soil Sci. Plant Nutr.* **2020**, *66*, 344–351.
79. Asadishad, B.; Chahal, S.; Akbari, A.; Cianciarelli, V.; Azodi, M.; Ghoshal, S.; Tufenkji, N. Amendment of Agricultural Soil with Metal Nanoparticles: Effects of Soil Enzyme Activity and Microbial Community Composition. *Environ. Sci. Technol.* **2018**, *52*, 1908–1918. [CrossRef]
80. Garcia-Gomez, C.; Fernandez, M.D.; Garcia, S.; Obrador, A.F.; Leton, M.; Babin, M. Soil pH Effects the Toxicity of Zinc Oxide Nanoparticles to the Soil Microbial Community. *Environ. Sci. Pollut. Res.* **2018**, *25*, 28140–28152. [CrossRef]
81. Liu, Y.; Li, Y.; Pan, B.; Zhang, H.; Steinberg, C.E.W.; Qiu, H.; Vijver, M.G.; Peijnenburg, W.J.G.M. Application of Low Dosage of Copper Oxide and Zinc Oxide Nanoparticles Boosts Bacterial and Fungal Communities in Soil. *Sci. Total Environ.* **2021**, *757*, 143807. [CrossRef]

82. Xu, J.B.; Luo, X.S.; Wang, Y.L.; Feng, Y.Z. Evaluation of Zinc Oxide Nanoparticles on Lettuce (*Lactuca sativa* L) Growth and Soil Bacterial Community. *Environ. Sci. Pollut. Res.* **2018**, *25*, 6026–6035. [CrossRef]
83. Alloway, B.L. *Zinc in Soils and Crop Nutrition*, 2nd ed.; IZA: Brussels, Belgium; IFA: Paris, France, 2008.
84. Boardman, R.; McGuire, D.O. The Role of Zinc in Forestry, I. Zinc in Forest Environments, Ecosystems and Tree Nutrition. *For. Ecol. Manag.* **1990**, *37*, 167–205. [CrossRef]
85. Macdonald, C.A.; Clark, I.M.; Zhao, F.J.; Hirsch, P.R.; Singh, B.K.; McGrath, S.P. Long-Term Impacts of Zinc and Copper Enriched Sewage Sludge Additions on Bacterial, Archaeal and Fungal Communities in Arable and Grassland Soils. *Soil Biol. Biochem.* **2011**, *43*, 932–941. [CrossRef]
86. Zaheer, A.; Malik, A.; Sher, A.; Qaisrani, M.M.; Mehmood, A.; Khan, S.U.; Ashraf, M.; Mirza, Z.; Karim, S.; Rasool, M. Isolation, Characterization, and Effect of Phosphate-Zinc-Solubilising Bacterial Strains on Chickpea (*Cicer arietinum* L.) Growth. *Saudi J. Biol. Sci.* **2019**, *26*, 1061–1067. [CrossRef]
87. Liu, P.; Yang, Y.S.; Xu, G.D.; Fang, Y.H.; Yang, Y.A.; Kalin, R.M. The effect of molybdenum and boron in soil on the growth and photosynthesis of three soybean varieties. *Plant Soil Environ.* **2005**, *51*, 197–205. [CrossRef]
88. Gupta, S.C.; Sahu, S. Response of Chickpea to Micronutrients and Biofertilizers in Vertisol. *Legume Res.* **2012**, *35*, 248–251.
89. Sun, T.; Wang, Y.P.; Wang, Z.Y.; Liu, P.; Xu, G.D. The Effects of Molybdenum and Boron on the Rhizosphere Microorganisms and Soil Enzyme Activities of Soybean. *Acta Physiol. Plant.* **2013**, *35*, 763–770. [CrossRef]
90. Mandal, M.; Naik, S.K.; Das, D.K. Effect of Boron and Sulphur Interaction on Some Important Biological Indices in an Inceptisol. *J. Plant Nutr.* **2018**, *41*, 197–209.
91. Abreu, I.; Cerda, M.E.; De Nanclares, M.P.; Baena, I.; Lloret, J.; Bonilla, I.; Bolanos, L.; Reguera, M. Boron Deficiency Affects Rhizobia Cell Surface Polysaccharides Important for Suppression of Plant Defence Mechanisms During Legume Recognition and for the Development of Nitrogen-Fixing Symbiosis. *Plant Soil* **2012**, *361*, 385–395. [CrossRef]
92. National Institutes of Health Office of Dietary Supplements: Boron. Available online: https://ods.od.nih.gov/factsheets/Boron-HealthProfessional/ (accessed on 3 August 2021).
93. Rana, M.S.; Hu, C.X.; Shaaban, M.; Imran, M.; Afzal, J.; Moussa, M.G.; Elyamine, A.M.; Bhantana, P.; Saleem, M.H.; Syaifudin, M.; et al. Soil Phosphorus Transformation Characteristics in Response to Molybdenum Supply in Leguminous Crops. *J. Environ. Manag.* **2020**, *268*, 110610. [CrossRef]
94. Rana, M.S.; Sun, X.C.; Imran, M.; Ali, S.; Shaaban, M.; Moussa, M.G.; Khan, Z.; Afzal, J.; Binyamin, R.; Bhantana, P.; et al. Hu CX Molybdenum-Induced Effects on Leaf Ultra-Structure Rhizosphere Phosphorus Transformation in *Triticum aestivum*, L. *Plant Physiol. Biochem.* **2020**, *153*, 20–29. [CrossRef]
95. Ge, X.X.; Vaccaro, B.J.; Thorgersen, M.P.; Poole, F.L.; Majumder, E.L.; Zane, G.M.; De Leon, K.B.; Lancaster, W.A.; Moon, J.W.; Paradis, C.J.; et al. Iron- and Aluminium- Induced Depletion of Molybdenum in Acidic Environments Impedes the Nitrogen Cycle. *Environ. Microbiol.* **2019**, *21*, 152–163. [CrossRef]
96. Rajakumari, R.; Oluwafemi, O.S.; Thomas, S.; Kalarikkal, N. Dietary Supplements Containing Vitamins and Minerals: Formulation, Optimization and Evaluation. *Powder Technol.* **2018**, *336*, 481–492. [CrossRef]
97. Hua, T.W.; Zhang, R.; Sun, H.W.; Liu, C.G. Alleviation of Boron Toxicity in Plants: Mechanisms and Approaches. *Crit. Rev. Environ. Sci. Technol.* **2020**, *51*, 2975–3015. [CrossRef]
98. Adhikari, L.; Missaoui, A.M. Nodulation Response to Molybdenum Supplementation in Alfalfa and its Correlation with Root and Shoot Growth in Low pH Soil. *J. Plant Nutr.* **2017**, *40*, 2290–2302. [CrossRef]
99. Kwiatkowski, C.A.; Harasim, E. Chemical Properties of Soil in Four-Field Crop Rotations under Organic and Conventional Farming Systems. *Agronomy* **2020**, *10*, 1045. [CrossRef]
100. Nadeem, F.; Farooq, M.; Nawaz, A.; Ahmad, R. Boron Improves Productivity and Profitability of Bread Wheat under Zero and Plough Tillage on Alkaline Calcareous Soil. *Field Crops Res.* **2019**, *239*, 1–9. [CrossRef]
101. Vera, A.; Moreno, J.L.; Siles, J.A.; Lopez-Mondejar, R.; Zhou, Y.; Li, Y.; Garcia, C.; Nicolas, E.; Bastida, F. Interactive Impacts of Boron and Organic Amendments in Plant-Soil Microbial Relationships. *J. Hazard. Mater.* **2021**, *408*, 124939. [CrossRef]
102. Bilen, S.; Bilen, M.; Bardhan, S. The Effects of Boron Management on Soil Microbial Population and Enzyme Activities. *Afr. J. Biotechnol.* **2011**, *10*, 5311–5319.
103. Mahmud, K.; Makaju, S.; Ibrahim, R.; Missaoui, A. Current Progress in Nitrogen Fixing Plants and Microbiome Research. *Plants* **2020**, *9*, 97. [CrossRef]
104. Parniske, M. Intracellular Accommodation of Microbes by Plants: A Common Developmental Program for Symbiosis and Disease? *Curr. Opin. Plant Biol.* **2000**, *3*, 320–328. [CrossRef]
105. Zehirov, G.; Georgiev, G. Relationships Between Cell Membrane Stability, Exudate Content and Infectivity of Bradyrhizobium japonicum Strain 639 to Boron Staved Soybean Plants. *Acta Physiol. Plant* **2006**, *28*, 171–179. [CrossRef]
106. Zhang, Y.; Gladyshev, V.N. Molybdoproteomes and Evolution of Molybdenum Utilisation. *J. Mol. Biol.* **2008**, *379*, 881–899. [CrossRef]
107. Kristek, S.; Resic, I.; Jovic, J.; Zmajic, K.; Lenart, L.; Kraljicak, Z.; Beslo, D.; Rasic, S. Effect of Various Rates of Boron on Yield and Quality of High-Grade Sugar Beet Varieties. *Lis. Cukrov. A Repar.* **2018**, *134*, 146–150.
108. Vera, A.; Moreno, J.L.; Garcia, C.; Morais, D.; Bastida, F. Boron in Soil: The Impacts on the Biomass, Composition and Activity of the Soil Microbial Community. *Sci. Total Environ.* **2019**, *685*, 564–573. [CrossRef]
109. Barron, A.R.; Wurzburger, N.; Bellenger, J.P.; Wright, S.J.; Kraepiel, A.M.L.; Hedin, L.O. Molybdenum Limitation of Asymbiotic Nitrogen Fixation in Tropical Forest Soils. *Nat. Geosci.* **2009**, *2*, 42–45. [CrossRef]

110. Wen, X.; Hu, C.X.; Sun, X.C.; Zhao, X.H.; Tan, Q.L.; Liu, P.J.; Xin, J.; Qin, S.Y.; Wang, P.C. Characterisation of Vegetable Nitrogen Uptake and Soil Nitrogen Transformation in Response to Continuous Molybdenum Application. *J. Plant Nutr. Soil Sci.* **2018**, *181*, 516–527. [CrossRef]
111. Franczuk, J.; Rosa, R.; Zaniewicz-Bajkowska, A.; Slonecka, D. Effects of Boron Application and Treatment with Effective Microorganisms on the Growth, Yield and some Quality Attributes of Broccoli. *J. Elem.* **2019**, *24*, 1335–1348.
112. Shcherbakova, E.N.; Shcherbakov, A.V.; Andronov, E.E.; Gonchar, L.N.; Kalenskaya, S.M.; Chebotar, V.K. Combined Pre-Seed Treatment with Microbial Inoculants and Mo Nanoparticles Changes Composition of Root Exudates and Rhizosphere Microbiome Structure of Chickpea (*Cicer arietinum* L.) Plants. *Symbiosis* **2017**, *73*, 57–59. [CrossRef]
113. Taran, N.Y.; Gonchar, O.M.; Lopatko, K.G.; Batsmanova, L.M.; Patyka, M.V.; Volkogon, M.V. The Effect of Colloidal Solution of Molybdenum Nanoparticles on the Microbial Composition in Rhizosphere of *Cicer arietinum* L. *Nanoscale Res. Lett.* **2014**, *9*, 289. [CrossRef]
114. Lehto, T.; Smolander, A.; Aphalo, P.J. Decomposition and Element Concentrations of Silver Birch Leaf Litter as Affected by Boron Status of Litter and Soil. *Plant Soil* **2010**, *329*, 195–208. [CrossRef]
115. Reed, S.C.; Cleveland, C.C.; Townsend, A.R. Relationships among Phosphorus, Molybdenum and Free-Living Nitrogen Fixation in Tropical Rainforests: Results from Observational and Experimental Analyses. *Biogeochemistry* **2013**, *114*, 135–147. [CrossRef]
116. Wichard, T.; Mishra, B.; Myneni, S.C.B.; Bellenger, J.P.; Kraepiel, A.M.L. Storage and Bioavailability of Molybdenum in Soils Increased by Organic Matter Complexation. *Nat. Geosci.* **2009**, *2*, 625–629. [CrossRef]
117. Liermann, L.J.; Guynn, R.L.; Anbar, A.; Brantley, S.L. Production of Molybdophore during Metal Targeted Dissolution of Silicates by Soil Bacteria. *Chem. Geol.* **2005**, *220*, 285–302. [CrossRef]
118. Zulfiqar, U.; Hussain, S.; Ishfaq, M.; Ali, N.; Ahmad, M.; Ihsan, F.; Sheteiwy, M.S.; Rauf, A.; Hano, C.; El-Esawi, M.A. Manganese Supply Improves Bread Wheat Productivity, Economic Returns and Grain Biofortification Under Conventional and No Tillage Systems. *Agriculture* **2021**, *11*, 142. [CrossRef]
119. Mukhopadhyay, M.J.; Sharma, A. Manganese in Cell Metabolism in Higher Plants. *Bot. Rev.* **1991**, *57*, 117–149. [CrossRef]
120. World Health Organisation: Regional Office of Europe (WHO) Chapter 6.8: Manganese. 2001. Available online: https://www.euro.who.int/__data/assets/pdf_file/0003/123078/AQG2ndEd_6_8Manganese.pdf (accessed on 11 August 2021).
121. Zulfiqar, U.; Hussain, S.; Ishfaq, M.; Ali, N.; Yasin, M.U.; Ali, M.A. Foliar Manganese Supply Enhances Crop Productivity, Net Benefits, and Grain Manganese Accumulation in Direct-Seeded and Puddled Transplanted Rice. *J. Plant Growth Regul.* **2021**, *40*, 1539–1556. [CrossRef]
122. Yang, W.H.; Zhang, Z.; Chen, H.; Liu, J.; Ali, M.; Liu, F.; Li, L. Population Structure of Manganese-Oxidising Bacteria in Stratified Soils and Properties of Manganese Oxide Aggregates Under Manganese-Complex Medium Enrichment. *PLoS ONE* **2013**, *8*, e73778.
123. Terzano, R.; Cuccovillo, G.; Gattullo, C.E.; Medici, L.; Tomasi, N.; Pinton, R.; Mimmo, T.; Cesco, S. Combined Effect of Soil Organic Acids and Flavonoids on the Mobilization of Major and Trace Elements from Soil. *Biol. Fertil. Soils* **2015**, *51*, 685–695. [CrossRef]
124. Zhang, F.S. Mobilization of Iron and Manganese by Plant-Borne and Synthetic Metal Chelators. *Plant Soil* **1993**, *155*, 111–114. [CrossRef]
125. Williams, C.H.; David, D.J. Effects of Pasture Improvement with Subterranean Clover and Superphosphate on the Availability of Trace Metals to Plants. *Aust. J. Soil Res.* **1976**, *14*, 85–93. [CrossRef]
126. Bowles, T.M.; Mooshammer, M.; Socolar, Y.; Calderon, F.; Cavigelli, M.A.; Culman, S.W.; Deen, W.; Drury, C.F.; Y Garcia, A.G.; Gaudin, A.C.M.; et al. Long-Term Evidence Shows that Crop-Rotation Diversification Increases Agricultural Resilience to Adverse Growing Conditions in North America. *One Earth* **2020**, *2*, 284–293. [CrossRef]
127. Ogut, M.; Er, F. Micronutrient Composition of Field-Grown Dry Bean and Wheat Inoculated with Azospirillum and Trichoderma. *J. Plant Nutr. Soil Sci.* **2006**, *169*, 699–703. [CrossRef]
128. Posta, K.; Marschner, H.; Romheld, V. Manganese Reduction in the Rhizosphere of Mycorrhizal and Nonmycorrhizal Maize. *Mycorrhiza* **1994**, *5*, 119–124. [CrossRef]
129. Kothari, S.K.; Marschner, H.; Romheld, V. Effect of Vesicular Arbuscular Mycorrhizal Fungus and Rhizosphere Microorganisms on Manganese Reduction in the Rhizosphere and Manganese Concentrations in Maize (*Zea-mays* L.). *New Phytol.* **1991**, *117*, 649–655. [CrossRef]
130. Piromyou, P.; Buranabanyat, B.; Tantasawat, P.; Tittabutr, P.; Boonkerd, N.; Teaumroong, N. Effect of Plant Growth Promoting Rhizobacteria (PGPR) Inoculation on Microbial Community Structure in Rhizosphere of Forage Corn Cultivated in Thailand. *Eur. J. Soil Biol.* **2011**, *47*, 44–54. [CrossRef]
131. Rana, A.; Saharan, B.; Nain, L.; Prasanna, R.; Shivay, Y.S. Enhancing Micronutrient Uptake and Yield of Wheat through Bacterial PGPR Consortia. *Soil Sci. Plant Nutr.* **2012**, *58*, 573–582. [CrossRef]
132. Marschner, P.; Ascher, J.S.; Graham, R.D. Effect of Manganese-Reducing Rhizosphere Bacteria on the Growth of Gaeumannomyces graminis var. tritici and on Manganese Uptake by Wheat (*Triticum aestivum* L.). *Biol. Fertil. Soil.* **1991**, *12*, 33–38. [CrossRef]
133. Innangi, M.; Schenk, M.K.; D'Alessandro, F.; Pinto, S.; Menta, C.; Papa, S.; Fioretto, A. Field and Microcosms Decomposition Dynamics of European Beech Leaf Litter: Influence of Climate, Plant Material and Soil with Focus on N and Mn. *Appl. Soil Ecol.* **2015**, *93*, 88–97. [CrossRef]
134. Jones, M.E.; LaCroix, R.E.; Zeigler, J.; Ying, S.C.; Nico, P.S.; Keiluweit, M. Enzymes, Manganese, or Iron? Drivers of Oxidative Organic Matter Decomposition in Soils. *Environ. Sci. Technol.* **2020**, *54*, 14114–14123. [CrossRef]

135. Sun, T.; Cui, Y.; Berg, B.; Zhang, Q.; Dong, L.; Wu, Z.; Zhang, L. A Test of Manganese Effects on Decomposition in Forest and Cropland Sites. *Soil Biol. Biochem.* **2019**, *129*, 178–183. [CrossRef]
136. Whalen, E.D.; Smith, R.G.; Grandy, S.; Frey, S.D. Manganese Limitation as a Mechanism for Reduced Decomposition in Soils under Atmospheric Nitrogen Deposition. *Soil Biol. Biochem.* **2018**, *127*, 252–263. [CrossRef]
137. Keiluweit, M.; Nico, P.; Harmon, M.E.; Mao, J.D.; Pett-Ridge, J.; Kleber, M. Long-Term Litter Decomposition Controlled by Manganese Redox Cycling. *Proc. Natl. Acad. Sci. USA* **2015**, *112*, E5253–E5260. [CrossRef]
138. Alguacil, M.D.; Torres, M.P.; Montesinos-Navarro, A.; Roldan, A. Soil Characteristics Driving Arbuscular Mycorrhizal Fungal Communities in Semiarid Mediterranean Soils. *Appl. Environ. Microbiol.* **2016**, *82*, 3348–3356. [CrossRef]
139. Collins, R.N.; Kinsela, A.S. Pedogenic Factors and Measurements of the Plant Uptake of Cobalt. *Plant Soil* **2011**, *339*, 499–512. [CrossRef]
140. Banerjee, P.; Bhattacharya, P. Investigating Cobalt in Soil-Plant-Animal-Human System: Dynamics, Impact and Management. *J. Soil Sci. Plant Nutr.* **2021**, *21*, 2339–2354. [CrossRef]
141. Ahmed, S.; Evans, H.J. Cobalt: A micronutrient Element for the Growth of Soybean Plants under Symbiotic Conditions. *Soil Sci.* **1960**, *90*, 205–210. [CrossRef]
142. Akbar, F.M.; Zafar, M.; Hamid, A.; Ahmed, M.; Khaliq, A.; Khan, M.R.; Rehman, Z.U. Interactive Effect of Cobalt and Nitrogen Growth on Growth, Nodulation, Yield and Protein Content of Field Grown Pea. *Hortic. Environ. Biotechnol.* **2013**, *54*, 465–474. [CrossRef]
143. Wilson, S.B.; Hallsworth, E.G. Studies on the Nutrition of Forage Legumes: IV The Effect of Cobalt on the Growth and Nodulated and Non-Nodulated *Trifolium Subterraneum*. *Plant Soil* **1965**, *2*, 260–279. [CrossRef]
144. Warren, M.J.; Raux, E.; Schubert, H.L.; Escalante-Semerena, J.C. The Biosynthesis of Adenosylcobalamin (Vitamin B12). *Nat. Prod. Rep.* **2002**, *19*, 390–412. [CrossRef]
145. Zhang, Y.; Rodionov, D.A.; Gelfand, M.S.; Gladyshev, V.N. Comparative Genomic Analyses of Nickel, Cobalt and Vitamin B12 Utilisation. *BMC Genom.* **2009**, *10*, 78. [CrossRef]
146. Rod, N.K.; Gudadhe, N.N.; Karmakar, N.; Mehta, P.V.; Narwade, A.V. Cobalt Chloride Enhances Crop Duration, Increases Production, and Productivity of Chickpea. *J. Plant Nutri.* **2019**, *42*, 40–57. [CrossRef]
147. Amir, H.; Pineau, R. Relationships between Extractable Ni, Co, and Other Metals and Some Microbiological Characteristics in Different Ultramafic Soils from New Caledonia. *Aust. J. Soil Res.* **2003**, *41*, 215–228. [CrossRef]
148. Amir, H.; Pineau, R. Release of Ni and Co by Microbial Activity in New Caledonian Ultramafic Soils. *Can. J. Microbiol.* **2003**, *49*, 288–293. [CrossRef]
149. Okamoto, S.; Eltis, L.D. The Biological Occurrence and Trafficking of Cobalt. *Metallomics* **2011**, *3*, 963–970. [CrossRef]
150. Gikas, P. Single and Combined Effects of Nickel (Ni(II)) and Cobalt (Co(II)) Ions on Activated Sludge and on Other Aerobic Microorganisms: A Review. *J. Hazard. Mater.* **2008**, *159*, 187–203. [CrossRef]
151. Quantin, C.; Becquer, T.; Rouiller, J.H.; Berthelin, J. Oxide Weathering and Trace Metal Release by bacterial Reduction in a New Caledonia Ferralsol. *Biogeochemistry* **2001**, *53*, 323–340. [CrossRef]
152. Chaudhari, D.; Rangappa, K.; Das, A.; Layek, J.; Basavaraj, S.; Kandpal, B.K.; Shouche, Y.; Rahi, P. Pea (*Pisum Sativuml.*) Plant Shapes its Rhizosphere Microbiome for Nutrient Uptake and Stress Amelioration in Acidic Soils of the North-East Region of India. *Front. Microbiol.* **2020**, *11*, 968. [CrossRef] [PubMed]
153. Inaba, S.; Takenaka, C. Changes in Chemical Species of Copper Added to Brown Forest Soil in Japan. *Water Air Soil Pollut.* **2005**, *162*, 285–293. [CrossRef]
154. Schoffer, J.T.; Sauve, S.; Neaman, A.; Ginocchio, R. Role of Leaf Litter on the Incorporation of Copper-Containing Pesticides into Soils under Fruit Production: A Review. *J. Soil Sci. Plant Nutr.* **2020**, *10*, 990–1000. [CrossRef]
155. Klevay, L.M. Is the Western Diet Adequate in Copper? *J. Trace Elem. Med. Biol.* **2011**, *25*, 204–212. [CrossRef]
156. Jenkins, S.N.; Murphy, D.V.; Waite, I.S.; Rushton, S.P.; O'Donnell, A.G. Ancient Landscapes and the Relationship with Microbial Nitrification. *Sci. Rep.* **2016**, *6*, 30733. [CrossRef]
157. Mishra, S.; Keswani, C.; Abhilash, P.C.; Fraceto, L.F.; Singh, H.B. Integrated Approach of Agri-Nanotechnology: Challenges and Future Trends. *Front. Plant Sci.* **2017**, *8*, 471. [CrossRef]
158. Simonin, M.; Colman, B.P.; Tang, W.Y.; Judy, J.D.; Andersono, S.M.; Bergemanno, C.M.; Rocca, J.D.; Unrine, J.M.; Cassar, N.; Bernhardt, E.S. Plant and Microbial Response to Repeated $Cu(OH)_2$ Nanopesticide Exposures Under Different Fertilization Levels in an Agro-Ecosystem. *Front. Microbiol.* **2018**, *9*, 1769. [CrossRef]
159. Zhao, S.Y.; Su, X.X.; Wang, Y.Y.; Yang, X.Y.; Bi, M.; He, Q.; Chen, Y. Copper Oxide Nanoparticles Inhibited Denitrifying Enzymes and Electron Transport System Activities to Influence Soil Denitrification and N_2O Emission. *Chemosphere* **2020**, *245*, 125394. [CrossRef]
160. Adak, A.; Prasanna, R.; Babu, S.; Bidyarani, N.; Verma, S.; Pal, M.; Shivay, Y.S.; Nain, L. Micronutrient Enrichment Mediated by Plant-Microbe Interactions and Rice Cultivation Practices. *J. Plant Nutr.* **2016**, *39*, 1216–1232. [CrossRef]
161. Keiblinger, K.M.; Schneider, M.; Gorfer, M.; Paumann, M.; Deltedesco, E.; Berger, H.; Jochlinger, L.; Mentler, A.; Zechmeister-Boltenstern, S.; Soja, G.; et al. Assessment of Cu Applications in Two Contrasting Soil-Effects on Soil Microbial Activity and Fungal Community Structure. *Ecotoxicology* **2018**, *27*, 217–233. [CrossRef]

soil systems

Article

Nitrogen Budget and Statistical Entropy Analysis of the Tiber River Catchment, a Highly Anthropized Environment

Alessandra De Marco [1], Maria Francesca Fornasier [2], Augusto Screpanti [1], Danilo Lombardi [3] and Marcello Vitale [3,*]

[1] Department for Sustainability, Italian National Agency for New Technologies, Energy and Sustainable Economic Development—Casaccia, Via Anguillarese, S. Maria di Galeria, 301-00123 Rome, Italy; alessandra.demarco@enea.it (A.D.M.); augusto.screpanti@enea.it (A.S.)
[2] Department for Environmental Assessment, Controls and Sustainability, Institute for Environmental Protection and Research (ISPRA), Via Vitaliano Brancati, 48-00144 Rome, Italy; mariafrancesca.fornasier@isprambiente.it
[3] Department of Environmental Biology, Sapienza University of Rome, Piazzale Aldo Moro, 5-00185 Rome, Italy; danilo.lombardi@uniroma1.it
* Correspondence: marcello.vitale@uniroma1.it; Tel.: +39-06-4991-2901

Abstract: Modern farming causes a decline in the recycling of the soil's inorganic matter due to losses by leaching, runoff, or infiltration into the groundwater. The Soil System Budget approach was applied to evaluate the net N budget at the catchment and sub-catchment levels of the Tiber River (central Italy) in order to establish the causes for different N budgets among the sub-catchments. Statistical Entropy Analysis (SEA) was used to evaluate the N efficiency of the Tiber River and its sub-catchments, providing information on the dispersion of different N forms in the environment. The total N inputs exceeded the total outputs, showing a low N retention (15.8%) at the catchment level, although some sub-catchments showed higher N retention values. The Utilized Agricultural Area was important in the determination of the N balance, as it was linked to zoo- and agricultural activities, although the Random Forest analysis showed that the importance ranking changed with the land use. The low N retention of the Tiber catchment was due to the soil characteristics (Cambisols and Leptosols), loads from atmospheric deposition, biological fixation, and the livestock industry. The SEA simulations showed a reduction of the N released into the atmosphere and groundwater compartments from 34% to 6% through a reduction of the N loads by 50%.

Keywords: agroecosystems; land use; nitrogen uptake efficiency; soil typologies; statistical entropy analysis

Citation: De Marco, A.; Fornasier, M.F.; Screpanti, A.; Lombardi, D.; Vitale, M. Nitrogen Budget and Statistical Entropy Analysis of the Tiber River Catchment, a Highly Anthropized Environment. *Soil Syst.* **2022**, *6*, 17. https://doi.org/10.3390/soilsystems6010017

Academic Editors: Antonella Lavini and Mohamed Houssemeddine Sellami

Received: 23 November 2021
Accepted: 20 January 2022
Published: 2 February 2022

1. Introduction

Freshwater is essential for agriculture, industry, human existence, and energy production [1,2], but it is also a limited resource on Earth [3]. Water quality is affected by human activities such as industrialization, urbanization, tourism, and garbage production, and by natural events such as rainfall, erosion, and climate change [4–6]. The reduction of inland water's availability as a resource constitutes one of the most important environmental problems of the last century [7]. Surface water and wastewater discharges are the main factors causing an increase of the inorganic nutrients in the rivers, lakes, and seas, inducing the eutrophication phenomenon and water pollution [2]. It is known that nitrogen (N) pollution is a global problem [8,9]. The N and phosphorus (P) concentrations play a key role in the characterization of the ecological status of water systems. As they are essential for the biogeochemical cycle, these elements can enter the waters through anthropogenic activities such as domestic sewage, industrial, and "unknown" source spills [10,11]. Excessive N and P loads are often considered to be the major cause of eutrophication, which is one of the most serious environmental threats to aquatic ecosystems. Because of eutrophication and algal blooms, many environmental problems occur, such as the reduction of oxygen in the water, which causes the death of aquatic organisms, taste and bad smell problems, and biodiversity loss [12–14].

Over the last 50 years, nitrogen cycling in watersheds has been heavily exploited by urbanization, and agriculture and animal farming have undergone major alterations because of multiple interplaying factors [15]. Sewage treatment plants, manure production and spreading, the excessive use of industrially fixed N-based fertilisers, fixation by crops, and atmospheric deposition have resulted in punctual and diffuse releases of reactive N species into the environment, greatly exceeding the crop uptake and other N-removal processes in both the soil and aquatic compartments [16–19].

Concurrently, intensive agricultural practices have simplified the landscape and removed the natural buffers such as vegetated riparian areas and wetlands [20]. The absence of these elements enhanced the lateral and vertical movements of nitrogen, and made the ground surface and groundwater more prone to N contamination [21]. This risk is augmented by the high use of water for irrigation, and by traditional practices based on soil flooding over permeable areas, thus enhancing the N losses through runoff and leaching [17,22]. Furthermore, high infiltration rates decrease the groundwater residence time, altering the rates of biogeochemical reactions [23]. High N concentrations in the surface waters may saturate both microbial processes and uptake by primary producers, making the N control by natural processes less effective [24,25]. This increased N loading has, therefore, a suite of negative consequences, including demonstrated health effects and contributions to global warming [26,27]. The recent literature reports numerous studies on N dynamics at the catchment scale [28,29], investigations of nitrate ion (NO_3^-) origin in surface watercourses [30–33], and evidence of the increase of the reactive nitrogen concentration in river basins [18].

Agricultural practices can have low nitrogen-use-efficiency (NUE), especially under increasing N inputs. As a result, losses of reactive N to the environment have increased greatly, including the nitrate pollution of watercourses and emissions of both ammonia and nitrous oxide into the atmosphere, with impacts on biodiversity and climate change [34]. In response to population growth and the associated activities, riverine nutrient loads have increased, leading to a deterioration in the ecological state of rivers, lakes, reservoirs, and marine areas in many regions of the world [35–37]. Consequently, great attention is paid to the surface–groundwater interaction [38,39], and to assimilation and biological removal processes [31,40]. Iit is important to understand the dynamics underlining the nutrient transportation by water flows in rivers and lakes because they are critical for resource management, the conservation of ecological processes and functions, and the prevention of eutrophication [41].

Open questions about the fate of the N surplus in impacted watersheds concern where and how long the excess N accumulates, and the processes and transformations that it undergoes [42,43]. Because of increasing problems concerning water quality and quantity, the European Commission has launched the Water Framework Directive (WFD) [44,45]. Member States that availed themselves of an extension beyond 2015 were required to achieve all of the WFD environmental objectives by the end of the second and third management cycles, which extended from 2015 to 2021, and from 2021 to 2027, respectively [46]. The most important objectives of the WFD were to protect the status of water bodies, and for all of the bodies of the European Union to achieve good ecological status by the target dates (2021 and 2027) through an integrated approach to water management [47]. However, fifteen years after the WFD was introduced, the achievement of its objectives remains a challenge, with 47% of EU surface waters not reaching a good ecological status in 2015, which was a central objective of the EU water legislation [48]. Furthermore, despite considerable effort, the accurate modelling of N sources, transformations, and sinks at the catchment scale remains a challenge. The improved knowledge on the uptake capacity of N of the watersheds and the response of biological reserves to climate change or N deposition, which are both pressing concerns, have also provided evidence for increasing anthropogenic N emissions [49].

In this context, a national initiative was launched in Italy at the beginning of 2014 (Italian Nitrogen Network, INN), which included limnologists, ecologists, biologists, agronomists, and hydrogeologists working with nitrogen and its consequences on ecosystems and ecosystem services. The initiative consisted in the sharing of a common methodology to evaluate the nitrogen budget at the catchment level. Data collection and budget calculation were also discussed and shared in the INN. In this frame, our research is an interesting case study which was intended to obtain information about nitrogen use efficiency, and to acquire better knowledge of nitrogen sources and fates in the catchment area of the Tiber River (Central Italy). Therefore, the main aim of this work concerns the integration of the nitrogen budget model, which was quantified based on the difference between inputs and outputs, with the system ability to dilute or concentrate nitrogenous substances within the entire river catchment and in its sub-catchments. These two complementary approaches allow the estimation of the impacts of nitrogen use change scenarios at the local level and the definition of the spatial distribution of the highest risk of nitrogen accumulation both at the catchment level and local areas (at the sub-catchment level), to suggest conceivable corrective actions.

2. Materials and Methods

2.1. Study Area

The Tiber River catchment (TRC) extends for 17,169 square kilometres in central Italy and represents the second largest river basin in Italy [50]. The catchment area covers a great part of Central Italy (mainly the Umbria and Latium regions) and 335 municipalities. The Tiber flows through important cities (Perugia, Terni, Rieti, and Rome), and it is the third greatest Italian river in terms of length (409 km) and volumetric flow rate (240 m^3/s measured at the mouth), with three left tributaries (Chiascio, Nera and Aniene) and three right tributaries (Nestore, Paglia, and Treja). The database built for the Tiber River catchment was parted into sub-catchments, such as the Tiber, Cerfone, Farfa, Treja, Paglia, Nestore, Nera, Chiascio, and Aniene (Figure 1).

From a biogeographic point of view, the Tiber catchment is part of the Mediterranean district of the middle-Tyrrhenian sector. From a hydro-morphological perspective, there are three hydro-morphological sectors (Tiber, Aniene, and Nera) in the main watercourses of the catchment. The upper sector starts in the mountainous belt, where watercourses are generally characterized by steep slopes, rocky substrata, and a rapid flow. The middle sector is characterized by a pebbly–gravelly substratum and changing slopes that affect the behaviour of the water flow from turbulent shallow waters to medium laminar flowing water (usually in the deeper tracts). Finally, the lower sector has a typical fluvial regime, slow-flowing waters, and calcareous sandy–muddy substrate [51]. A Temperate climate characterizes the upper and middle sectors, whereas a Mediterranean climate is typical in the lower sector. The main litho-type of the Tiber catchment is calcareous, whereas natural vegetation is generally well preserved in the mountain and sub-mountain areas. On the other hand, a great part of the original characteristics of the riparian system has been lost in the bottom of valleys and plains, with some exceptions, due to intensive agricultural practices and urban sprawl. These land uses are the main cause of the general mineralization and eutrophication of the waters in the lower sector [52]. In Table 1, some important characteristics of the Tiber River sub-catchments are reported.

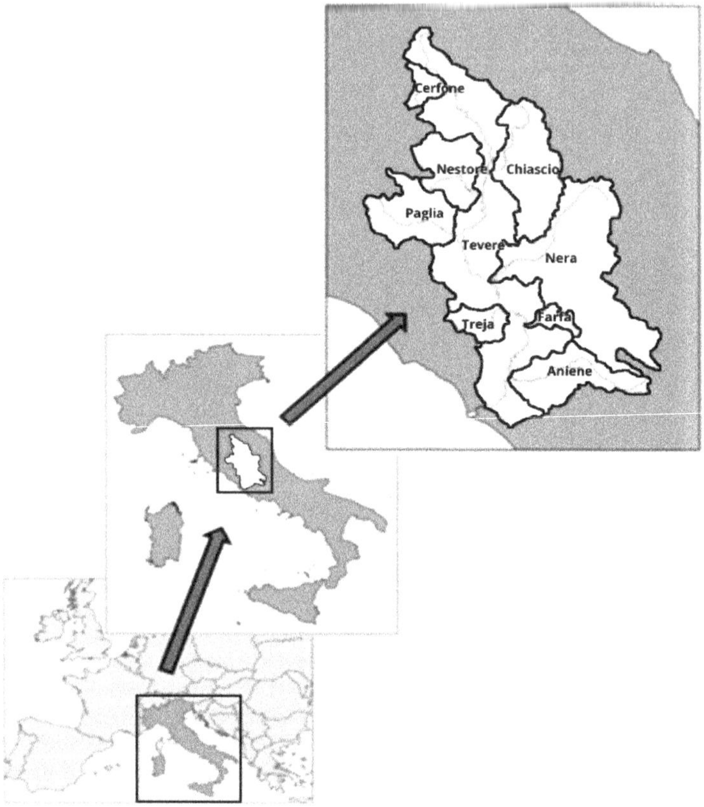

Figure 1. The Tiber River catchment area and the tributaries' sub-catchments.

Table 1. Data regarding the Tiber River catchment (TRC) and its sub-catchments. All of the data come from the database of the Italian National Institute of Statistics (ISTAT), which was produced as part of the sixth agricultural census in 2010.

Sub-Catchment	Area (ha)	Population ($\times 10^6$)	Utilised Agricultural Area (UAA) (ha)	Total N Input (t N/yr)	Total N Input/UAA t N/(ha yr)	Total N Output (t N/yr)	Total N Output/UAA t N/(ha yr)
Tiber	560,502	2.874	267,821	35,351	0.132	28,455	0.106
Nera	411,429	0.389	133,315	13,321	0.100	9520	0.071
Chiascio	163,052	0.215	76,475	10,395	0.136	7279	0.095
Paglia	129,746	0.079	50,356	5249	0.104	6179	0.123
Aniene	114,560	0.358	30,904	2459	0.080	1157	0.037
Nestore	87,137	0.086	36,087	4972	0.138	3716	0.103
Treja	40,223	0.059	18,980	2094	0.110	2196	0.116
Cerfone	29,794	0.013	11,161	1011	0.091	865	0.077
Farfa	18,011	0.017	8562	1109	0.130	432	0.050
TRC	1,554,454	4.090	633,662	75,961	0.120	59,798	0.094

2.2. Data Collection

In order to evaluate the N losses from agricultural and zoo-technical sources, the Soil System Budget approach [53] was applied to the Tiber River and its sub-catchments. The soil system budget records all of the nutrient inputs and nutrient outputs, including the nutrient gains and losses within and from the soil. The system approach also allows partitioning between the various nutrient loss pathways and the storage and/or depletion of nutrients in nutrient compartments within the system. A surplus/deficit is a measure of the net depletion (output > input) or enrichment (output < input) of the system. Most of the necessary data at the spatial resolution of municipalities (i.e., agricultural productions in terms of surfaces, typologies, and amounts of farmed animals, etc.) were downloaded from the Italian National Institute of Statistics (ISTAT) and formed part of the sixth Agricultural Census database for 2010. All of the data were converted into nitrogen units (the actual amount of nitrogen included in agricultural products—horticultural, crops, and industrial crops—and synthetic fertilizers) by employing appropriate site- and product-specific coefficients, which are available in the ISTAT database. An inventory of inputs (livestock manure, synthetic fertilisers, atmospheric deposition, biological fixation) and outputs (crop uptake, denitrification in soil, nitrogen leaching, and ammonia volatilization) was produced, and the net budget was calculated. The N budget and the inclusion of accessory information (e.g., population density, land use, slopes, the presence of wetlands, soil permeability) were derived by the ISTAT database, allowing inferences to be made regarding the system's ability to metabolize nitrogen loads, and allowing the planning of appropriate management actions.

The N contribution by livestock manure was derived from the number of cattle for each animal category (thirty-six animal categories were considered, according to typologies, age, and use; see File S1). The average weight and N production coefficients for each animal category were provided by the Rural Development Programs of the Lombardy [54] and Emilia Romagna [55] Regions, and the ISTAT database. Nitrogen inputs due to chemical fertilization were calculated considering the extension of each fertilised crop type, the annual sales of mineral fertilisers, and the N content of each type of mineral fertiliser. These data were subdivided by a percentage based on the utilized agricultural area (UAA) of each municipality falling inside the Tiber River catchment. The amount of N biological fixation was derived from the literature by considering woods, arable lands, permanent grasses, and N-fixing crop (Fabaceae) areas, and the N-fixing rates for woods, permanent grass and pastures, alfalfa, soy, and legumes [56–60].

The reduced and oxidised N depositions for the year 2010 were derived from the Greenhouse Gas Air Pollution Interactions and Synergies model—GAINS (The GAINS' features http://www.iiasa.ac.at/web/home/research/researchPrograms/air/GAINS.html (accessed on 26 January 2022)). Details about GAINS and the references are reported in File S2.

The N uptake by crops was calculated by multiplying the area of each culture by its own yield, as reported in the sixth Agricultural Census for 2010, whereas the N content of each crop was derived from the Rural Development Programmes of the Lombardy and Emilia Romagna Regions [54,55]. The N output due to ammonia volatilization was evaluated according to the literature [61,62], excluding the locally re-deposited fraction corresponding to 60% [63]. The denitrification in the soil was calculated as 10% of the sum of the N derived from livestock manure and chemical fertilisers [64]. The data regarding the water flow and N concentration at the mouth of the Tiber River were provided by the Regional Environmental Agency of the Lazio Region.

Data regarding the soil type were obtained from the European Soil Data Centre (ESDAC) for Europe at a 1 km resolution [65]. The portion of nitrogen lost due to the leaching and runoff were calculated taking into consideration the national average values [66]. The fraction of N lost per runoff was 4% of the nitrogen resulting from manure and synthetic fertilization, whereas the fraction of N lost due to leaching was 18% of the estimated surplus, corrected by the ammonia (NH_3) losses and runoff. The quantities of nitrogen derived from manure, synthetic fertilization and NH_3 losses were provided by the sixth Agricultural Census for 2010 [66]. The inputs and outputs of N were assigned to each sub-catchment, and in turn, the N retentions were calculated as the difference between the inputs and outputs both at the catchment and sub-catchment levels.

2.3. Statistical Analysis

Significant differences of N retention (N input–N output, a response variable) among the sub-catchments (discriminant factor) were quantified by one-way analysis of variance (one-way ANOVA, $p \leq 0.05$). The Neumann–Keuls post hoc test ($p \leq 0.05$) was applied for the definition of which sub-catchments were statistically different. The Random Forest (RF) analysis [67,68] was applied to determine the most important predictors that significantly affected the response variable. RF is a machine learning method that builds an ensemble of classification or regression trees [69]. The RF algorithm estimates the importance of a variable by looking at the extent to which the prediction error increased when out-of-bag (OOB) data (OOB data are those samples that are not included in the bootstrap samples, and the final prediction is the average or majority vote among all of the predictors, or any bootstrap-aggregated methods). When this process is repeated, such as when building a random forest, many bootstrap samples and OOB sets are created for the variable that was permuted while all of the others were left unchanged. The calculations were carried out tree by tree as the random forest was constructed. The final predictor importance values were computed such that the highest average was assigned a value of 1, and the importance of all of the other predictors was expressed in terms of the relative magnitude of the average values of the predictor statistics, relative to the most important predictor [70]. Note that (a) OOB samples are unique to Bagging, which is a variance reduction technique based on a collection of predictors trained on bootstrap samples, and (b) the bootstrap is any test or metric that uses random sampling with replacement, and falls under the broader class of resampling methods.

In this analysis, the most important predictors affecting the N retention were selected until their percentage difference from the most important one was 30–35%. The RF analysis has been performed both at the catchment and sub-catchment areas, taking into consideration all of the N inputs and outputs, and the environmental parameters (area, Utilised Agricultural Area (UAA), resident's number, etc.; see File S3 for a complete list of the predictors). All of the statistical analyses were carried out using the software package STATISTICA 12 (StatSoft Inc., Tulsa, OK, USA).

2.4. Statistical Entropy Analysis

Statistical Entropy Analysis (SEA) is a tool that evaluates the ability of a system to dilute or concentrate a substance (known as the system's power). Due to Shannon's concept of statistical entropy, SEA allows the calculation of the probability of the appearance of an element through its concentration in each system [71]. Therefore, SEA was used to analyse the performances of the Tiber River both at the catchment and sub-catchment levels by quantifying the dispersion of N in several environmental compartments (soil, water, and air), unlike the nitrogen use efficiency (NUE), which only operated on the total N input and N output quantities [72]. In this paper, SEA was used to analyse the N budget related to agricultural, zoo-technical, and other human-based activities occurring in the TRC and its sub-catchments. The input of the N load was due to synthetic and natural fertilizers in the form of urea ($CO(NH_2)_2$), ammonium nitrate (NH_4NO_3) and organic nitrogen (N_{org}). Some chemical species of N (NO_3^-, NH_4^+ and nitrogen entering as N_2) were derived from atmospheric depositions and ground fixation processes. Outgoing nitrogen was lost in the atmosphere in a gaseous form mainly as ammonia (NH_3) and nitrogen (N_2), and in smaller quantities as nitrous oxide (N_2O) and more generally nitrogen oxides (NO_x). The N emissions in groundwater were in the form of NO_3^-, NH_4^+ and N_{org} chemical species (Figure 2a). When all of the incoming nitrogen forms were diluted into groundwaters, the maximum value of the entropy (H_{max}) arose (Figure 2b). When, instead, all of the nitrogen was in the N_{org} form, which could be found in the products, the system had the minimum (H_{min}) or optimum (H_{opt}) entropy value (Figure 2c).

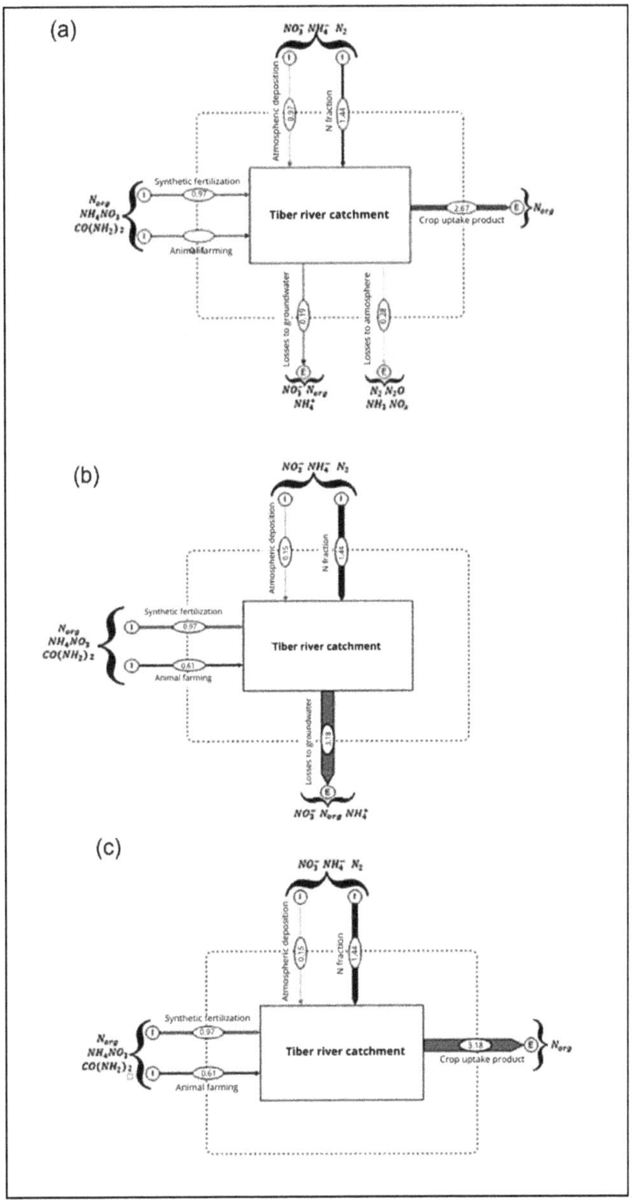

Figure 2. Tiber river catchment system's N budget. The four input fluxes were synthetic fertilization, farming animals, atmospheric deposition, and N fixation. Three fluxes were the outputs: ammonia volatilization, denitrification in soil, and crop uptake product) (**a**); worst entropy H_{max} scenario (**b**); minimal entropy scenario H_{opt} (**c**). All of the data were normalized to 1 Kg N/ha year anthropogenic input (animal farming, synthetic fertilization, and N fixation from seeds) (modified from [72]).

The basic formula for the calculation of the statistical entropy (*H*) was:

$$H(c_{iN}, m_i) = -\sum_{i=1}^{k} m_i c_{iN} \log_2(c_{iN}) \geq 0 \quad (1)$$

with

$$m_i = \frac{M_i}{\sum_{i=1}^{k} X_{iN}} \quad (2)$$

$$x_{iN} = M_i c_{iN} \quad (3)$$

where c_{iN} (expressed in mass per mass) was the N concentration in the mass flow of goods *I*, such as the amount of dry and humid depositions, fertilisers, and products; x_{iN} was the flow of N in the goods flow *i*, which was the product of M_i (equal to the material flow *i*, expressed in mass *per* time) and c_{iN}. The mass fraction of the material flow *i* (m_i) was calculated using Equation (2). *K* was the maximum number of material flows, where for the imported material flows *K* = 4 (such as synthetic fertilisation, farming animals, atmospheric deposition, and N-fixation), and for the exported material flows *K* = 3 (ammonia volatilization, denitrification in soil, and crop uptake product) (Figure 2a). The relative values of H (dimensionless) for the input and output scenario were also calculated. The variation of entropy ΔH indicated the nitrogen behaviour in the system: if the system concentrated nitrogen, then $\Delta H < 0$, otherwise $\Delta H > 0$ (the system was able to dilute nitrogen). The dilution and/or concentration concerned the natural levels of the environmental concentrations of the element or compounds examined. In general, entropy increased when the nitrogen emitted in a compartment was at a higher concentration than the natural background [71–73].

$$\Delta H = \frac{H_{out} - H_{in}}{H_{in}} \% \quad (4)$$

2.5. *Simulations of N Performance Variation in the TRC System*

After the entropy values' calculation for a realistic N efficiency of the TRC system, five scenarios were simulated, aiming to study the N performance variation in the hydrological system. In the first scenario (S1), we assumed a 50% reduction in the nitrogen load coming from synthetic fertilization. In the second (S2), the nitrogen load from the livestock manure was reduced by 50%. In the third and fourth scenarios (S3–S4) we simulated concurrent reductions in nitrogen loads from synthetic fertilization and livestock manure by 25% and 50%, respectively. The fifth scenario (S5) simulated an increase of 50% of the N inputs derived from synthetic fertilization and livestock manure.

3. Results

3.1. *Nitrogen Budget*

The Tiber catchment area is populated by 4.06 million people. Overall, the total N inputs amounted to 7.60×10^4 tons (average 227.43 ± 28.72 tons S.E., N = 334). The global N balance yielded a value of 1.62×10^4 tons, which was retained within the catchment area, and 5.98×10^4 tons came out of the hydrographic system. The main input was biological N-fixation, with a 45% contribution, followed by synthetic fertilisers (31%) and livestock (19%). The remaining 5% depended on atmospheric N deposition. The main output was given by food, horticulture, and industrial crop uptake (79.43%), with a small contribution of other budget components, as reported in Table 2.

The low nitrogen retention (12,053 t N/yr; 15.83%) of the whole catchment area was relevant. However, some sub-catchments showed higher nitrogen retentions when they were related to their own N inputs, such as Farfa (677 t N/yr; 61.06%), Aniene (1303 t N/yr; 52.97%), Chiascio (3116 t N/yr; 29.98%) and Nera (3801 t N/yr; 28.53%). The Tiber basin showed an N retention of 19.51% (6897 t N/yr), although its area is the highest of the whole catchment (Table 1). The lowest N retention values were found for the Paglia (−930 t N/yr) and Treja (−103 t N/yr) sub-catchments.

Table 2. The Soil System Budget Model for the Tiber River Catchment (TRC).

INPUT	t N/year	% of the total
Animal farming	14,649.73	19.25%
Synthetic fertilization	23,400.36	30.75%
Biological fixation	34,496.55	45.33%
Atmospheric deposition	3552.67	4.67%
Σinput	76,099.31	100%
OUTPUT	**t N/year**	**% of the total**
Food, horticulture and industrial crops uptake	50,874.46	79.43%
Nitrogen leaching	2645.05	4.13%
Ammonia volatilization	5203.13	8.12%
Denitrification in soil	3805.01	5.94%
Runoff	1522.00	2.38%
Σoutput	64,046.66	100%
Σinput − Σoutput	12,052.65	

The Tiber and Nera were characterised by high values of N loads (35,351 t N/yr and 13,321 t N/yr, respectively), representing 62% of the catchment area and 79.8% of the total residential population (Figure 3a,b). It could be argued that the N loads were likely due to an excessive amount of nutrients coming from the discharge of domestic sewage. However, as will become clearer later, the role of the nitrogen load derived from urban, peri-urban, and rural wastewater is practically nil in influencing the nitrogen retention capacity, except in the Paglia sub-catchment. In this basin an imbalance of N was evident, with a surplus of N output compared to inputs, although showing a low percentage of residents (1.94%) compared to the total resident population (Figure 3a) and a small extension of the sub-catchment area (Figure 3b).

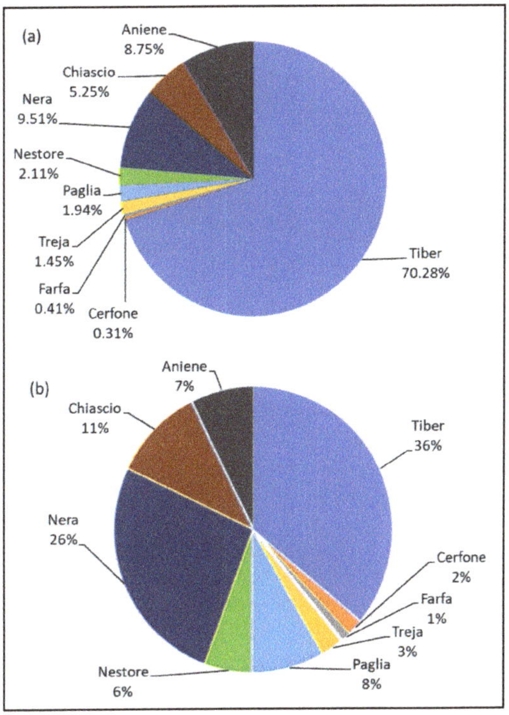

Figure 3. Relative percentages of residential people (**a**) and land extensions (**b**) for each sub-catchment.

3.2. Random Forest Analysis

In order to clarify these dynamics, it was essential to estimate the role of each predictor and its importance in affecting N retention. The Random Forest highlighted different ranks of importance for the catchment and each sub-catchment area. The important predictors affecting N retention at the catchment and sub-catchment levels were reported in Table 3, including predictors' changes to the most important one within a threshold set at up to 35%; the others included in the RF analysis were not considered because they exceeded the threshold of 35%.

Table 3. The most important variables affecting the N retention are sorted by their importance in the Tiber catchment and its sub-catchments. The Random Forest analysis was not able to fulfil the minimal threshold of the monitoring sites (equal to or above 20 sites) for the sub-catchments Nestore, Cerfone, Farfa, and Treja; therefore, these were not reported in this analysis.

Importance Rank	Tiber River Catchment	Sub-Catchments				
		Tiber	Aniene	Paglia	Nera	Chiascio
1	Total N loss by denitrification (Livestock + Agriculture)	Total N loss by denitrification (Livestock + Agriculture)	N load by atmospheric deposition	Animal husbandry load density	N load by atmospheric deposition	Total N load by biological fixation
2	N load by atmospheric deposition	Total N load by biological fixation	Total N load by biological fixation	Total N load due to livestock industry	Total N load due to livestock industry	
3	Utilised Agricultural Area	N load by atmospheric deposition	Total N loss by denitrification (Livestock + Agriculture)	Total N load by biological fixation	Utilised Agricultural Area	
4	Total N load due to livestock industry	Utilised Agricultural Area		N load by residents	Total N loss by denitrification (Livestock + Agriculture)	
5	Total N loss (sum of fractions effectively lost)	Total N load due to livestock industry				
6	Nitrogen losses from non-nitrogen fixing plants	Total N loss (sum of fractions effectively lost)				
7	Total N load by biological fixation					

The total N losses by denitrification with the contribution of livestock and agriculture, total N loads due to atmospheric deposition, utilised agricultural area, total N loads due to livestock industry, total N losses due to the sum of fractions effectively lost (i.e., fractions of N of livestock origin, N-ammonium nitrate, N-ternary, organic compounds, and N-urea ammonium sulphate totally lost by volatilisation), N losses from non-nitrogen fixing plants (i.e., N losses from cereal crops; industrial and textile crops, including aromatic plant crops; open-field horticultural crops; floral crops and ornamental plants; seedlings, alternating forage crops; agricultural tree crops, i.e., vineyards, olive groves, and orchards; and family gardens), and total N loads by biological fixation (symbiotic and non-symbiotic N fixation) were considered to be main predictors contributing to and affecting the N retention at the catchment level (Table 3). It was meaningful that the N load due to civil sewages and the number of residents living in the catchment area did not significantly affect the N retention, with this being in the intermediate and/or final positions in the importance ranking (not reported here).

Looking at Random Forest results concerning each sub-catchment (Table 3), the negative value of the nitrogen balance of the Paglia sub-catchment seemed to be due to N loads from the livestock industry and its territorial density, biological fixation, and the nitrogen contributed by the resident people (Table 3). The output N surplus in the balance of this sub-catchment likely originated from a non-equilibrium between the loads and losses; therefore, the total N loads exceed the total N losses by removal through N-fixing species (alfalfa, soy, and legumes). The biological fixation (symbiotic and non-symbiotic N fixation) may be altered in its components when nitrogen sources, such as ammonium ions and nitrates, are available in soils, inducing a strong reduction of the non-symbiotic N fixation [74] by prokaryotes such as *Azotobacter* and *Clostridium* living in the soil and waters. With the N outputs being greater than the N inputs, the significant and negative correlation between N retention and non-symbiotic N fixation (r = minus 0.379, $p = 0.046$) suggested that the N outputs exceed the N inputs when non-symbiotic N fixation was reduced by the high availability of N in the soil (that is, the total N losses were low, at 61% N retention). This high availability of N in the soil should be due to the agricultural practices that enriched the soil through synthetic N fertilization. The Paglia sub-catchment was typified by 54.8% non-irrigated arable land and agricultural land (Figure 4). Similar considerations could be made for the Treja sub-catchment (data not shown), where the N retention was negative, and correlated in a negative and significant way with non-symbiotic N fixation (r = −0.789, $p << 0.001$); its territory was represented by non-irrigated arable land (44.7%) and fruit tree and berry plantations (11.7%; Figure 4).

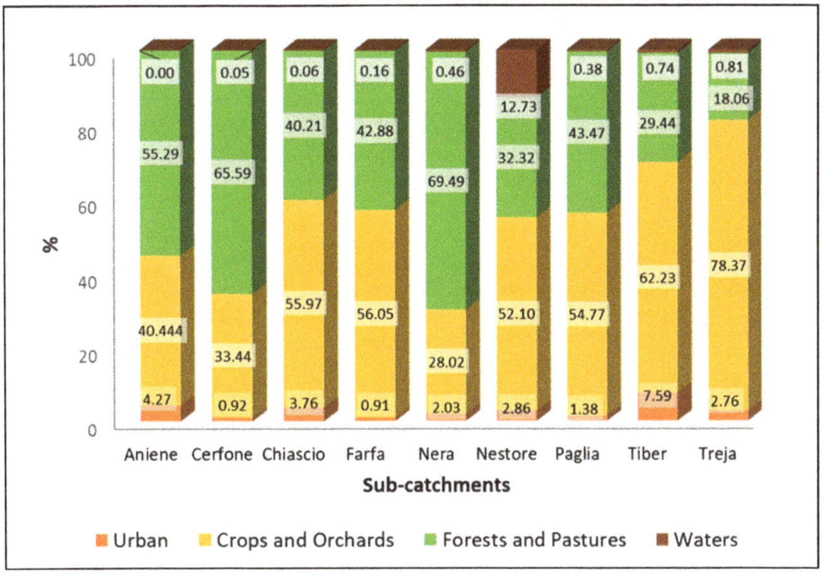

Figure 4. Repartition of the land-use categories among the sub-catchments.

The equilibrium between the loads and losses also characterised the other sub-catchments (Table 3); the total N losses by denitrification, and as the sum of fractions effectively lost (N from ammonia volatilization, i.e., losses due to the volatilization of nitrogen of zootechnical origin, N urea-ammonium sulphate, N ammonium nitrate, and N ternary + various organic), were opposed to the N load by biological fixation, atmospheric deposition, and livestock industry (e.g., the Tiber, Table 3). The total N loss by denitrification was opposed to atmospheric N deposition and the total N load due to livestock industry loads (Nera,

Table 3). Generally, it was significant that the N loads were not due to domestic sewage but agricultural and livestock practices because, looking at the land use assessment, the Tiber and Nera represented 62% of the entire catchment area, where the most typical categories were crops and orchards, forests, and pastures (Figure 4). Furthermore, similar considerations could be made for the Aniene sub-catchment (Table 3). Atmospheric N deposition could be due to the contribution of some of the sub-catchments where this variable was one of the most important ones, such as the Tiber, Nera, and Aniene sub-catchments (Table 3), which covered 68% of the whole catchment area. Because industrial areas were scarcely distributed in the territory, N deposition could arrive from surrounding areas or local industrial ensembles. De Marco et al. [75] reported a differential action of oxidised and reduced nitrogen depositions to the net primary production of Italian forests that was linked to the geographical location of the N sources, which were mainly situated in northern Italy. The Utilised Agricultural Area (UAA) was another important variable that was generally associated with the Total N load due to livestock industry, or the Total N loss (denitrification) by livestock and agricultural practices. In fact, UAA was tightly correlated with these parameters ($r = 0.854$, $p << 0.01$ and $r = 0.944$, $p << 0.01$ for the Total N load due to livestock industry and the Total N loss (denitrification) by livestock and agricultural practices, respectively). There appeared to be a balance between these two processes that prevented (or limited) the release of N compounds outside the catchment area..

3.3. Soil Type Distribution at the Catchment and Sub-Catchment Levels

Looking at the distribution of the soil types existing in the Tiber catchment, it was clear that Cambisol was the most representative type (84.9%), followed by Leptosol (Rendzina) (8.7%), Luvisol (2.6%), Regosol and Andosol (1.6 and 1.5%, respectively), and Lithosol (<1%). As reported by the World Reference Base for Soil Resources [76], Cambisol is related to mineral soils of which the formation is conditioned by their limited age, that is, soils that are only moderately developed on account of their limited paedogenic age or because of the rejuvenation of the soil material. On the other hand, Leptosols are very shallow soils over a hard rock or highly calcareous material but also deeper soils that are extremely gravelly and/or stony. Leptosols are azonal soils with an incomplete solum and/or without clearly expressed morphological features. They are particularly common in mountain regions. Leptosols correlate with the 'Lithosols' taxa of many international classification systems (USA, FAO), and with 'Lithic' subgroups of other soil groupings. In many systems, Leptosols on the calcareous rock are denoted as 'Rendzinas'. From these definitions, Cambisols and Leptosols (covering 93.6% of the Tiber catchment area) were not able to retain nitrogen in the soil, although in some sub-catchments (e.g., the Farfa) the nitrogen retention was high (61%). In this sub-catchment, the prevalent soil typology was Chromic Cambisols (65.2%), and in hilly (mainly colluvial) terrain Eutric, Calcaric, and Chromic Cambisols were planted with a variety of annual and perennial crops or were used as grazing land. These soils are medium-textured and have good structural stability, a high porosity, a good water holding capacity, and good internal drainage, thus allowing us to retain much more nitrogen than in the other sub-catchment. Note that, in the Farfa sub-catchment, crops and forests and pastures extended for 56% and 43% of the area, respectively (Figure 4).

3.4. The Relative Statistical Entropy Analysis: Actual Management Scenario

The N performances of the Tiber River Catchment (TRC) system and its sub-catchments were evaluated by relative statistical entropy (RSE) analysis and by Nitrogen Use Efficiency (NUE) [77], the values of which are shown in Table 4. The N_{in} and N_{out} values for the TRC showed a satisfactory N efficiency ($\Delta H = 33\%$), and all of the sub-catchments, excluding the Aniene and Farfa, performed close to their optimum values (Table 4).

Table 4. Input and output entropy (H_{in} and H_{out}, respectively), RSE, and NUE values for the assessment of the N performance concerning the Tiber River Catchment (TRC) system and its sub-catchments. Relative Statistical Entropy, $RSE_x = H_x/H_{max}$ (x = in, out), $\Delta H = (H_{out} - H_{in}/H_{in}) \times 100$ and $NUE = N_{out}/N_{in}$. The values of Nin and Nout for the TRC refer to those in Table 1. The values of N_{in} and N_{out} for each sub-basin are not reported here.

	H_{in}	H_{out}	H_{max}	ΔH	RSE_{in}	RSE_{out}	NUE
Aniene	0.843	7.861	10.914	833%	0.077	0.720	24%
Cerfone	1.688	2.323	9.814	46%	0.171	0.251	88%
Chiascio	1.721	5.158	13.085	200%	0.132	0.394	84%
Farfa	0.587	9.436	9.503	1508%	0.062	0.993	23%
Nera	1.967	3.455	13.416	76%	0.147	0.257	60%
Nestore	2.210	3.830	11.896	73%	0.186	0.322	96%
Paglia	1.607	1.698	12.616	6%	0.127	0.135	119%
Tiber	1.917	2.692	14.842	40%	0.129	0.181	92%
Treja	1.132	2.361	11.282	109%	0.100	0.209	65%
Tiber River Catchment	1.906	2.885	15.967	51%	0.119	0.181	84%

It is important to note that the NUE values did not give any information about the system's ability to metabolise different nitrogen forms, which, in turn, could be assessed by RSE_{out} and ΔH values. The Paglia sub-catchment showed a very high efficiency of N (ΔH = 6%), where the incoming N was mainly transformed into products (70.67%); the others (the Cerfone, Nera, Nestore, and the Tiber) showed average middle-efficiency values of N of about 59 ± 18% (Table 4). The Chiascio and Treja sub-catchments showed high values of ΔH, at 200% and 109%, respectively. Very high values of ΔH were instead found in the Aniene (833%) and Farfa (1508%) sub-catchments. The largest contributions to entropy found in the Aniene, Chiascio, Farfa and Treja sub-catchments were due to the significant nitrogen dilution in the atmosphere, surface water, and groundwater. In the Aniene sub-catchment, the major contribution was due to the nitrogen dilution in the groundwater (77%) and the atmosphere (51%) (Figure 5). In the Chiascio sub-catchment, the contribution to the high value of ΔH is related mainly to the dilution of the nitrogen load in the groundwater (105%); the same is true for the Treja sub-catchment (ΔH_groundwater = 30%). For the Farfa sub-catchment, the contribution to ΔH is due to the high N dilution in all of the environmental compartments (69% for groundwater, 31% for surface water and 48% for atmosphere) (Figure 5). Their high entropy values could be a consequence of high human activity pressures in the sub-catchments positioned in the southern areas of the TRC system, where more than 70% of the total population lives [78]. Besides this, the high contribution to entropy due to the N dilution in the groundwater—especially for the Farfa area—could be justified by the soil type, Chromic Cambisol, which is characterized by high N retention.

Although N converted into a product cannot be considered to be diluted, it contributed to the generation of entropy because the primarily nitrogen contained in the synthetic fertilizers was more concentrated than the nitrogen contained in the products. Therefore, an agricultural production system will always have an $H > 0$ [72]. The entropy related to the transformation of N_{org} in the product contributed to 63% of the total entropy of the TRC system. The remaining entropy contribution to the N losses in the groundwater, surface water and atmosphere were 14%, 9% and 20%, respectively (Figure 5).

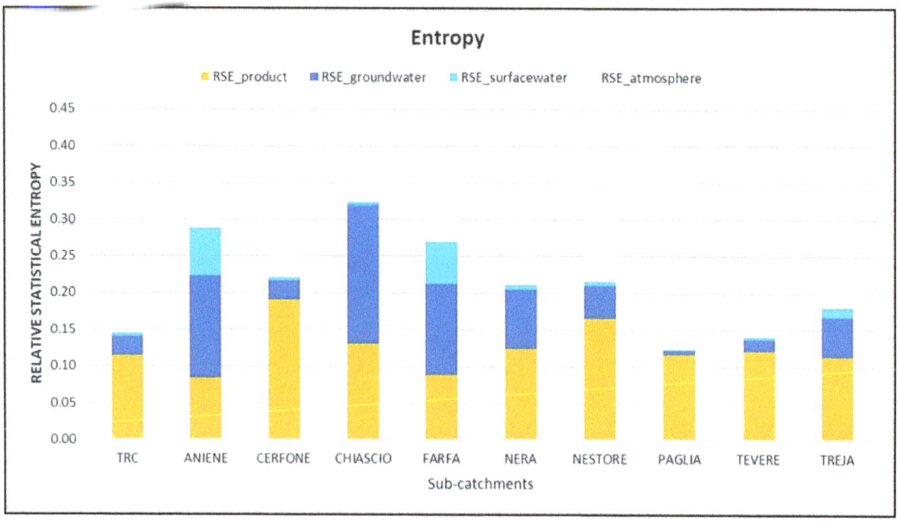

Figure 5. RSE output contributions for the TRC system and its sub-catchments. The RSE product represents nitrogen converted mainly into biomass (crop uptake, in our case). It corresponds to the system's ability to concentrate the input nitrogen into the product. RSE_groundwater, RSE_surfac water and RSE_atmosphere indicate the loss of nitrogen by the system in the various environmental matrices. The total relative statistical entropy of each sub-basin and the whole TRC is given by the sum of the single components. A high system efficiency regarding the handling of nitrogen is indicated by low RSE values. Such values indicate that most of the nitrogen is converted into a product with minimal dispersion ("dilution") in the environment.

3.5. The Relative Statistical Entropy Analysis: Different Management Scenarios

Looking at the simulations (Figure 6), in the scenarios where a reduction in the input nitrogen loads is achieved, a lower N_{org} amount in the products was obtained by about −3% for the S1, S2 and S3 scenarios, and −6% for the S4 scenario. With the 50% increase in nitrogen loading from both synthetic fertilization and livestock manure (S5 scenario), we estimated an increase of 2% in the N_{org} amount in the products. The differences between the S1 and S2 scenarios in entropy were characterised by an increase of N efficiency (ΔH) from 27% to 29%, respectively. With the reduction of chemical fertilisers (S1 scenario), the entropy contribution to the groundwater and surface water decreased by 68% and 54% compared to the current scenario, but it had a low impact on the entropy contribution to the atmosphere (−7%). The reduction by 50% of the nitrogen load (S2 scenario) involved a reduction of the relative entropy (ΔH = 29%) concerning TRC, with this being mainly due to a reduction of the N losses to groundwater, surface water and atmosphere by 38%, 37% and 80%, respectively (Figure 6). In the S3 and S4 scenarios, the reduction of nitrogen loads by acting on chemical fertilisation and livestock manure yielded a further enhancement of the N efficiency (ΔH = 15% and 9% for S3 and S4, respectively) of the overall system (Figure 6). In these scenarios, the nitrogen included was almost entirely converted into the products as N_{org}, with minimal N losses towards the groundwater and atmosphere compartments. The contribution of N_{org} to entropy in the S3 scenario was higher than that in the S4 scenario (RSE_product = 0.111 for S3 and 0.107 for S4), although S4 showed a minor dilution capacity for N into the atmosphere, groundwater, and surface water compartments (Figure 6). On the other hand, an increase in the nitrogen loads (S5 scenario) implied a reduction of the N efficiency (ΔH = 101%) in the TRC system, although we observed an increase in N_{org} (2%) in the products compared to the actual scenario (Figure 6). This could be explained by a strong entropy contribution due to the increase of N forms in the groundwater (170%) and in the surface water (103%), despite the best yield. Finally, the S4 scenario was the best one. Hence, a reduction of 50% in the N loads within the TRC would allow for a better system performance, which would be able to maximize the yields through the significant reduction of nitrogen losses in the environment.

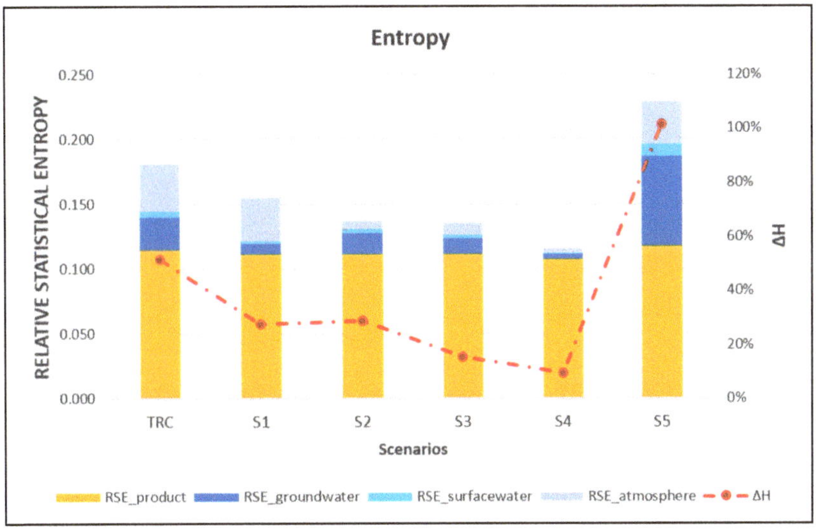

Figure 6. Simulated relative statistical entropy output contributions for the Tiber River Catchment system (RTC) with the relative efficiency values ΔH, expressed as a percentage, in different management scenarios compared to the actual TRC scenario.

4. Discussion

The Tiber catchment showed low nitrogen retention. This phenomenon could be due to several reasons, including (1) the type of soil having a low N retention capacity (prevalently Cambisols and Leptosols), and (2) nitrogen loads coming from atmospheric deposition, biological fixation, and the livestock industry. The use of conventional agricultural practices makes soils more compressed, decreasing the mineral component uptakes, such as the nitrogen uptake. An additional consequence of the conventional agricultural practices is to increase the run-off and enhance the nitrogen load in the surface- and in drainage waters. UAA is a very important variable in the determination of the N balance in the sub-catchments. UAA is linked to zoo-technical and agricultural activities, such as the Total N loads due to the livestock industry and the Total N losses (denitrification) by livestock and agricultural practices, which, in turn, control the N release out of the catchment or among the sub-catchments. Nitrogen loads due to wastewater, which are directly linked to the residents' number, did not seem to affect the nitrogen budget. The RSE analysis showed a good N efficiency for the Tiber River catchment, despite its low nitrogen retention value.

The best efficiency observed in the Paglia sub-catchment was due to its ability to concentrate almost all of the N loads into the products (yields), unlike the Aniene and Farfa sub-catchments, which instead highlighted powerful N losses in the environmental matrices. Importantly, the worst N efficiency of the Farfa area was due to the lithological nature of the soil, which made it a system with a high N retention. In order to increase the N retention capacity and NUE, agricultural coverage should be increased in the area and managed through best practices oriented to the reduction of impacts through soft-or-not-soil tillage. The reduction of the N-based chemical fertilisers would be the best practice to aim to increase the percentage of non-symbiotic—and therefore biological—N fixation, and to reduce N loads by a decrease of the livestock number, in order to further practice crop rotation in a more spread-out way, and to maximize the use of crop species which are able to increase the efficiency of nitrogen uptake. A reduction of 25% or 50% of the N loads increased the N efficiency of the Tiber River catchment system, as demonstrated by the RSE simulations. The first intercomparison exercise was carried out by comparing the values obtained for the Tiber River Catchment with those obtained for the Austrian catchments of Wulka and Ybbs [73]. We observed that the TRC system has a greater nitrogen management efficiency, with a delta surplus = 15.83% (19 kg N/ha yr), NUE = 84% and ΔH = 51%, despite the greater surface area of the catchment and UAA. The Wulka catchment showed delta surplus values of 37% (21 kg N/ha yr), and NUE and ΔH values of 67% and 180%, respectively, while the Ybbs catchment had delta surplus, NUE and ΔH values equal to 57% (79 kg N/ha yr), 43% and 335%, respectively. This intercomparison highlights the homogeneity of the values and the usefulness of the RSE analysis, which is a method to evaluate the system's efficiency and dispersion of various nitrogen compounds into the systemic environmental matrices. The recommendation reported in [73] concerning the use of SEA as an agri-environmental indicator to be integrated into the decision-making processes regarding nitrogen management strategies is here renewed.

The surplus calculation may be considered a good indicator of the environmental impacts due to agricultural practices, but its integration with the information about the type of N that is released into the environmental matrices, and in nitrogen compounds and load variations, allows a more detailed analysis of N input and output fluxes, enabling more effective management actions.

In conclusion, the important points of this study are related to the importance of the environmental matrix in the determination of the nitrogen budget at the river basin level. This matrix is mainly characterized by the different types of soil, by the mosaic of agricultural and forest coverings, and therefore, by the current land use. In this environmental complexity, the approach based on the integration of nitrogen budget calculation—based on the soil budget system, with the random forest and Statistical Entropy Analysis—provides a powerful analysis and monitoring tool for future land-use changes. It will soon be possible to carry out a historical analysis aimed at comparing the management and land-use changes, and the effects on the nitrogen budget on the Tiber River Catchment with the release of the seventh general agricultural census provided by the Italian National Institute of Statistics (ISTAT) in March 2022.

Supplementary Materials: The following supporting information can be downloaded at: https://www.mdpi.com/article/10.3390/soilsystems6010017/s1, File S1: Animal categories that are considered in the ISTAT database for calculating the N contribution by livestock manure; File S2: The GAINS model: an overview; File S3: Predictors used in the Random Forest (RF) analysis.

Author Contributions: Conceptualization, A.D.M.; methodology, M.V. and D.L.; software, D.L.; formal analysis, M.V. and D.L.; resources, A.D.M.; data curation, M.F.F. and A.S.; writing—original draft preparation, A.D.M.; writing—review and editing, M.V.; visualization, M.F.F. and A.S.; supervision, M.V. and A.D.M. All authors have read and agreed to the published version of the manuscript.

Funding: This research received no external funding.

Institutional Review Board Statement: Not applicable.

Informed Consent Statement: Not applicable.

Data Availability Statement: The data utilized in this work came from the sixth Agricultural Census for 2010 (http://dati-censimentoagricoltura.istat.it/Index.aspx), which was supplied by The Italian National Institute of Statistics.

Acknowledgments: The authors wish to thank the International Nitrogen Management System (INMS) project for its support, and all of the Italian personnel involved in the national network for nitrogen studies (Italian Nitrogen Network, INN).

Conflicts of Interest: The authors declare no conflict of interest.

References

1. Alkan, A.; Serdar, S.; Fidan, D.; Akbaş, U.; Zengin, B.; Kılıç, M.B. Physico-Chemical Characteristics and Nutrient Levels of the Eastern Black Sea Rivers Ali. *Turkish J. Fish. Aquat. Sci.* **2013**, *13*, 847–859.
2. Kumar, R.N.; Solanki, R.; Nirmal Kumar, J.I. An assessment of seasonal variation and water quality index of Sabarmati River and Kharicut canal at Ahmedabad, Gujarat. *Electron. J. Environ. Agric. Food Chem.* **1970**, *10*, 2771–2782.
3. Venkatesharaju, K.; Ravikumar, P.; Somashekar, R.; Prakash, K. Physico-Chemical and Bacteriological Investigation on the River Cauvery of Kollegal Stretch in Karnataka. *Kathmandu Univ. J. Sci. Eng. Technol.* **1970**, *6*, 50–59. [CrossRef]
4. Ananthan, G.; Sampathkumar, P.; Palpandi, C.; Kannan, L. Distribution of heavy metals in Vellar estuary, Southeast coast of India. *J. Ecotoxicol. Environ. Monit.* **2006**, *16*, 501–506.
5. Shin, J.Y.; Artigas, F.; Hobble, C.; Lee, Y.S. Assessment of anthropogenic influences on surface water quality in urban estuary, northern New Jersey: Multivariate approach. *Environ. Monit. Assess.* **2013**, *185*, 2777–2794. [CrossRef]
6. Valença, A.P.M.C.; Santos, P.J.P. Macrobenthic community for assessment of estuarine health in tropical areas (Northeast, Brazil): Review of macrofauna classification in ecological groups and application of AZTI Marine Biotic Index. *Mar. Pollut. Bull.* **2012**, *64*, 1809–1820. [CrossRef]
7. Tanriverdi, Ç.; Alp, A.; Demirkiran, A.R.; Üçkardeş, F. Assessment of surface water quality of the Ceyhan River basin, Turkey. *Environ. Monit. Assess.* **2010**, *167*, 175–184. [CrossRef]
8. Hadjikakou, M. Modelling nitrogen in the Yeilirmak River catchment in Northern Turkey: Impacts of future climate and environmental change and implications for nutrient management. *Sci. Total Environ.* **2011**, *409*, 2404–2418. [CrossRef]
9. Fowler, D.; Coyle, M.; Skiba, U.; Sutton, M.A.; Cape, J.N.; Reis, S.; Sheppard, L.J.; Jenkins, A.; Grizzetti, B.; Galloway, J.N.; et al. The global nitrogen cycle in the twenty-first Century. *Philos. Trans. R. Soc. B Biol. Sci.* **2013**, *368*, 20130164. [CrossRef]
10. Billen, G.; Thieu, V.; Garnier, J.; Silvestre, M. Modelling the N cascade in regional watersheds: The case study of the Seine, Somme, and Scheldt rivers. *Agric Ecosyst. Environ.* **2009**, *133*, 234–246. [CrossRef]
11. Campbell, J.L.; Hornbeck, J.W.; Mitchell, M.J.; Adams, M.B.; Castro, M.S.; Driscoll, C.T.; Kahl, J.S.; Kochenderfer, J.N.; Likens, G.E.; Lynch, J.A.; et al. Input-output budgets of inorganic nitrogen for 24 forest watersheds in the Northeastern United States: A review. *Water Air Soil Pollut.* **2004**, *151*, 373–396. [CrossRef]
12. Passy, P.; Gypens, N.; Billen, G.; Garnier, J.; Thieu, V.; Rousseau, V.; Callens, J.; Parent, J.-Y.; Lancelot, C. A model reconstruction of riverine nutrient fluxes and eutrophication in the Belgian Coastal Zone since 1984. *J. Mar. Syst.* **2013**, *128*, 106–122. [CrossRef]
13. Liu, C.; Kroeze, C.; Hoekstra, A.Y.; Gerbens-Leenes, W. Past and future trends in grey water footprints of anthropogenic nitrogen and phosphorus inputs to major world rivers. *Ecol. Indic.* **2012**, *18*, 42–49. [CrossRef]
14. Paerl, H.W. Controlling eutrophication along the freshwater-Marine continuum: Dual nutrient (N and P) reductions are essential. *Estuaries Coasts* **2009**, *32*, 593–601. [CrossRef]
15. Vitousek, P.M.; Aber, J.D.; Howarth, R.W.; Likens, G.E.; Matson, P.A.; Schindler, D.W.; Schlesinger, W.H.; Tilman, D.G. Human alteration of the global nitrogen cycle: Sources and consequences. *Ecol. Appl.* **1997**, *7*, 737–750. [CrossRef]
16. Puckett, L.J. Identifying the Major Sources of Nutrient Water Pollution: A national watershed-based analysis connects nonpoint and point sources of nitrogen and phosphorus with regional land use and other factors. *Environ. Sci. Technol.* **1995**, *29*, 408A–414A. [CrossRef]
17. Cassman, K.G.; Dobermann, A.; Walters, D.T. Agroecosystems, nitrogen-use efficiency, and nitrogen management. *Ambio* **2002**, *31*, 132–140. [CrossRef]
18. Galloway, J.N.; Townsend, A.R.; Erisman, J.W.; Bekunda, M.; Cai, Z.; Freney, J.R.; Martinelli, L.A.; Seitzinger, S.P.; Sutton, M.A. Transformation of the nitrogen cycle: Recent trends, questions, and potential solutions. *Science* **2008**, *320*, 889–892. [CrossRef]
19. Hertel, O.; Skjøth, C.A.; Reis, S.; Bleeker, A.; Harrison, R.M.; Cape, J.N.; Fowler, D.; Skiba, U.; Simpson, D.; Jickells, T.; et al. Governing processes for reactive nitrogen compounds in the European atmosphere. *Biogeosciences* **2012**, *9*, 4921–4954. [CrossRef]
20. Somma, F. *River Basin Network on Water Framework Directive and Agriculture: Practical Experiences and Knowledge Exchange in Support of the WFD Implementation (2010–2012)*; European Commission-Joint Research Centre-Institute for Environment and Sustainability, Reference JRC78538 Luxembourg; Publications Office of the European Union: Ispra, Italy, 2013.
21. Balestrini, R.; Arese, C.; Delconte, C.A.; Lotti, A.; Salerno, F. Nitrogen removal in subsurface water by narrow buffer strips in the intensive farming landscape of the Po River watershed, Italy. *Ecol. Eng.* **2011**, *37*, 148–157. [CrossRef]
22. Böhlke, J.K.; Verstraeten, I.M.; Kraemer, T.F. Effects of surface-water irrigation on sources, fluxes, and residence times of water, nitrate, and uranium in an alluvial aquifer. *Appl. Geochem.* **2007**, *22*, 152–174. [CrossRef]
23. Maher, K. The dependence of chemical weathering rates on fluid residence time. *Earth Planet. Sci. Lett.* **2010**, *294*, 101–110. [CrossRef]
24. Böhlke, J.K.; O'Connell, M.E.; Prestegaard, K.L. Groundwater stratification and delivery of nitrate to an incised stream under varying flow conditions. *J. Environ. Qual.* **2008**, *36*, 664–680. [CrossRef] [PubMed]
25. Mulholland, P.J.; Helton, A.M.; Poole, G.; Hall, R.O.; Hamilton, S.K.; Peterson, B.J.; Tank, J.L.; Ashkenas, L.R.; Cooper, L.; Dahm, C.N.; et al. Stream denitrification across biomes and its response to anthropogenic nitrate loading. *Nature* **2008**, *452*, 202–205. [CrossRef]
26. Ward, M.H.; Dekok, T.M.; Levallois, P.; Brender, J.; Gulis, G.; Nolan, B.T.; VanDerslice, J. Workgroup report: Drinking-water nitrate and health-Recent findings and research needs. *Environ. Health Perspect.* **2005**, *113*, 1607–1614. [CrossRef]

27. Davidson, E.A. The contribution of manure and fertilizer nitrogen to atmospheric nitrous oxide since 1860. *Nat. Geosci.* **2009**, *2*, 659–662. [CrossRef]
28. Rock, L.; Mayer, B. Nitrogen budget for the Oldman River Basin, southern Alberta, Canada. *Nutr. Cycl. Agroecosystems* **2006**, *75*, 147–162. [CrossRef]
29. Wollheim, W.M.; Vörösmarty, C.J.; Bouwman, A.F.; Green, P.; Harrison, J.; Linder, E.; Peterson, B.J.; Seitzinger, S.P.; Syvitski, J.P.M. Global N removal by freshwater aquatic systems using a spatially distributed, within-basin approach. *Global Biogeochem. Cycles* **2008**, *22*, 1–14. [CrossRef]
30. Neal, C.; Jarvie, H.P.; Neal, M.; Hill, L.; Wickham, H. Nitrate concentrations in river waters of the upper Thames and its tributaries. *Sci. Total Environ.* **2006**, *365*, 15–32. [CrossRef]
31. Neal, C.; Jarvie, H.P.; Love, A.; Neal, M.; Wickham, H.; Harman, S. Water quality along a river continuum subject to point and diffuse sources. *J. Hydrol.* **2008**, *350*, 154–165. [CrossRef]
32. Ribbe, L.; Delgado, P.; Salgado, E.; Flügel, W.A. Nitrate pollution of surface water induced by agricultural non-point pollution in the Pocochay watershed, Chile. *Desalination* **2008**, *226*, 13–20. [CrossRef]
33. Lassaletta, L.; García-Gómez, H.; Gimeno, B.S.; Rovira, J.V. Agriculture-induced increase in nitrate concentrations in stream waters of a large Mediterranean catchment over 25 years (1981–2005). *Sci. Total Environ.* **2009**, *407*, 6034–6043. [CrossRef] [PubMed]
34. Flohre, A.; Fischer, C.; Aavik, T.; Bengtsson, J.; Berendse, F.; Bommarco, R.; Ceryngier, P.; Clement, L.W.; Dennis, C.; Eggers, S.; et al. Agricultural intensification and biodiversity partitioning in European landscapes comparing plants, carabids, and birds. *Ecol. Appl.* **2011**, *21*, 1772–1781. [CrossRef] [PubMed]
35. Carstensen, J.; Andersen, J.H.; Gustafsson, B.G.; Conley, D.J. Deoxygenation of the Baltic Sea during the last century. *Proc. Natl. Acad. Sci. USA* **2014**, *111*, 5628–5633. [CrossRef]
36. Mockler, E.M.; Deakin, J.; Archbold, M.; Gill, L.; Daly, D.; Bruen, M. Sources of nitrogen and phosphorus emissions to Irish rivers and coastal waters: Estimates from a nutrient load apportionment framework. *Sci. Total Environ.* **2017**, *601*, 326–339. [CrossRef]
37. Morales-Marín, L.A.; Wheater, H.S.; Lindenschmidt, K.E. Assessment of nutrient loadings of a large multipurpose prairie reservoir. *J. Hydrol.* **2017**, *550*, 166–185. [CrossRef]
38. Reichard, J.S.; Brown, C.M. Detecting groundwater contamination of a river in Georgia, the USA using baseflow sampling. *Hydrogeol. J.* **2009**, *17*, 735–747. [CrossRef]
39. Ouyang, Y. Estimation of shallow groundwater discharge and nutrient load into a river. *Ecol. Eng.* **2012**, *38*, 101–104. [CrossRef]
40. Alexander, R.B.; Bohlke, J.K.; Boyer, E.; David, M.B.; Harvey, J.; Mulholland, P.J.; Seitzinger, S.P.; Tobias, C.R.; Tonitto, C.; Wollheim, W. Dynamic modeling of nitrogen losses in river networks unravels the coupled effects of hydrological and biogeochemical processes. *Biogeochemistry* **2009**, *93*, 91–116. [CrossRef]
41. Luo, G.; Bu, F.; Xu, X.; Cao, J.; Shu, W. Seasonal variations of dissolved inorganic nutrients transported to the Linjiang Bay of the Three Gorges Reservoir, China. *Environ. Monit. Assess.* **2011**, *173*, 55–64. [CrossRef]
42. Van Breemen, N.V.; Boyer, E.W.; Goodale, C.L.; Jaworski, N.A.; Paustian, K.; Seitzinger, S.P.; Lajtha, K.; Mayer, B.; Van Dam, D.; Howarth, R.W.; et al. Where did all the nitrogen go? Fate of nitrogen inputs to large watersheds in the north-eastern USA. *Biogeochemistry* **2002**, *57*, 267–293. [CrossRef]
43. Puckett, L.J.; Tesoriero, A.J.; Dubrovsky, N.M. Nitrogen contamination of surficial aquifers-A growing legacy. *Environ. Sci. Technol.* **2011**, *45*, 839–844. [CrossRef] [PubMed]
44. European Commission. Common implementation strategy for the water framework directive (2000/60/EC). In *Guidance document no. 7—Monitoring under the Water Framework Directive*; Office for Official Publications of the Europe: Luxembourg, 2003.
45. Teodosiu, C.; Barjoveanu, G.; Teleman, D. Sustainable water resources management 1. River basin management and the EC water framework directive. *Environ. Eng. Manag. J.* **2003**, *2*, 377–394. [CrossRef]
46. European Commission. *Report from the Commission to the European Parliament and the Council on the implementation of the European Statistical Programme 2013–2017*; European Commission: Brussels, Belgium; Luxembourg, 2015.
47. Sánchez-Montoya, M.D.M.; Arce, M.I.; Vidal-Abarca, M.R.; Suárez, M.L.; Prat, N.; Gómez, R. Establishing physico-chemical reference conditions in Mediterranean streams according to the European Water Framework Directive. *Water Res.* **2012**, *46*, 2257–2269. [CrossRef]
48. European Commission. Report from the commission to the European parliament and the council on the implementation of the Water Framework Directive (2000/60/EC). In *River Basin Management Plans*; European Commission: Brussels, Belgium; Luxembourg, 2012.
49. Galloway, J.N.; Hiram Levy, I.I.; Kasibhatla, P.S. Year 2020: Consequences of population growth and development on deposition of oxidized nitrogen. *Ambio* **1994**, *23*, 120–123.
50. Rizzoni, M.; Gustavino, B.; Ferrari, C.; Gatti, L.G.; Fano, E.A. An integrated approach to the assessment of the environmental quality of the Tiber River in the urban area of Rome: A mutagenesis assay (micronucleus test) and an analysis of macrobenthic community structure. *Sci. Total Environ.* **1995**, *162*, 127–137. [CrossRef]
51. Istituto di ricerca sulle acque Roma. *Indagini Sull'inquinamento del Fiume Tevere*; Istituto di ricerca sulle acque Roma: Roma, Italy, 1978.
52. Casini, M.; Giussani, E. *Il Tevere a Roma. Autorità di Bacino del Fiume Tevere—Citera*; Edizioni Ambiente: Milano, Italy, 2006.
53. Oenema, O.; Kros, H.; De Vries, W. Approaches, and uncertainties in nutrient budgets: Implications for nutrient management and environmental policies. *Eur. J. Agron.* **2003**, *20*, 3–16. [CrossRef]

54. ERSAF Lombardia. ERSAF Lombardia (2016) Allegato A—"Programma d'Azione regionale per la protezione delle acque dall'inquinamento provocato dai nitrati provenienti da fonti agricole nelle zone vulnerabili ai sensi della Direttiva nitrati 91/676/CEE". Available online: https://www.ersaf.lombardia.it/it/file/2103/08bbc92c/Allegato+A+Programma+d%27 Azione+Nitrati+2016-2019.pdf (accessed on 29 January 2022).
55. Regione Emilia Romagna, Deliberazione dell'Assemblea Legislativa della Regione Emilia-Romagna del 16 gennaio 2007, n. 96-Attuazione del decreto del Ministro delle Politiche agricole e forestali 7 aprile 2006. Programma d'azione per le zone vulnerabili ai nitrati da fonte agricola. *Bollettino Ufficiale Regione Emilia-Romagna.* 2007. Available online: https://bur.regione.emilia-romagna. it/archivio/bollettino_download?anno=2007&num_boll=16. (accessed on 27 January 2022).
56. McKee, L.J.; Eyre, B.D. Nitrogen and phosphorus budgets for the sub-tropical Richmond River catchment, Australia. *Biogeochemistry* **2000**, *50*, 207–239. [CrossRef]
57. De Falco, E.; Landi, G.; Sassano, A.; De Franchi, A.S. Harvest management effects on lucerne (Medicago sativa L.) forage production and quality in a hilly environment in southern Italy. In *Optimal Forage Systems for Animal Production and the Environment, Proceedings of the 12th Symposium of the European Grassland Federation, Pleven, Bulgaria, 26–28 May 2003*; Bulgarian Association for Grassland and Forage Production (BAGFP): Bulgaria, Balkans, 2003; pp. 348–351.
58. Herridge, D.F.; Peoples, M.B.; Boddey, R.M. Global inputs of biological nitrogen fixation in agricultural systems. *Plant Soil* **2008**, *311*, 1–18. [CrossRef]
59. Salvagiotti, F.; Cassman, K.; Specht, J.; Walters, D.; Weiss, A.; Dobermann, A. Nitrogen uptake, fixation, and response to fertilizer N in soybeans: A review. *F Crop Res.* **2008**, *108*, 1–13. [CrossRef]
60. Jordan, T.E.; Weller, D.E. Human contributions to terrestrial nitrogen flux. *Bioscience* **1996**, *46*, 655–664. [CrossRef]
61. Van Der Weerden, T.J.; Jarvis, S.C. Ammonia emission factors for N fertilizers applied to two contrasting grassland soils. *Environ. Pollut.* **1997**, *95*, 205–211. [CrossRef]
62. Bussink, D.W.; Oenema, O. Ammonia volatilization from dairy farming systems in temperate areas: A review. *Nutr. Cycl. Agroecosystems* **1998**, *51*, 19–33. [CrossRef]
63. Asman, W.A.H. Factors influencing local dry deposition of gases with special reference to ammonia. *Atmos Env.* **1998**, *32*, 415–421. [CrossRef]
64. Smil, V. Nitrogen in crop production: An account of global flows. *Glob. Biogeochem Cycles* **1999**, *13*, 647–662. [CrossRef]
65. Panagos, P.; Van Liedekerke, M.; Jones, A.; Montanarella, L. European Soil Data Centre: Response to European policy support and public data requirements. *Land Use Policy* **2012**, *29*, 329–338. [CrossRef]
66. Velthof, G.; Oudendag, D.; Witzke, H.; Asman, W.; Klimont, Z.; Oenema, O. Integrated Assessment of Nitrogen Losses from Agriculture in EU-27 using MITERRA-EUROPE. *J. Environ. Qual.* **2009**, *38*, 402–417. [CrossRef]
67. Breiman, L. Bagging Predictors. *Mach. Learn.* **1996**, *24*, 123–140. [CrossRef]
68. Breiman, L. ST4_Method_Random_Forest. *Mach. Learn.* **2001**, *45*, 5–32. [CrossRef]
69. Breiman, L.; Cutler, A. Manual–Setting Up, Using, And Understanding Random Forests V4.0, Technical report, University of California, Berkeley. 2003. Available online: https://www.stat.berkeley.edu/~{}breiman/Using_random_forests_v4.0.pdf (accessed on 15 December 2021).
70. Vitale, M.; Proietti, C.; Cionni, I.; Fischer, R.; De Marco, A. Random forests analysis: A useful tool for defining the relative importance of environmental conditions on crown defoliation. *Water. Air. Soil Pollut.* **2014**, *225*, 1992. [CrossRef]
71. Rechberger, H.; Brunner, P.H. A new, entropy-based method to support waste and resource management decisions. *Environ. Sci. Technol.* **2002**, *36*, 809–816. [CrossRef] [PubMed]
72. Sobańtka, A.P.; Zessner, M.; Rechberger, H. The extension of statistical entropy analysis to chemical compounds. *Entropy* **2012**, *14*, 2413–2426. [CrossRef]
73. Sobańtka, A.P.; Thaler, S.; Zessner, M.; Rechberger, H. Extended statistical entropy analysis for the evaluation of nitrogen budgets in Austria. *Int. J. Environ. Sci. Technol.* **2014**, *11*, 1947–1958. [CrossRef]
74. Bürgmann, H.; Widmer, F.; Sigler, W.V.; Zeyer, J. mRNA extraction and reverse transcription-PCR protocol for detection of nifH gene expression by Azotobacter vinelandii in soil. *Appl. Environ. Microbiol.* **2003**, *69*, 1928–1935. [CrossRef]
75. De Marco, A.; Screpanti, A.; Attorre, F.; Proietti, C.; Vitale, M. Assessing ozone and nitrogen impact on net primary productivity with a Generalised Non-Linear Model. *Environ. Pollut.* **2013**, *172*, 250–263. [CrossRef]
76. IUSS Working Group WRB. World Reference Base for Soil Resources 2014, update 2015. International soil classification system for naming soils and creating legends for soil maps. In *World Soil Resources Reports No. 106*; FAO: Rome, Italy, 2015.
77. Oenema, O.; Brentrup, F.; Lammel, J.; Bascou, P.; Billen, G.; Dobermann, A.; Erisman, J.W.; Garnett, T.; Hammel, M.; Haniotis, T.; et al. *Nitrogen Use Efficiency (NUE)—An Indicator for the Utilization of Nitrogen in Agriculture and Food Systems*; EU Nitrogen Expert Panel; Wageningen University: Wageningen, The Netherlands, 2015; p. 47.
78. Gozzi, C.; Filzmoser, P.; Buccianti, A.; Vaselli, O.; Nisi, B. Statistical methods for the geochemical characterisation of surface waters: The case study of the Tiber River basin (Central Italy). *Comput. Geosci.* **2019**, *131*, 80–88. [CrossRef]

soil systems

MDPI

Article

Manganese Uptake to Wheat Shoot Meristems Is Differentially Influenced by Arbuscular Mycorrhiza Fungal Communities Adapted to Acidic Soil

Jorge M. S. Faria [1,2,*], Dora Martins Teixeira [3,4], Diana Ferreira [4], Pedro Barrulas [3], Isabel Brito [2,4], Ana Paula Pinto [2,4] and Mário Carvalho [2,4]

1 INIAV, I.P., National Institute for Agrarian and Veterinarian Research, Quinta do Marquês, 2780-159 Lisbon, Portugal
2 MED, Mediterranean Institute for Agriculture, Environment and Development & CHANGE—Global Change and Sustainability Institute, Institute for Advanced Studies and Research, Évora University, Pólo da Mitra, Ap. 94, 7006-554 Évora, Portugal; ibrito@uevora.pt (I.B.); app@uevora.pt (A.P.P.); mjc@uevora.pt (M.C.)
3 HERCULES Laboratory, Évora University, Largo Marquês de Marialva 8, 7000-809 Évora, Portugal; dmt@uevora.pt (D.M.T.); pbarrulas@uevora.pt (P.B.)
4 Chemistry and Biochemistry Department, Science and Technology School of Évora University, Rua Romão Ramalho n°59, 7000-671 Évora, Portugal; l26304@alunos.uevora.pt
* Correspondence: fariajms@gmail.com

Citation: Faria, J.M.S.; Teixeira, D.M.; Ferreira, D.; Barrulas, P.; Brito, I.; Pinto, A.P.; Carvalho, M. Manganese Uptake to Wheat Shoot Meristems Is Differentially Influenced by Arbuscular Mycorrhiza Fungal Communities Adapted to Acidic Soil. *Soil Syst.* **2022**, *6*, 50. https://doi.org/10.3390/soilsystems6020050

Academic Editors: Antonella Lavini and Mohamed Houssemeddine Sellami

Received: 15 April 2022
Accepted: 24 May 2022
Published: 26 May 2022

Abstract: Soil acidity is a strong promoter of the bioavailability of Al, Fe, and Mn, whose concentrations can sometimes reach toxic levels for plants. In agricultural soils, the use of arbuscular mycorrhizal fungi (AMF) has shown a protective influence on wheat growth under Mn toxicity. The intact extraradical mycelium (ERM) promotes faster AMF colonization, leading to a higher wheat shoot weight, lower Mn uptake, and changes in antioxidant enzyme activity. Its effect on the uptake and distribution of plant nutrients according to the developmental stage of shoot organs has seldomly been analyzed. In the present study, Mn, Mg, Ca, and K were quantified by ICP–MS in leaves and apical meristems of wheat grown in soil with two different ERM consortia, developed from the native plants *Lolium rigidum* (LOL) and *Ornithopus compressus* (ORN). The ORN treatment induced the highest wheat shoot weights and the lowest Mn levels. In the leaves, no significant differences were detected for Mg, Ca, or K, but in the apical meristems, the ORN treatment slightly lowered the Ca concentration. The AMF associated with ORN was seen to enhance wheat weight partly by protecting the zones of active growth against high Mn levels in Mn toxic soils. The use of ORN in acidic soils with Mn toxicity provides a sustainable alternative and an efficient complement to current farming practices to lower the negative impacts of farming on the environment.

Keywords: arbuscular mycorrhizal fungi; intact extraradical mycelium; Mn toxicity; soil acidity; sustainable agriculture

1. Introduction

Soil microorganisms play key roles in the ecological functions of living soil. Their activity can change the soil's biochemical, chemical, and physical properties and influence the structure of plant and animal communities [1,2]. The latest environmental policies encourage the ecological restoration of degraded lands. This has prompted researchers to develop more sustainable farming practices to apply in agroecological contexts [3]. Arbuscular mycorrhizal fungi (AMF) are at the forefront of agricultural research for the eco-friendly improvement of agricultural systems. In the last four years (2018 to 2021), a high number of reports citing the keywords "arbuscular mycorrhizal fungi" and "agricultural" in the abstract have been published, totaling ca. 40% of publications since 1977 [4]. Additionally, 2018 to 2021 was the period with the highest registered number of citations on this subject.

Taken together, these indicators point to a growing interest of the research community in sustainable agriculture based on AMF.

Arbuscular mycorrhiza result from the mutualistic symbiosis between AMF and ca. 80% of plant species; their origins are believed to span back to the colonization of land by modern plants' first ancestors [5]. These fungi belong to the phylum Glomeromycota, with 342 species described at the moment, distributed through 43 genera, 12 families, and 4 orders [6]. AMF are obligate symbionts, i.e., dependent on the presence of a plant host to complete their life cycle. Once in symbiosis, the fungi can be supplied with up to 20% of the plant's photosynthates and other important compounds, while the plant gains increased access to water and essential plant nutrients, particularly the ones less mobile in the soil solution, e.g., phosphorus [7]. These natural symbiotic systems can be used to improve sustainability and protect modern agroecosystems, given that they (1) improve the soil's structure and biochemical characteristics, (2) increase crop growth and productivity, and (3) enhance plant resistance to environmental stress and pest attacks [8].

Wheat is the second-most-produced global staple crop after maize, and the cereal is mainly used in developed countries. When grown in acidic soils, wheat is mainly affected by toxic levels of bioavailable Al and/or Mn [9–11]. High levels of Mn influence the uptake and distribution of other essential nutrients, e.g., Mg, Ca, and P. Under Mn toxicity, wheat tends to decrease element uptake, retain Ca in the root apoplast, and redirect excessive Mn, along with Mg and P, to the vacuole to avoid oxidative stress [12,13]. The development of natural stress-adapted native plant–AMF symbiotic systems in acidic soils was found to alter their biological composition and influence the growth and mineral composition of subsequently planted crops [14–16]. In wheat, this improvement can be more significant depending on the symbiotic soil microbiota associated with the native plant that preceded it and on maintaining a reduced disturbance to soil structure [14,17]. The intact extraradical mycelium (ERM) of AMF is known to provide faster root colonization than spores or colonized root fragments if kept undisturbed in the soil [18]. By granting earlier plant colonization, the intact ERM also allows an earlier capitalization of the benefits granted by mycorrhization, such as growth improvement and extended protection of the plant under stressful conditions. Under Mn toxicity resulting from soil acidity, wheat grown after the previous development of the native plants *Lolium rigidum* L. (LOL) or *Ornithopus compressus* L. (ORN) and colonized by an intact ERM will harbor different AMF communities on its roots and show reduced levels of Mn, effectively decreasing its toxic effects inside wheat tissues [14]. Additionally, the colonization of roots by the intact ERM of these symbiotic systems can alter wheat nutrient levels, namely, increasing P and S and decreasing Mg and Ca, and change plant redox status by altering the activity of antioxidant enzymes [14,17,19,20]. Depending on the previously developed native mycotrophic plant and the specific composition of the AMF community colonizing wheat roots from the intact ERM, different biochemical mechanisms are induced on wheat's internal tissues [15,21].

The AMF functional diversity underlying the influence of these symbiotic systems on the elemental composition of wheat shoots is currently being analyzed by assessing wheat redox status and mapping the uptake, translocation, and subcellular distribution of essential nutrients [15,17–21]. The present study analyses, for the first time, the transport of essential micro and macronutrients, with different mobilities *in planta*, to wheat leaves and meristems in acidic soil with different AMF consortia. This study aims to uncover the impact of the soil AMF community on the distribution of (1) plant mobile K and Mg and (2) plant immobile Ca and Mn on (a) leaves and (b) apical meristems of 3-week-old wheat grown under the influence of the intact ERM from LOL or ORN symbiotic systems in acidic soil.

2. Materials and Methods

2.1. Physicochemical and Biological Soil Characterization

The acidic soil used for plant growth was collected from the top 20 cm of a natural pasture site on the Mitra experimental farm of Évora University, Portugal (38°32′ N; 08°00′ W) [14]. This site is characterized by Eutric Cambisols of granitic origin, extensively used for research on the effect of acidic soils with Mn toxicity on crop development, particularly using wheat [12,22–24]. The air-dried and sieved (2 mm) soil is characterized as sandy loam with 11 g SOM (soil organic matter, chromic acid wet oxidation)/kg, a cation exchange capacity (CEC) of 4.5 centimoles of charge per kg (cmol(+)/kg), a base saturation of 60%, and a pH of 5.6 (soil:water = 1:2.5 (*w*/*v*)) [9,12]. Chemical analysis quantified P at 26 mg/kg (Egner-Rhiem), $N\text{-}NO_3$ at 0.4 mg/kg, K at 67 mg/kg (Egner-Rhiem), Mg at 112 mg/kg (1 M ammonium acetate, pH 7), and Mn at 41 mg/kg (Lakanen) [17]. AMF abundance was previously quantified at 180 viable propagules per g of dry acidic soil by Brito et al. [14]. Composition of the AMF consortia associated with each native plant and wheat after each treatment used in the present study was previously described by Brígido et al. [15]. Extraction of the soil solution for each soil treatment was performed according to Faria et al. [9]. Briefly, ca. 40 g samples were centrifuged at 2500× *g* for 60 min at 4 °C in 50 mL tubes fitted with 0.45 µm polyethersulfone filters to isolate the interstitial soil solution. The volume was measured, and the aqueous solutions were kept at −20 °C until analysis.

2.2. Treatments and Experimental Protocol

Experiments were performed in a greenhouse under controlled conditions using pot trials with the acidic soil characterized above, according to Faria et al. [17]. Briefly, dark plastic 8 L pots were packed with the homogenized acidic soil and kept hydrated to ca. 70% of maximum water holding capacity by weight for 1 week for soil stabilization. Then, six ORN or LOL seedlings in similar development stages were selected from seeds previously germinated in hydrated filter paper and planted at equally distanced positions in each of four replicate pots for each treatment, with a total of eight pots. These native plants were allowed to develop naturally in the acidic soil and establish symbiotic relationships with their respective AMF consortium for 7 weeks. Pots were kept in the greenhouse, minimum and maximum air temperatures were recorded daily, and temperature control was set to a maximum of 30 °C. Naturally germinated weed plants were manually excised to maintain the AMF consortium composition characteristic of each planted native species. Soil was maintained hydrated, as described above. After 7 weeks, the native plants were eliminated by cut, and six wheat seedlings (*Triticum aestivum* L. cv. Ardila) were planted per pot at equally distanced positions in a total of eight pots. In all instances, pots were kept fully randomized in the greenhouse. After 3 weeks, the wheat was in stage 1 of principal growth, and the shoots were excised and weighed. Three random plants per pot were used whole for shoot element quantification, and the remaining three were each divided into leaves and apical meristems, with a total of 48 plants and 72 samples. All samples were immediately frozen in liquid nitrogen and stored at −80 °C until analysis.

2.3. Quantification of Mn, Mg, Ca, and K in the Soil Solution and Wheat Shoot Tissues

Levels of Mn, Mg, Ca, and K were determined through inductively coupled plasma mass spectrometry (ICP–MS), according to Faria et al. [17]. Briefly, samples were ground in liquid nitrogen, and 0.5 g of fresh ground shoot tissues were dried in an oven at 60 °C for 3 days and then weighed for the assessment of dry weight (DW). Afterwards, 3 mL of HNO_3 (Suprapur, 67–69%, Fisher Chemicals, Hampton, NH, USA) and 2 mL of HCl (Trace Metal Analysis, 37%, Fisher Chemicals) were added to closed Teflon vessels and left overnight at room temperature. Then, microwave-assisted acid digestion was performed after adding 3 mL of HNO_3 and 2 mL of HCl to this solution by heating to 240 °C for 40 min, with a final cooling step of 15 min, using a Mars 6 microwave digestion system (CEM, Matthews, NC, USA). Digested shoot tissue solutions and soil solutions were filtered through a 0.45 μm pore PTFE filter, and ultrapure water was added to a final volume of 20 mL. Diluted samples (40- and 1000-fold) were analyzed in an 8800 Triple Quadrupole ICP–MS (Agilent, Santa Clara, CA, USA) equipped with a Micromist nebuliser. Agilent ICP–MS tuning solution of 2% HNO_3 containing 10 μg/L each of Ce, Co, Li, Tl, and Y (Agilent Technologies, Palo Alto, CA, USA) was used for instrument optimization. External calibration was performed with the multi-element certificate standard solution ICP–MS-68B-A (100 mg/L) from High-Purity Standards (Charleston, SC, USA). Matrix effects and instrumental drifts were corrected on the basis of the internal standards ruthenium (Ru), rhodium (Rh), and iridium (Ir). The collision/reaction cell was set to "He mode" for elemental quantification. The plasma gas flow rate was 15 mL/min, and the collision and reaction gas (He) flow rate was 4 mL/min. Analyses were optimized at 1550 W forward power and 1.1 L/min Ar carrier gas flow, with no dilution or makeup gas. Sampling depth (10 mm) and lens parameters were optimized for the highest signal and optimum peak shape while maintaining low oxides and doubly charged species. MS/MS scan type was used in all the operation modes. Element levels were expressed per shoot dry weight.

2.4. Data Analysis

The bioconcentration factor for each element was determined according to Faria et al. [12] using the formula $BCF = C_{tissue}/C_{soil\ solution}$, where C_{tissue} is the element concentration in wheat shoots, leaves, or meristems and $C_{soil\ solution}$ is the element concentration in the soil solution. Statistical analysis was performed with SPSS statistics software version 27, according to Faria et al. [12]. The statistical significance of data was determined with one-way ANOVA, using Tukey's post hoc test for means comparison at a 95% significance level ($p < 0.05$). The Shapiro–Wilk test and the Browns–Forsythe test were used for the evaluation of normality and homoscedasticity, respectively. The results presented correspond to the average and standard error of four (n = 4) biological replicates.

3. Results

3.1. Wheat Growth

A higher shoot DW ($p < 0.05$) was obtained for wheat grown in soil from ORN (271 ± 6 mg per plant) than for wheat grown in soil from LOL (176 ± 3 mg per plant) (Figure 1a). Wheat plants did not show symptoms of Mn toxicity, e.g., chlorosis in older leaves, but plants grown in LOL treatment showed impaired development when compared to those grown in ORN treatment (Figure 1b).

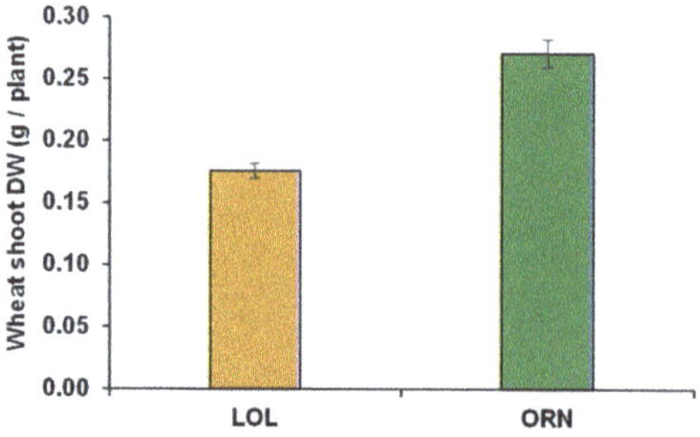

(a)

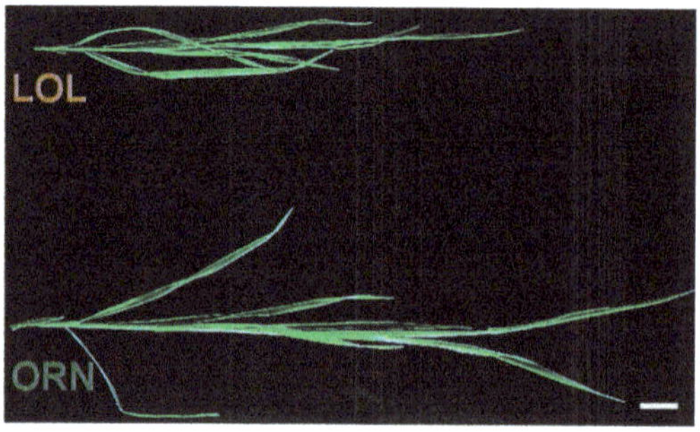

(b)

Figure 1. Shoots of wheat plants grown for 3 weeks in acidic soil after the previous development of *Lolium rigidum* (LOL) or *Ornithopus compressus* (ORN). (**a**) Growth assessed through shoot dry weight (DW) per plant; (**b**) visual aspect of shoots of wheat grown in each soil treatment. Scale bar = 1 cm.

3.2. Mn, Mg, Ca, and K Concentration

3.2.1. Soil Solution

Except for K, element levels were generally higher in the soil solution of the ORN treatment than in LOL ($p < 0.05$) (Table 1). For K, levels in the LOL soil solution were higher than for standard acidic soil ($p < 0.05$). However, the Mg/Mn ratio was higher for the standard acidic soil (0.015) than for either LOL or ORN soils (both 0.011).

Table 1. Concentrations of Mn, Mg, Ca, and K (mg/L) in the soil solution of standard acidic soil and of acidic soil collected in the vicinity of roots of wheat grown for 3 weeks after the previous development of *Lolium rigidum* (LOL) or *Ornithopus compressus* (ORN). Different letters indicate statistically significant differences ($p < 0.05$) on the basis of Tukey's test in each element.

Elements (mg/L)	Mn	Mg	Ca	K
Acidic soil	110.7 ± 3.1 [a]	7566.3 ± 721.4 [a]	23,247.6 ± 1351.7 [a]	3060.5 ± 8.4 [a]
LOL/Wheat	74.0 ± 7.5 [b]	6700.2 ± 478.4 [a]	22,518.3 ± 1783.1 [a]	5171.7 ± 515.4 [b]
ORN/Wheat	138.8 ± 8.7 [a]	12,028.4 ± 1290.0 [b]	37,445.7 ± 849.0 [b]	7900.4 ± 115.1 [c]

3.2.2. Wheat Shoot Tissues

To understand the effect of the acidic soil with intact ERM from the native plants on the internal distribution of Mn, Mg, Ca, and K in wheat shoots, the levels of these elements were quantified in leaves and meristems. The levels of Mn in wheat shoots rose to 50.0 ± 1.7 mg/kg plant DW for the LOL treatment and to 39.5 ± 3.0 mg/kg plant DW for the ORN treatment (Figure 2a). In the leaves, Mn reached similar levels for both treatments but was lower in meristems ($p < 0.05$), namely, 51.3 ± 2.0 and 31.9 ± 3.1 mg/kg plant DW, respectively, for the LOL treatment and 43.8 ± 3.2 and 20.8 ± 2.9 mg/kg plant DW, respectively, for the ORN treatment. In the LOL treatment, the Mg/Mn shoot ratios were similar to those found in leaves but were lower for the meristems. In the ORN treatment, wheat shoots showed the lowest Mg/Mn ratio value, while the meristems showed the highest (Figure 2a). The levels of Mg were lower in wheat shoots ($p < 0.05$) and leaves (not statistically significant) from the ORN treatment than those from LOL. In the meristems, the values were lower ($p < 0.05$) than in the shoots or leaves, but no substantial difference was found between treatments (Figure 2b). Calcium showed no major variations except for the lower levels detected in the meristems of wheat grown in soil after ORN (although not statistically significant) (Figure 2c). Potassium levels showed no statistical difference between plant tissues; however, a 19% lower K % was seen in wheat shoots from the ORN treatment when compared to LOL. Additionally, K levels in meristems were lower than in shoots or leaves in both treatments (Figure 2d).

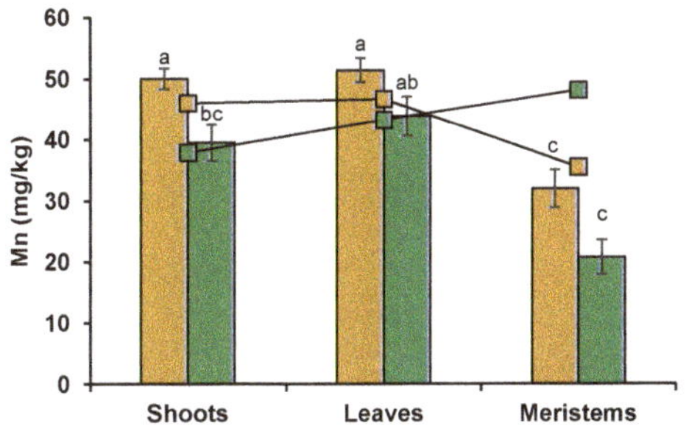

(a)

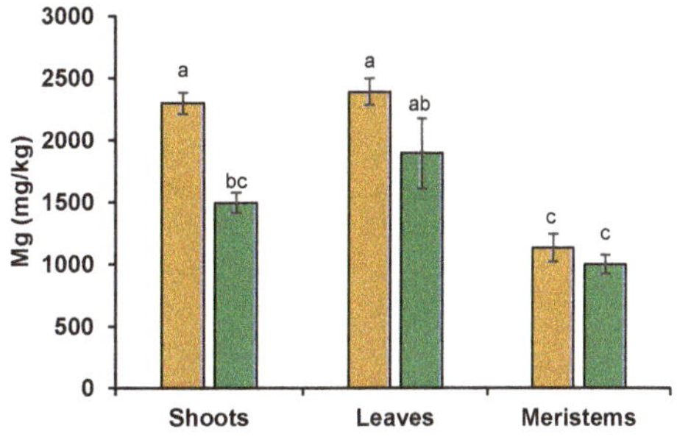

(b)

Figure 2. *Cont.*

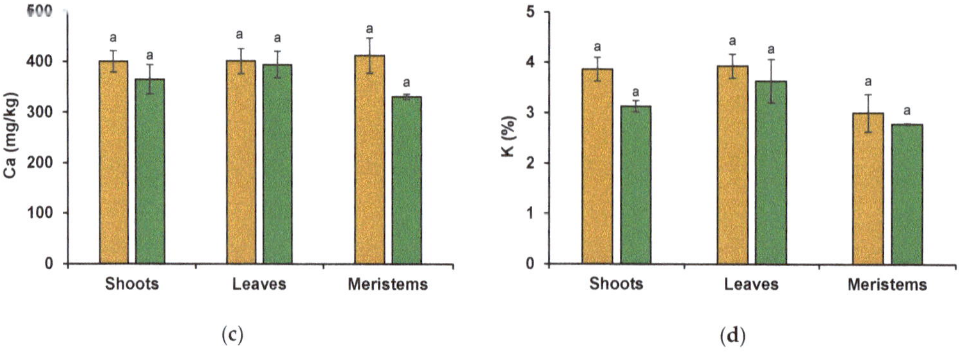

Figure 2. Concentrations of (**a**) Mn, (**b**) Mg, (**c**) Ca, (**d**) K, and (**a**) Mg/Mn ratio (line) in the shoots, leaves, and meristems of wheat grown for 3 weeks in acidic soil after the previous development of *Lolium rigidum* (LOL) (light brown) or *Ornithopus compressus* (ORN) (green). Different letters indicate statistically significant differences ($p < 0.05$) on the basis of Tukey's test.

For each element, the bioconcentration factor values determined were lower in the shoot tissues of wheat grown in ORN soil than in LOL soil (Table 2). For Mn, the bioconcentration factor values were 2.4-, 2.2-, and 2.9-fold lower; for Mg, 2.7-, 2.2-, and 2.0-fold lower; for Ca, 1.8-, 1.7-, and 2.1-fold lower; and for K, 1.9-, 1.7-, and 1.7-fold lower in shoots, leaves, and meristems, respectively.

Table 2. Bioconcentration factors of Mn, Mg, Ca, and K for shoots, leaves, and meristems of wheat grown for 3 weeks in acidic soil after the previous development of *Lolium rigidum* (LOL) or *Ornithopus compressus* (ORN). Different letters indicate statistically significant differences ($p < 0.05$) on the basis of Tukey's test for each element.

Plant Part/Bioconcentration Factor		Mn × 10^{-1}	Mg × 10^{-1}	Ca × 10^{-2}	K
Shoot	LOL	6.9 ± 0.3 [a]	3.5 ± 0.1 [a]	1.8 ± 0.1 [a]	7.6 ± 0.4 [a]
	ORN	2.9 ± 0.1 [b,c]	1.3 ± 0.1 [a,b]	1.0 ± 0.0 [b]	4.0 ± 0.1 [b]
Leaves	LOL	7.1 ± 0.3 [a]	3.6 ± 0.1 [a]	1.8 ± 0.1 [a]	7.7 ± 0.4 [a]
	ORN	3.2 ± 0.2 [c]	1.6 ± 0.1 [b]	1.1 ± 0.0 [b]	4.6 ± 0.3 [b]
Meristems	LOL	4.4 ± 0.3 [c]	1.7 ± 0.1 [c]	1.9 ± 0.1 [a]	5.9 ± 0.5 [c]
	ORN	1.5 ± 0.1 [b]	0.8 ± 0.0 [d]	0.9 ± 0.0 [b]	3.5 ± 0.1 [b]

4. Discussion

Stress-adapted native plants modify the microbiome in the rhizosphere by stimulating the development of fungal consortia that can be specific to each environment–soil–plant combination. These communities are composed of AMF symbiotic with the native plants and the associated microbiota, providing a beneficial environment to plant growth under stressful conditions [18]. This soil microbial ecosystem has been previously used to improve the development and productivity of crop plants [15,25,26]. In acidic soils with problems of Mn toxicity, the use of soil with intact ERM associated with LOL or ORN showed strong effects on wheat growth by differentially (a) shifting the structure of the communities of colonizing AMF [15,21], (b) inducing the transcription of genes in the host root involved in growth or stress evasion [27], (c) influencing the activity of antioxidant enzymes [20], and (d) altering internal tissue levels of essential nutrients and their subcellular distribution [17,19].

By applying LOL and ORN microbial communities separately, in the present study, we found differences in wheat growth and the translocation patterns of Mn, Mg, Ca, and K in shoot tissues of wheat grown under Mn toxicity imposed by soil acidity. Soil from the LOL treatment was less successful in promoting wheat shoot growth, and the average shoot DW was 35% lower than that of wheat grown in soil from the ORN treatment. Similar differences were also previously reported for these systems (with values ranging from 26% to 38%) [14,21], yet, they were considerably higher (up to 54% for the ORN treatment) when compared to wheat grown on soils where no native plant was previously developed [14]. Although the differences observed in wheat tissues as a response to these distinct native plants could be attributed to altered nutrient uptakes, the levels of some of the main elements, e.g., P, S, or Mg, were not seen to change in previous experiments [14]. This suggests that different previous native plants can influence nutrient distribution in wheat shoots at the organ or cellular levels under Mn toxicity. In the present work, in wheat shoots from both soil treatments, the levels of Mn were below the concentrations considered toxic for most cereals (100–200 mg/kg) [14,28]. Both in the leaves and apical meristems of wheat grown in ORN soil, Mn levels were maintained at lower levels than those in LOL soil. This was also observed for the Mg/Mn ratios in shoot tissues, a reported marker for Mn toxicity stress [22,23,29]. In wheat meristems that are actively growing tissues, the ORN soil treatment induced lower Mg/Mn values than LOL, even though in ORN soils, wheat shoots showed lower Mg levels (Figure 2a,b). Additionally, Ca was slightly lower in the meristems of wheat grown in ORN soil.

The AMF communities associated with ORN appear to be more beneficial for wheat in terms of both growth and managing element levels in different plant parts. In a previous study, these communities were analyzed and detailed through a metagenomics approach [15]. The AMF communities identified in symbiosis with roots of wheat grown in LOL or ORN soil showed only small differences in terms of diversity. However, some major differences were detected in the abundance of some genera; for example, several operational taxonomic units (OTUs) of *Rhizophagus* spp. were seen with greater abundance in LOL or ORN treatments; *Paraglomus* spp. was more abundant in ORN treatments, and *Claroideoglomus* spp. was more abundant in LOL treatment [15]. These authors also identified similar differences in AMF abundance in the roots of the source native plants. This suggests that the changes in the biochemical mechanisms regulating wheat growth and element distribution in response to the development in acidic soil from different native plants may be related to a cumulative activity of AMFs that are common to these native plants rather than to the influence of a specific AMF diversity. An additional study reports on the influence of this differential AMF abundance on the transcriptome of wheat roots [27]. Within 1 week of root colonization, genes encoding for putative transporter proteins (importins, potassium transporter, calcium-transporting ATPase, plasma membrane $Na+/H+$ antiporter) were upregulated in the ORN treatment, with high fold changes ($logFC > 10$), while in the LOL treatment, four ATP-binding cassette (ABC) transporters (transport of a wide variety of substrates across membranes in cells), three potassium transporters, and one AKT2 potassium channel were reportedly upregulated. By the fifth week after AMF colonization, wheat roots were differentially expressing ABC binding protein genes from the ABC transport system, one of the largest transporter protein families of compounds and elements in plants, localized in most plant membranes, namely, the plasma membrane, tonoplasts, chloroplasts, mitochondria, and peroxisomes. In contrast, in the ORN treatment, only two were found to be upregulated in the LOL soil treatment, where over 30 genes were upregulated. Additionally, genes for five phosphate transporters were upregulated in the LOL treatment, as well as two ETHYLENE-INSENSITIVE 2 (EIN2) genes and the NRAMP5 genes, which belong to the NRAMP family of metal transporters. Differential transporter gene expression points towards distinct management of the uptake and translocation of plant nutrients influenced by the differential AMF abundance associated with each native plant. The regulation of element transport and translocation in plants influenced by AMF abundance is a new concept that requires further research; it may contribute to the understanding of the inconsistency in the reported effects of AMF on plant nutrient uptake, which was either absent, positive, or negative for the same element in different reports [30].

Although reports on the management of these plant nutrients by natural AMF communities are very sparse, some research has been performed on one or more artificial communities of a small number of AMF species. For Mg, in trifoliate orange under Mg deficiency, the AMF *Funneliformis mosseae* increased shoot chlorophyll levels and stimulated the activity of antioxidant enzymes and leaf soluble protein, suggesting an influence on Mg nutrition [31]. For Mn, the introduction of AMF propagules substantially altered the soil microbiome. For example, the addition of *Funneliformis mosseae* (syn. *Glomus mosseae*) to a natural soil decreased the abundance of microbes able to reduce Mn and Fe, leading mycorrhizal plants to accumulate less Mn [32]. Additionally, plant responses to AMF activity are often specific to the fungal symbiont. For instance, the application of *Glomus macrocarpum* enhanced soybean growth under Mn-deficient conditions but induced Mn toxicity symptoms in soil with abundant Mn availability; in contrast, in the same study, *G. etunicatum* and *Rhizophagus intraradices* (syn. *G. intraradices*) increased plant performance in both Mn soil conditions [33]. Ultimately, the sustainable improvement of agricultural practices based on AMF species or natural AMF developers must first consider the intricate ecological relationships between each fungi–plant symbiosis and the chemical and biochemical mechanisms induced in the host.

5. Conclusions

The growth of ORN native plants, known to influence the abundance of specific Glomeromycota taxa in acidic soils, appears to favor growth and tolerance to Mn toxicity in wheat plantlets by managing the nutritional composition of its shoot tissues. Under the influence of an intact ERM, the actively growing wheat apical meristems show lower Mn levels and higher Mg/Mn ratios, indicative of reduced Mn toxicity stress. The screening of native plants, as developers of AMF, with a functional application contributes to the design of environmentally friendlier agronomical approaches to enhance crop productivity in acid soils. Uncovering and understanding the potential Mn detoxification mechanisms induced by an intact ERM structure on wheat is key to implementing a wide range of usable benefits granted by AMF in the frame of sustainable agriculture, contributing to research against the degradation of soil ecosystems.

Author Contributions: Conceptualization, J.M.S.F., D.M.T., A.P.P., I.B., P.B., M.C.; data curation, J.M.S.F., D.F., P.B.; formal analysis, J.M.S.F., D.F., P.B.; funding acquisition, D.M.T., A.P.P., I.B., M.C.; investigation, J.M.S.F.; methodology, J.M.S.F., D.M.T., A.P.P., I.B., P.B., M.C.; project administration, M.C.; resources, D.M.T., A.P.P., I.B., M.C.; writing—original draft, J.M.S.F.; writing—review and editing, J.M.S.F., D.M.T., D.F., A.P.P., I.B., P.B., M.C. All authors have read and agreed to the published version of the manuscript.

Funding: This study was partially funded by Fundo Europeu de Desenvolvimento Regional (FEDER), Programa Operacional Regional Alentejo 2020, under research contract ALT20-03-0145-FEDER-000039.

Institutional Review Board Statement: Not applicable.

Informed Consent Statement: Not applicable.

Data Availability Statement: The raw data supporting the findings of this study are available from the corresponding author (Jorge M. S. Faria) upon reasonable request.

Conflicts of Interest: The authors declare no conflict of interest.

References

1. Trinchera, A.; Ciaccia, C.; Testani, E.; Baratella, V.; Campanelli, G.; Leteo, F.; Canali, S. Mycorrhiza-mediated interference between cover crop and weed in organic winter cereal agroecosystems: The mycorrhizal colonization intensity indicator. *Ecol. Evol.* **2019**, *9*, 5593–5604. [CrossRef] [PubMed]
2. Rillig, M.C. Arbuscular mycorrhizae, glomalin, and soil aggregation. *Can. J. Soil Sci.* **2011**, *84*, 355–363. [CrossRef]
3. Basu, S.; Rabara, R.C.; Negi, S. *AMF: The Future Prospect for Sustainable Agriculture*; Elsevier Ltd.: Amsterdam, The Netherlands, 2018; Volume 102, pp. 36–45.
4. Web of Science. Available online: www.webofscience.com (accessed on 29 January 2022).

5. Humphreys, C.P.; Franks, P.J.; Rees, M.; Bidartondo, M.I.; Leake, J.R.; Beerling, D.J. Mutualistic mycorrhiza-like symbiosis in the most ancient group of land plants. *Nat. Commun.* **2010**, *1*, 103. [CrossRef] [PubMed]
6. Glomeromycota PHYLOGENY. Available online: http://amf-phylogeny.com/ (accessed on 29 January 2022).
7. Thirkell, T.J.; Pastok, D.; Field, K.J. Carbon for nutrient exchange between arbuscular mycorrhizal fungi and wheat varies according to cultivar and changes in atmospheric carbon dioxide concentration. *Glob. Chang. Biol.* **2020**, *26*, 1725–1738. [CrossRef]
8. Thirkell, T.J.; Charters, M.D.; Elliott, A.J.; Sait, S.M.; Field, K.J. Are mycorrhizal fungi our sustainable saviours? Considerations for achieving food security. *J. Ecol.* **2017**, *105*, 921–929. [CrossRef]
9. Faria, J.M.S.; Teixeira, D.M.; Pinto, A.P.; Brito, I.; Barrulas, P.; Carvalho, M. Aluminium, Iron and Silicon Subcellular Redistribution in Wheat Induced by Manganese Toxicity. *Appl. Sci.* **2021**, *11*, 8745. [CrossRef]
10. Khabaz-Saberi, H.; Rengel, Z.; Wilson, R.; Setter, T.L. Variation of tolerance to manganese toxicity in Australian hexaploid wheat. *J. Plant Nutr. Soil Sci.* **2010**, *173*, 103–112. [CrossRef]
11. Bojórquez-Quintal, E.; Escalante-Magaña, C.; Echevarría-Machado, I.; Martínez-Estévez, M. Aluminum, a friend or foe of higher plants in acid soils. *Front. Plant Sci.* **2017**, *8*, 1767. [CrossRef]
12. Faria, J.M.S.; Teixeira, D.M.; Pinto, A.P.; Brito, I.; Barrulas, P.; Alho, L.; Carvalho, M. Toxic levels of manganese in an acidic Cambisol alters antioxidant enzymes activity, element uptake and subcellular distribution in Triticum aestivum. *Ecotoxicol. Environ. Saf.* **2020**, *193*, 110355. [CrossRef]
13. Faria, J.; Pinto, A.P.; Teixeira, D.; Brito, I.; Dias, L.; Barrulas, P.; Alho, L.; Carvalho, M. Elemental composition and antioxidant enzyme activity of roots and shoots of wheat grown in manganese spiked Montado alentejano soil. *Free Radic. Biol. Med.* **2018**, *120*, S151. [CrossRef]
14. Brito, I.; Carvalho, M.; Alho, L.; Goss, M.J. Managing arbuscular mycorrhizal fungi for bioprotection: Mn toxicity. *Soil Biol. Biochem.* **2014**, *68*, 78–84. [CrossRef]
15. Brígido, C.; van Tuinen, D.; Brito, I.; Alho, L.; Goss, M.J.; Carvalho, M. Management of the biological diversity of AM fungi by combination of host plant succession and integrity of extraradical mycelium. *Soil Biol. Biochem.* **2017**, *112*, 237–247. [CrossRef]
16. Alho, L.; Carvalho, M.; Brito, I.; Goss, M.J. The effect of arbuscular mycorrhiza fungal propagules on the growth of subterranean clover (*Trifolium subterraneum* L.) under Mn toxicity in ex situ experiments. *Soil Use Manag.* **2015**, *31*, 337–344. [CrossRef]
17. Faria, J.M.S.; Teixeira, D.M.; Pinto, A.P.; Brito, I.; Barrulas, P.; Carvalho, M. The Protective Biochemical Properties of Arbuscular Mycorrhiza Extraradical Mycelium in Acidic Soils Are Maintained throughout the Mediterranean Summer Conditions. *Agronomy* **2021**, *11*, 748. [CrossRef]
18. Brito, I.; Goss, M.J.; Alho, L.; Brígido, C.; van Tuinen, D.; Félix, M.R.; Carvalho, M. Agronomic management of AMF functional diversity to overcome biotic and abiotic stresses—The role of plant sequence and intact extraradical mycelium. *Fungal Ecol.* **2019**, *40*, 72–81. [CrossRef]
19. Faria, J.M.S.; Teixeira, D.M.; Pinto, A.P.; Brito, I.; Barrulas, P.; Carvalho, M. Arbuscular Mycorrhiza Inoculum Type Influences Phosphorus Subcellular Distribution in Shoots of Wheat Grown in Acidic Soil under Sustainable Agricultural Practices. *Biol. Life Sci. Forum* **2020**, *4*, 62. [CrossRef]
20. Faria, J.M.S.; Pinto, A.P.; Teixeira, D.; Brito, I.; Carvalho, M. Diversity of Native Arbuscular Mycorrhiza Extraradical Mycelium Influences Antioxidant Enzyme Activity in Wheat Grown Under Mn Toxicity. *Bull. Environ. Contam. Toxicol.* **2021**, *108*, 451–456. [CrossRef]
21. Campos, C.; Carvalho, M.; Brígido, C.; Goss, M.J.; Nobre, T. Symbiosis Specificity of the Preceding Host Plant Can Dominate but Not Obliterate the Association Between Wheat and Its Arbuscular Mycorrhizal Fungal Partners. *Front. Microbiol.* **2018**, *9*, 2920. [CrossRef]
22. Le Bot, J.; Goss, M.J.; Carvalho, M.J.G.P.R.; Van Beusichem, M.L.; Kirkby, E.A. The significance of the magnesium to manganese ratio in plant tissues for growth and alleviation of manganese toxicity in tomato (*Lycopersicon esculentum*) and wheat (*Triticum aestivum*) plants. *Plant Soil* **1990**, *124*, 205–210. [CrossRef]
23. Carvalho, M.; Goss, M.J.; Teixeira, D. Manganese toxicity in Portuguese Cambisols derived from granitic rocks: Causes, limitations of soil analyses and possible solutions. *Rev. Ciênc. Agrár.* **2015**, *38*, 518–527. [CrossRef]
24. Goss, M.J.; Carvalho, M.J.G.P.R.; Cosimini, V.; Fearnhead, M.L. An approach to the identification of potentially toxic concentrations of manganese in soils. *Soil Use Manag.* **1992**, *8*, 40–43. [CrossRef]
25. Brito, I.; Goss, M.J.; de Carvalho, M.; Chatagnier, O.; van Tuinen, D. Impact of tillage system on arbuscular mycorrhiza fungal communities in the soil under Mediterranean conditions. *Soil Tillage Res.* **2012**, *121*, 63–67. [CrossRef]
26. Brito, I.; de Carvalho, M.; Goss, M.J. Summer survival of arbuscular mycorrhiza extraradical mycelium and the potential for its management through tillage options in Mediterranean cropping systems. *Soil Use Manag.* **2011**, *27*, 350–356. [CrossRef]
27. Campos, C.; Nobre, T.; Goss, M.J.; Faria, J.; Barrulas, P.; Carvalho, M. Transcriptome Analysis of Wheat Roots Reveals a Differential Regulation of Stress Responses Related to Arbuscular Mycorrhizal Fungi and Soil Disturbance. *Biology* **2019**, *8*, 93. [CrossRef] [PubMed]
28. Pinto, E.; Aguiar, A.A.R.M.; Ferreira, I.M.P.L.V.O. Influence of soil chemistry and plant physiology in the phytoremediation of Cu, Mn, and Zn. *Crit. Rev. Plant Sci.* **2014**, *33*, 351–373. [CrossRef]
29. Goss, M.J.; Carvalho, M.J.G.P.R. Manganese toxicity: The significance of magnesium for the sensitivity of wheat plants. *Plant Soil* **1992**, *139*, 91–98. [CrossRef]
30. Clark, R.B.; Zeto, S.K. Mineral acquisition by arbuscular mycorrhizal plants. *J. Plant Nutr.* **2000**, *23*, 867–902. [CrossRef]

31. Zhang, F.; Du, P.; Song, C.; Wu, Q. Alleviation of Mycorrhiza to Magnesium Deficiency in Trifoliate Orange: Changes in Physiological Activity. *Emirates J. Food Agric.* **2015**, *27*, 763–769. [CrossRef]
32. Kothari, S.K.; Marschner, H.; Römeheld, V. Effect of a vesicular—Arbuscular mycorrhizal fungus and rhizosphere micro-organisms on manganese reduction in the rhizosphere and manganese concentrations in maize (*Zea mays* L.). *New Phytol.* **1991**, *117*, 649–655. [CrossRef]
33. Nogueira, M.A.; Cardoso, E.J.B.N. Mycorrhizal effectiveness and manganese toxicity in soybean as affected by soil type and endophyte. *Sci. Agric.* **2003**, *60*, 329–335. [CrossRef]

soil systems

Article

Phytoremediating a Wastewater-Irrigated Soil Contaminated with Toxic Metals: Comparing the Efficacies of Different Crops

Iftikhar Ahmad [1], Saeed Ahmad Malik [1], Shafqat Saeed [2], Atta-ur Rehman [3] and Tariq Muhammad Munir [4,5,*]

1 Department of Botany, Institute of Pure and Applied Biology, Bahauddin Zakariya University, Multan 60800, Pakistan
2 Department of Entomology, Muhammad Nawaz Sharif University of Agriculture, Multan 66000, Pakistan
3 Soil and Water Testing Laboratory, Vehari, Multan 61100, Pakistan
4 Department of Geography and Planning, Global Institute for Water Security, University of Saskatchewan, Saskatoon, SK S7N 5C8, Canada
5 Department of Geography, University of Calgary, Calgary, AB T2N 1N4, Canada
* Correspondence: tmmunir@ucalgary.ca; Tel.: +1-403-971-5693

Abstract: A formidable challenge in suburban agriculture is the sustainability of soil health following the use of wastewater for irrigation. The wastewater irrigation likely toxifies the crop plants making them unconsumable. We used a multivariate, completely randomized design in a greenhouse, comparing the phytoextraction capacities of *Brassica juncea*, *Eruca sativa*, *Brassica rapa*, and *Brassica napus*—all grown on silt loam soil irrigated with industrial wastewater, canal water, and a 1:1 mixture, during 2018. The studied *Brassica* plants were generally closely efficient in remediating toxic metals found in wastewater irrigated soil. Substantial differences between *Brassica* and *Eruca* plants/parts were recorded. For example, *B. napus* had significantly higher metal extraction or accumulation compared to *E. sativa* for Zn (71%), Cu (69%), Fe (78%), Mn (79%), Cd (101%), Cr (57%), Ni (92%). and Pb (49%). While the water and plant were the main predictors of metal extraction or accumulation, an interaction between the main effects substantially contributed to Cu, Mn, and Fe extractions from soil and accumulations in plants. Significant correlations between biological accumulation coefficient and biological transfer coefficient for many metals further supported the metal extraction or accumulation efficiencies as: *B. napus* > *B. juncea* > *B. rapa* > *E. sativa*. Root-stem mobility index correlation with stem-leaf mobility index indicated the metal translocation along the root-stem-leaf continuum. Therefore, we suggest that these crops may not be used for human or animal consumption when grown with industrial wastewater of toxic metal concentrations ≥ permissible limits. Rather these plants may serve as effective remediators of toxic metal-polluted soil.

Keywords: *Brassica*; canola; cauliflower; heavy metal; phytoremediation; pollution; soil health; toxic; wastewater

Citation: Ahmad, I.; Malik, S.A.; Saeed, S.; Rehman, A.-u.; Munir, T.M. Phytoremediating a Wastewater-Irrigated Soil Contaminated with Toxic Metals: Comparing the Efficacies of Different Crops. *Soil Syst.* **2022**, *6*, 77. https://doi.org/10.3390/soilsystems6040077

Academic Editors: Antonella Lavini and Mohamed Houssemeddine Sellami

Received: 14 September 2022
Accepted: 28 September 2022
Published: 10 October 2022

Publisher's Note: MDPI stays neutral with regard to jurisdictional claims in published maps and institutional affiliations.

1. Introduction

Worldwide, surface or groundwater supply for food crop production is limited. The available surface water is irrigating 17% of the arable land with crop production of 34% of the world's food demand. In Pakistan (5th most climate-change-impacted country on Earth), 85% of the food is produced by the lands irrigated with river water, canal water, and/or groundwater [1,2]. Climate warming combined with an increasing population (11.0 billion in 2100 [3]) is frequently predicted to result in diminishing surface and groundwater supplies [4,5]; therefore, irrigation water supplies may become increasingly constrained for food crop production. Hence, there is a tremendous potential to explore alternate but safe irrigation waters as well as their compatible crops to cope with the shortage of irrigation water and uncertainty in food security. While experiments have been conducted to explore alternate irrigation waters and compatible crops for successful phytoextraction or food production, controlled experimentation to compare the efficiencies of these waters and crops remains limited.

Canal water is the surface water which is obtained from an engineered distributary of a river—the distributaries can be thought of as small, constructed rivers. Irrigation water quality of canal water is regarded as the best (Table 1) compared to groundwater or wastewater (sewerage, industrial wastewater). Wastewater is one of the alternate waters drained from industries, for example, leather, tannery, packaging, textile, and pesticide. In addition to nutrients, the wastewater may contain one or more of the essential or non-essential heavy metals, i.e., zinc (Zn), copper (Cu), iron (Fe), manganese (Mn), cadmium (Cd), chromium (Cr), nickel (Ni), and lead (Pb) in excessive (toxic) amounts and may be toxic for soils, plants and/or humans [6–8]. Toxicity thresholds for many micronutrients or metals are reached at considerably lower doses [9]. The wastewater may also contain a large assortment of pollutants including dissolved salts, organic matter, hosts and/or pathogens [10], and unknown human or animal health product residues [11]. The constituents when encountering soil particles, occupy exchange sites, accumulate in the soil over time, and deteriorate soil health, resulting in cropland of low productivity. Crops grown on these lands, uptake the toxic ions and accumulate in roots, stems, leaves or fruits in different proportions.

Soil texture is one of the dominant controls over heavy metal adsorption on soil particle exchange sites and inter-lattice spaces [12]. Light textured soil (sandy, sandy loam, and silt loam) particles exhibit lower surface area, cation exchange capacity (CEC), and surface charge density than those of heavy-textured soils (clay, clay loam, silty clay) [13]. Other soil properties such as pH, redox potential, hydraulic conductivity, and tendency to uptake, accumulate and transport heavy metals to various plant parts also differ with texture [14].

Heavy metals are bonded to particulate organic and inorganic matter in the wastewater [14] and become in equilibrium with the soil solution following the soil is irrigated with wastewater. Heavy metals are mobilized to or accumulate in plant tissues via plant-root-driven mechanisms: (1) plant root secretion of metal-chelating compounds (called phytosiderophores) into the rhizosphere, (2) plant root biochemical reduction of ions bonded to soil particles, and/or (3) plant root rhizosphere acidification by humic substances [15] followed by the exchange of metals with protons [16]. Different plants show significantly different capacities for heavy metal toxicity tolerance using exclusion, inclusion, or selective absorption of metals in amounts equivalent to nutrient requirements [17,18]. These metals are bioaccumulated in plant edible portions—roots, stem, leaves, and/or fruits [19]. Different metals show different threshold levels to become toxic to soil properties or plant tissues. Zn toxicity in plants is commonly indicated by leaf chlorosis [20,21] while Cu toxicity additionally causes suppressed root growth.

Plant species have been used to remediate heavy metal polluted soils—a technique called phytoremediation which involves the use of plants for the extraction of contaminants from soils and accumulating them in their above- and below-ground parts [22]. While phytoextraction rate is dependent on the bioavailability of heavy metals, different phytoextractants (*Brassica juncea*, *Eruca sativa*, *Brassica rapa*, *Brassica napus*) have different efficiencies owing to their distinct growth rates, depth of root systems, and accumulation and translocation capabilities, e.g., [23]. These phytoremediators are amongst the efficient growing and high biomass yielding crops [24], though their efficiencies may differ. For example, *B. juncea*, *B. rapa* and *B. napus* are better at removing Zn than they can remove Cu. *Brassica* taxa are the best at removing a range of heavy metals (Zn, Cu, Fe, Mn, Cd, Cr, Ni, and Pb), though *B. juncea* removed the highest amounts of these metals and accumulated in the stem [23,24]. *E. sativa* is also grown on heavy metals-laden soils and has been found to have good tolerance and adaptability to wastewater irrigated soils [25].

To compare the phytoremediation efficiencies of the test crops (*B. juncea*, *E. sativa*, *B. rapa*, *B. napus*) irrigated with wastewater, we compared ratios of metal concentrations between soil and plant parts as biological accumulation coefficient (BAC); biological transfer coefficient (BTC); mobility index at soil-root (MISR); mobility index at root-stem (MIRS);

mobility index at stem-leaf (MISL) [19]. Ratios of ≥1 indicate effective movement and/or accumulation from soil to root or root to stem or stem to leaf. The higher the coefficient value or ratio, the greater would be the remediation or accumulation efficiency of a crop [19,26–28], and the greater would be the risk to the health of humans consuming these crops.

We hypothesized that (1) the type of irrigation water will influence the heavy metal extractions or concentrations in the studied crop plants/parts, and (2) heavy metal BAC, BTC, MISR, MIRS, and MISL of the test crops will be ≥1. Specific research objectives were to (1) compare the phytoextraction capacities of four common crops in response to irrigation with canal water, wastewater, and their 1:1 mix, and (2) compare the different indices for heavy metal uptake (BAC, BTC, MISR, MIRS, and MISL) by these crops.

2. Materials and Methods

2.1. Experimental Design

A pot experiment was conducted in a greenhouse at Cotton Research Station, Multan, Pakistan to quantify and compare the phytoextraction or phytoaccumulation capacities, BAC, BTC, MISR, MIRS, and MISL of four high biomass-producing oil-seed or vegetable species: *B. juncea* (Raya), *E. sativa* (Taramira), *B. rapa* (Toria) and *B. napus* (Canola)—all grown on a homogenous silt loam soil. The authors compared the role of soil texture in phytoextraction or phytoaccumulation by comparing sandy loam and clay loam textures (and their characteristics) in their previous research at this site [29], therefore, in this experiment, the factor already studied at this site is not included to focus on the remaining factors and sustain clarity and brevity of results. Three types of irrigation water were used: canal water (control), industrial wastewater (treatment) and, a 1:1 (v/v) mix of canal water, and wastewater (treatment).

On 19 September 2018, a total of 36, 20 kg capacity pots were taken—each was filled with 16 kg silt loam soil collected from 0–20 cm deep cultivated/polluted land irrigated with wastewater from Industrial Estate Wastewater Disposal Station (IEWDS), Multan (IEWDS; 30°11′66″ N, 71°32′67″ E). The IEWDS collects wastewater from approximately 190 industrial units—mostly of leather, tannery, packaging, textile, and pesticides. Use of the wastewater is preferred by urban/suburban farmers owing to its high concentrations of nutrients and, free and stable year-round availability, which support the livelihood of local farmers [30,31]. The infill soil was uniformly air-dried before filling the pots. For irrigation, canal water was collected from a distributary originating from the Chenab River, and irrigating the Government Agricultural Farms, Multan. Wastewater was obtained from IEWDS, Multan. Treatment of canal + wastewater mixture was prepared by mixing the waters in a 1:1 (v/v) ratio. The three water types were carried and stored in 25 liter plastic containers. Following Strack et al. 2019 [32], the waters were passed through Whatman 40 and then a series of filters: first through a 1.5 μm glass fiber and then through a 0.4 μm glass fiber filter the same day (continued through the next or following days) to retain any dissolved organic matter and, particulate organic and inorganic matter which may otherwise provide exchange sites for precipitation and cross-precipitation of the organic or inorganic metals and change salinity (ECiw, pHiw). Further, to minimize microbial degradation of heavy metals, the waters were stored in a laboratory at 4 °C before being used for irrigation. Before each irrigation, waters were allowed to warm to normal temperature and tested for any significant changes in salinity. No substantial changes in salinity were recorded ($p \geq 0.05$). Therefore, the waters we used for pot irrigation may have minimal organic matter/metals, changes in salinity, and subsequent precipitation or cross-precipitation—thus, no significant experimental source of uncertainty in results/interpretation was expected [33].

The pot experiment was organized in a completely randomized design [(4 crops × 3 water types) × 3 replications = 36 pots]. All pots were maintained at field capacity for approximately one month from 19 September 2018 to 14 October 2018. On 15 October 2018, a recommended, basal dose of NPK (60:30:30) was mixed with the topsoil, and then,

10 government-certified (disease-free), viable seeds of a study crop were sown in each pot as per the research design. The seeds were obtained from Punjab Seed Corporation, Government of the Punjab, Pakistan. Only the three most viable seedlings were allowed to grow in each pot. The three replicates from each treatment (3 waters × 4 crops) were subjected to standard agricultural practices during germination and growth. Overall, the plants received 8–9 h of direct or full-spectrum sunlight every day throughout the experiment. The greenhouse had transparent panels which may show an adequate or higher transpiration level [34]. On 30 December 2018, the crop plants, or parts from 36 pots were harvested, placed separately in paper bags (36 bags), and transported to the soil and water testing laboratory, Multan for sample preparation and chemical analyses.

2.2. Sampling and Analysis

Triplicate bulk surface soil samples (0–20 cm) were collected randomly from the Industrial Estate Wastewater Disposal Station-fed area of three acres. The soil was ensured to be enough for filling 16 kg soil in each of the 36 pots. The soil samples were air-dried, crushed, sieved to <2 mm, thoroughly mixed, and stored at room temperature before analyzing for some physicochemical properties and chosen heavy metal concentrations. Soil samples were analyzed for textural class, saturation pastes electrical conductivity (EC_s), and pH (pH_s) following methods described by the US Salinity Laboratory Staff [35]. Based on our laboratory limitations, only Zn, Cu, Fe, Mn, Cd, Cr, Ni, and Pb concentrations in soil and plants could be estimated. To quantify the planned water-soluble heavy metal concentrations in soil, 20 g of dry soil was extracted with 40 mL of deionized water [36] and the extractants were stored for chemical analyses.

Irrigation wastewater/samples were also collected around the soil sampling campaigns and prepared for chemical analyses and irrigation as explained in the 2nd paragraph of this section. Within a week of preparing irrigation water, a 50 mL sample was digested with 10 mL of concentrated HNO_3 at 80 °C until the solution turned clear [37]. The clear solution was then filtered through Whatman No. 42, diluted back to 50 mL using distilled water, and stored for chemical analyses.

Root, stem, and leaf parts of the four test crops were thoroughly cleaned for any dust material, washed sequentially with 1% HCl and double deionized water, air dried in shade for 24 h, and then oven dried at 70 °C until a constant weight was reached. The dry matter was ground to a powder form and sieved to <1 mm. A 1.0 g of the powder was digested with a di-acid mixture of $HClO_4$ and HNO_3 in a 1:2 ratio, respectively. The clear digest was filtered and diluted to 50 mL using distilled water and stored at $\leq$4 °C for chemical analyses.

Plant total and soil and wastewater soluble Zn, Cu, Fe, Mn, Cd, Cr, Ni, and Pb concentrations (mg L^{-1}) were determined from the stored extracts using an atomic absorption spectrophotometer (Model AAS Vario 6, Analytik Jena AG, Jena, Germany). Following Baker (1981); [19]), we calculated the BAC, BTC, MISR, MIRS, and MISL of heavy metals:

$$Biological\ Accumulation\ Coefficient\ (BAC) = \frac{Stem + Leaf\ concentration}{Soil\ concentration} \quad (1)$$

$$Biological\ Transfer\ Coefficient\ (BTC) = \frac{Stem + Leaf\ concentration}{Root\ concentration} \quad (2)$$

$$Mobility\ Index\ at\ Soil - Root\ (MISR) = \frac{Root\ concentration}{Soil\ concentration} \quad (3)$$

$$Mobility\ Index\ at\ Root - Leaf\ (MIRS) = \frac{Stem\ concentration}{Root\ concentration} \quad (4)$$

$$Mobility\ Index\ at\ Shoot - Leaf\ (MISL) = \frac{Leaf\ concentration}{Stem\ concentration} \quad (5)$$

2.3. Statistical Analyses

All data analyses were performed using the SPSS 26.0 package (SPSS, Chicago, IL, USA). We used a two-way (water and crop treatments) multivariate (Zn, Cu, Fe, Mn, Cd, Cr, Ni, Pb) ANOVA for quantifying the individual and interactive effects of the water and crop factors on the response variables of Zn, Cu, Fe, Mn, Cd, Cr, Ni, and Pb concentrations in crops. Separate models were run for plant parts. Regressions and correlations were also performed where meaningful. Data were normalized to $\log_{10}$ values when not normally distributed. Differences were significant when $p < 0.05$.

3. Results

Thoroughly mixed, pot infill soil had uniform texture, salinity (ECs = 3.84 dSm^{-1}, pH = 8.6; Table 1), and toxic levels (mg L^{-1}) of studied heavy metals (Table 1). The original irrigation water types (canal water, wastewater) were different in the EC_{iw} (ANOVA: $F_{1,5} = 121.50$, $p < 0.001$) and pH_{iw} (ANOVA: $F_{1,5} = 121.50$, $p < 0.001$) and heavy metal concentrations. Our study focused on some key drivers of phytoremediation and bioaccumulation, i.e., water type and crops. Other key drivers of phytoremediation or bioaccumulation, such as soil texture have been studied at the experimental site [29].

Table 1. Textural class of soil and, salinity and heavy metal concentrations of the soil and irrigation waters used in the study.

Soil Texture or Water Source	Salinity		Heavy Metal Concentration (mg L^{-1})							
	EC_s/EC_{iw} (dSm^{-1})	pH_s/pH_{iw}	Zn	Cu	Fe	Mn	Cd	Cr	Ni	Pb
Silt loam soil	3.84	8.64	4.52	1.74	15.36	4.62	1.02	0.18	0.20	2.76
Canal water	1.02	7.20	0.01	0.01	0.03	0.04	0.07	0.02	0.02	0.06
Wastewater	4.18	7.41	0.26	0.12	1.28	0.24	0.19	0.11	0.29	1.14
1:1 mix water	2.72	7.34	0.13	0.08	0.67	0.15	0.15	0.07	0.16	0.63
Permissible limits (water) †	1.5	6.5–8.5	2.00	0.20	5.00	0.20	0.01	0.01	0.20	0.50

† Recommended maximum concentration in irrigation water for crops [38,39]. ECs and pHs, and ECiw and pHiw denote electrical conductivity and pH of soil and irrigation water, respectively. For all pots, surface soil (0–20 cm) was collected from Industrial Estate Disposal Station, Multan.

Overall, the main effects of water and crop and their interaction were significant (two-way MANOVA, Wilks' lambda test: $p < 0.001$; Table 2) on heavy metal concentrations in the studied crops irrigated with three water types. However, the interaction was significant only for Cu, Fe, and Mn contents in plants in response to various waters. The Zn, Cd, Cr, Ni, and Pb concentrations did not differ.

More specifically, the irrigation water type was the dominant control over most metal contents in each of the test crops. It reflects that different water types produced different responses in a plant species, with overall highest metal concentrations recorded in response to wastewater followed by the 1:1 mix and canal water in that order. When compared, the four plant species differed in remediation or accumulation capacities, however, similarities were not uncommon, for example, *B. napus* was not different from *E. sativa* for the extraction of Cd, and *B. juncea* for Mn uptake and accumulation. Likewise, *B. juncea* and *B. rapa* had similar responses for their Cr concentrations. Bottom-line, extraction capacities were in the order of *B. napus* > *B. juncea* > *B. rapa* > *E. sativa* in plants and root > stem > leaf in plant parts.

Individually, water type had a significant effect (two-way MANOVA: Table 2; Figure 1A–D) on plant part metal concentrations, except: root Zn ($p = 0.067$), stem Zn ($p = 1.000$), root Cu ($p = 1.000$), stem Cu ($p = 0.678$), leaf Cu ($p = 0.211$), leaf Fe ($p = 1.000$), root Cd ($p = 0.001$), root Cr ($p = 0.424$), and leaf Cr ($p = 1.000$). Wastewater had the highest impact on extraction of all metals by roots > stem > leaves, followed by the impact of 1:1 combined water mix in the same order on plant parts.

Table 2. Results of a two-way MANOVA for phytoextraction of heavy metals by root, stem, and leaf of study crops (*B. juncea*, *E. sativa*, *B. rapa*, and *B. napus*) irrigated with canal water, wastewater, and canal + wastewater (1:1; v/v) in Multan.

Source	F/p	Zn	Cu	Fe	Mn	Cd	Cr	Ni	Pb
Water ($df_{2,24}$)	$p\leq$	0.000	0.000	0.000	0.000	0.000	0.000	0.000	0.000
root	F	13.81	64.06	44.35	134.21	4.71	8.17	3.68	12.47
	$p\leq$	0.000	0.000	0.000	0.000	0.020	0.000	0.040	0.000
stem	F	57.28	207.92	186.86	268.56	8.91	23.26	16.04	61.93
	$p\leq$	0.000	0.000	0.000	0.000	0.000	0.000	0.000	0.000
leaf	F	170.84	179.71	236.73	398.76	7.59	31.51	4.91	33.17
	$p\leq$	0.000	0.000	0.000	0.000	0.000	0.000	0.010	0.000
Crop ($df_{2,24}$)	$p\leq$	0.000	0.000	0.000	0.000	0.000	0.000	0.000	0.000
root	F	121.30	429.21	861.94	80.14	86.18	17.21	11.29	65.25
	$p\leq$	0.000	0.000	0.000	0.000	0.000	0.000	0.000	0.000
stem	F	243.51	305.18	640.30	333.54	31.82	33.66	9.05	114.00
	$p\leq$	0.000	0.000	0.000	0.000	0.000	0.000	0.000	0.000
leaf	F	497.12	189.28	537.08	273.62	15.78	23.48	8.59	263.15
	$p\leq$	0.000	0.000	0.000	0.000	0.000	0.000	0.000	0.000
Water × Crop ($df_{6,24}$)	$p\leq$	ns	0.000	0.000	0.00	ns	ns	ns	ns
root	F	0.99	14.85	3.29	9.71	0.80	0.95	0.06	0.32
	p	ns	0.00	0.02	0.00	ns	ns	ns	ns
stem	F	2.48	27.80	13.61	16.51	0.27	1.70	0.24	0.81
	p	0.05	0.00	0.00	0.00	ns	ns	ns	ns
leaf	F	6.83	24.04	1.18	26.50	0.43	1.45	0.08	0.35
	p	0.00	0.00	ns	0.00	ns	ns	ns	ns

F_{stat} = variation between sample means/variation within samples; Higher the F value greater is the variation between sample means relative to the variation within the samples. df represents the degree of freedom to choose from the number of observations available for a variable. $p \leq 0.05$ means we are 95% confident to reject the null hypothesis which states no difference among the group means (for example, Zn extraction in response to three water types). n.s represents non-significant. Overall, the effects of water and crop and their interaction were significant ($p < 0.001$; Wilks' lambda test) on heavy metal concentrations in the root, stem, and leaf of the study crops irrigated with three different water types. Exceptions are debated in the discussion section.

The bioaccumulations by plant parts were also significantly different between crops (two-way MANOVA: Table 2; Figure 1A–D), except: *B. juncea* did not differ from *B. napus* for leaf Zn ($p = 1.000$), stem Cu ($p = 0.773$), and leaf Fe ($p = 0.609$), from *E. sativa*, *B. rapa* and *B. napus* for root Cd ($p = 1.000, 1.000, 0.711$, respectively) and from *E. sativa*, *B. rapa* for leaf Cr concentrations ($p = 0.338, 0.817$, respectively); *E. sativa* did not differ from *B. rapa* for *B. rapa* for stem and leaf Zn ($p = 0.130$), root and stem Cu ($p = 0.424, 1.000$, respectively), stem Cd ($p = 0.314$) and root Cr concentrations ($p = 1.000$).

Overall, the effects of water and crop treatments (individually and interactively) on BAC, BTC, MISR, MIRS and MISL were significant (two-way MANOVA, Wilks' lambda test: $F_{26,22} = 283.97$, $p < 0.001$; $F_{39,33} = 898.25$, $p < 0.001$; $F_{78,66} = 14.61$, $p < 0.001$, respectively). Comparisons between the water treatments showed that all the wastewater irrigated crops had the highest values of BAC and BTC followed by the lower coefficient values of the 1:1 mixed water irrigated crops while the canal irrigated crops had the lowest BAC and BTC values (Table 3). Although water treatment had a significant effect on overall MI ($p < 0.001$), however, crop to crop comparisons for MISR, MIRS, and MISL had mixed results; however, many of the indexes had values higher than 1.0 (Figure 2).

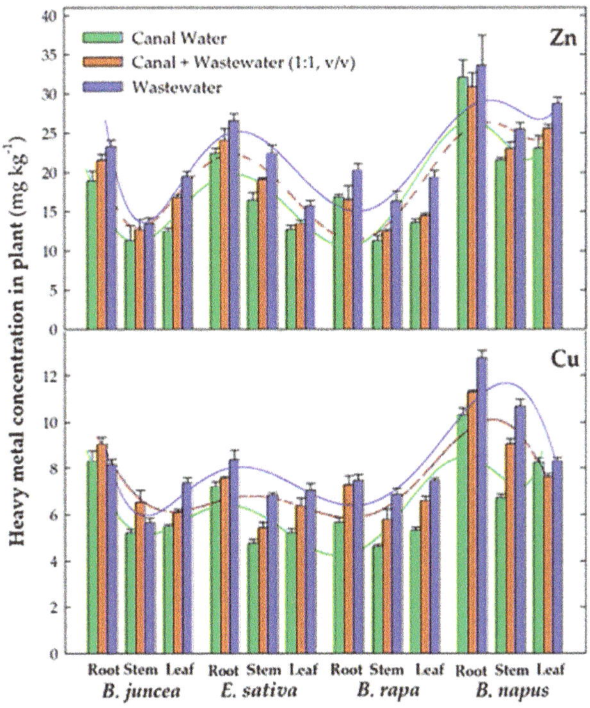

(**A**)

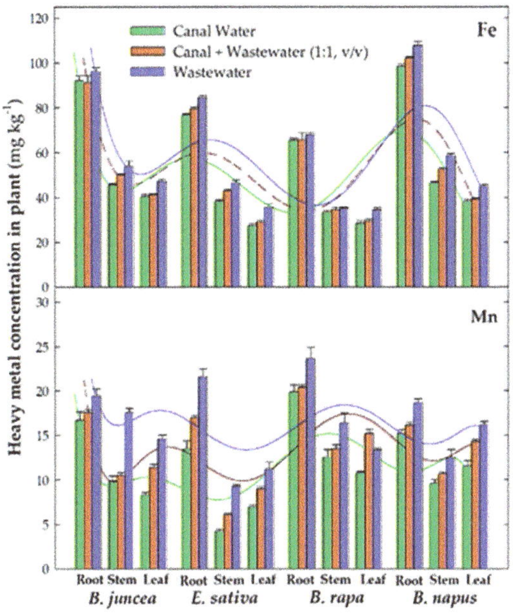

(**B**)

Figure 1. *Cont.*

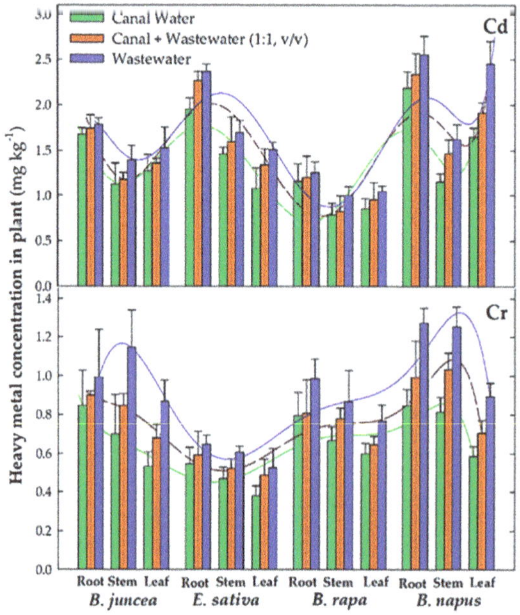

(**C**)

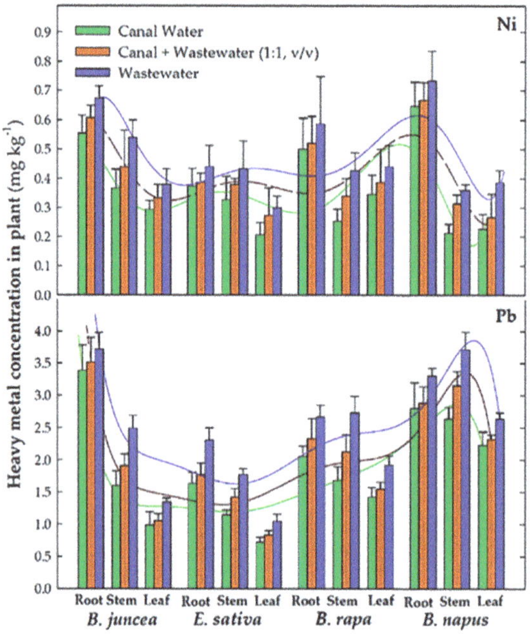

(**D**)

Figure 1. Impact of irrigation water type on Zn and Cu (**A**), Fe and Mn (**B**), Cd and Cr (**C**), and Ni and Pb (**D**) concentrations (mg kg^{-1} dry weight) in the root, stem, and leaf of *B. juncea*, *E. sativa*, *B. rapa* and *B. napus* in a pot experiment. Bars and line of same color represent one water type. The statistical significances or *p* values (instead of significance letters on bars) are provided in Table 2.

Table 3. Results of a two-way MANOVA for biological accumulation coefficient (BAC) and biological transfer coefficient (BTC) of studied crops, and mobility index (MI; soil-root, root-stem, and stem-leaf) of these crops irrigated with canal water, wastewater, and canal + wastewater in Multan.

Source	F/p	Zn	Cu	Fe	Mn	Cd	Cr	Ni	Pb
Water ($df_{2,24}$)									
BAC	F	164.65	322.04	427.26	690.84	14.89	34.19	20.59	98.36
	p	0.000	0.000	0.007	0.001	0.007	0.000	0.002	0.001
BTC	F	11.41	41.91	65.83	83.42	1.33	1.13	3.68	4.562
	p	0.000	0.006	0.007	0.007	ns	ns	0.042	0.021
MI (soil-root)	F	22.20	60.70	20.10	118.08	1.73	5.69	2.79	16.72
	p	0.000	0.005	0.005	0.009	ns	0.010	0.081	0.000
MI (root-stem)	F	3.69	29.86	54.87	94.57	1.84	1.16	4.71	10.53
	p	0.040	0.004	0.001	0.007	ns	ns	0.022	0.000
MI (stem-leaf)	F	0.57	22.43	22.89	40.80	0.24	0.00	0.34	2.19
	p	ns	0.000	0.001	0.001	ns	ns	ns	ns
Crop ($df_{3,24}$)									
BAC	p	0.002	0.000	0.000	0.000	0.004	0.000	0.000	0.000
BTC	p	0.001	0.000	0.001	0.000	0.029	ns	0.000	0.000
MI (soil-root)	p	0.000	0.000	0.000	0.001	0.002	0.001	0.000	0.000
MI (root-stem)	p	0.000	0.001	0.009	0.000	0.068	ns	0.001	0.005
MI (stem-leaf)	p	0.001	0.008	0.005	0.000	0.008	0.007	0.011	0.088
Water × Crop ($df_{6,24}$)									
BAC	p	ns	0.000	0.000	0.004	ns	ns	ns	ns
BTC	p	ns	0.000	ns	0.002	ns	ns	ns	ns
MI (soil-root)	p	ns	0.005	ns	0.000	ns	ns	ns	ns
MI (root-stem)	p	ns	0.010	0.010	0.000	ns	ns	ns	ns
MI (stem-leaf)	p	0.0100	0.001	0.000	0.007	ns	ns	ns	ns

Overall, the effects of water and crop and their interaction were significant ($p < 0.001$; Wilks's lambda test) on BAC, BTC, and MI of the four crops irrigated with three different water types. n.s represents non-significant.

The BAC-BTC Pearson correlations were found significant for Zn, Cu, Fe, Mn, and Cd contents of *B. napus*, Zn, Fe, and Mn contents of *E. sativa*, Mn, and Cr contents of *B. juncea*, and only Fe contents of *B. rapa*. (Figure 2). While most MIRS–MISL correlations (R^2) for the studied metals and crops were significant and positive, we found some relationships were negative but meaningful, for example, Mn MIRS–MISL relationship in *E. sativa* and Ni MIRS–MISL relationship in *B. juncea* (Figure 3).

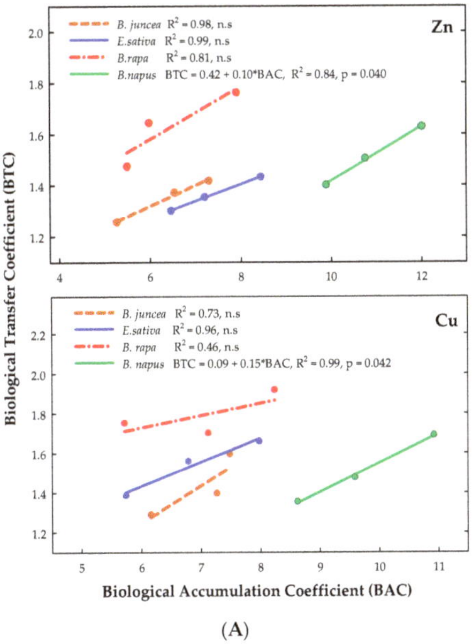

Figure 2. *Cont.*

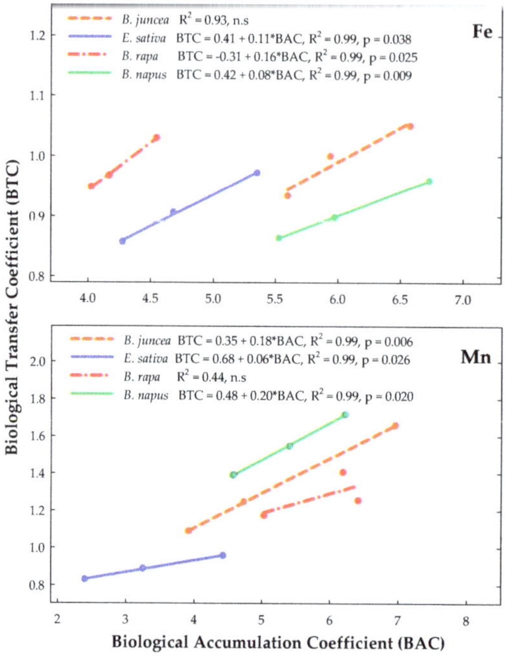

(B)

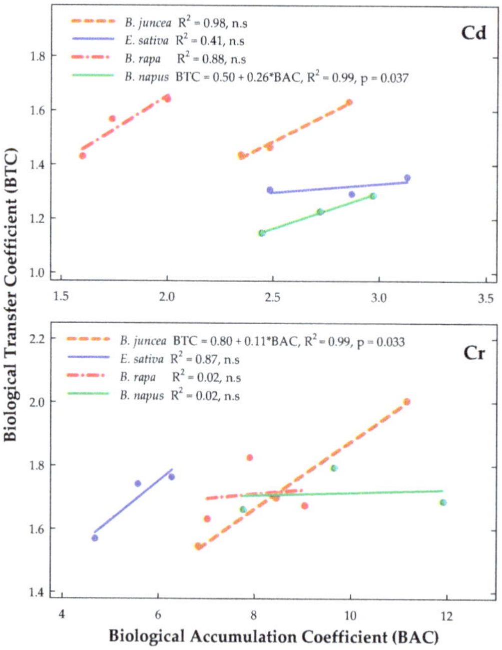

(C)

Figure 2. *Cont.*

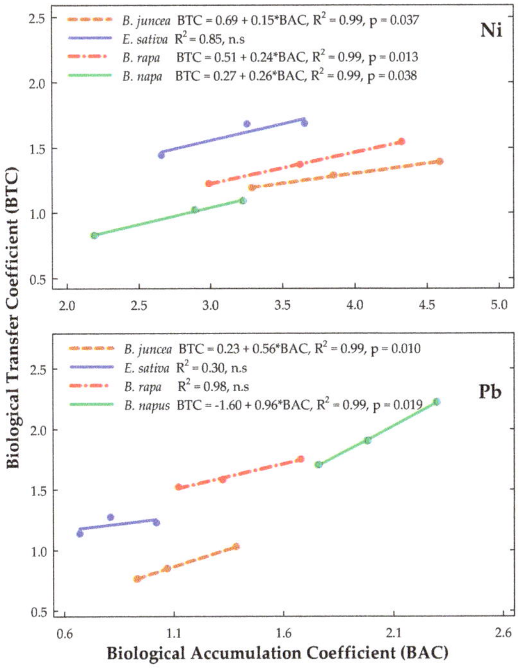

(D)

Figure 2. Relationships between biological accumulation coefficient (BAC) and biological transfer coefficient (BTC) of test crops for Zn and Cu (**A**), Fe and Mn (**B**), Cd and Cr (**C**), and Ni and Pb (**D**) concentrations (mg kg^{-1} dry weight). Relationship is significant at $p < 0.05$. n.s represents non-significant ($p > 0.05$).

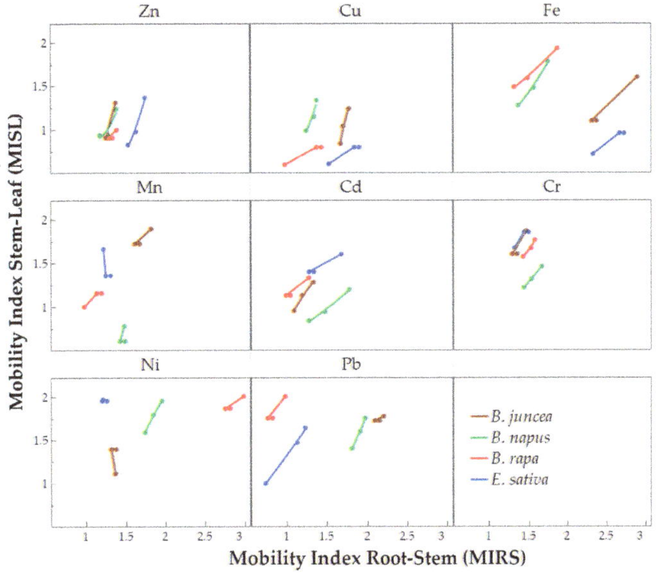

Figure 3. Relationships between mobility index at root-stem (MIRS) and mobility index at stem-leaf (MISL) of crops for heavy metals concentrations (mg kg^{-1} dry weight).

4. Discussion

Unlike most previous work, we conducted multivariate research—it involved wastewater-fed soil, polluted with toxic mineral heavy metals. Four different crops variably remediated the metals that accumulated in the root, stem, and leaf parts. While the three *Brassica* and one *Eruca* plant species and their parts were generally effective for phytoextraction and phytoaccumulation (respectively), significant differences between species and among parts for metal extraction or accumulation were not uncommon—though, *B. juncea*, *B. rapa*, and *B. napus* were closer for metal removal efficiencies compared to *Eruca* plants. While the disparity in metal extraction or accumulation function is generally attributed to the type of irrigation water, nature of the toxic metals, plant species/physiology, and/or soil/plant redox potentials [40], why the different water types produced different phytoremediation responses in the studied plants is discussed in detail in the following paragraphs. Our findings have important implications for (1) urban and suburban farmers who use industrial wastewater for cropping and/or planning to remediate their polluted soils, and (2) industrial ecologists who seek to remediate toxic heavy metal contaminated soils or treat contaminated waters using treatment wetlands in industrial settings.

4.1. Types of Irrigation Water and Phytoremediators

Based on the heavy metal toxicity criteria for irrigation water (Table 1), the wastewater concentrations of Cd, Cr, Ni, and Pb were well above the permissible limits and significantly higher than those in canal water. The 1:1 combined water still had higher than normal concentrations, except Ni. We did not manipulate the excessive metal concentrations in the 1:1 mix, but rather adopted a holistic approach and used the 1:1 mix as such to make this work practically useful for growers. Using the 1:1 mix without further dilution provided us with an opportunity to assess the extraction or accumulation responses of test plants to waters of varying toxic metal concentrations. Additionally, the metal accumulations by roots, stems, and leaves in response to these abnormal waters were also measured. Four important crop plants belonging to the *Brassicaceae* family (mustards) namely *Juncea* (Raya), *rapa* (Toria), *napus* (Canola), and *sativa* (Taramira) were used as phytoremediators.

4.2. Effect of Wastewater on Phytoremediation or Phytoaccumulation

Overall, different water types produced different responses in plants. The plant metal concentrations responded to wastewater > 1:1 mix > canal water in all plants with the corresponding concentrations of Cd (1.6, 1.4, 1.3 mg kg^{-1}), Cr (0.9, 0.7, 0.6 mg kg^{-1}), Ni (0.5, 0.4, 0.3 mg kg^{-1}), and Pb (2.5, 2.1, 1.8 mg kg^{-1}), all respectively. *B. napus* extracted or accumulated the highest quantity of Cd, Cr, Ni, and Pb in response to wastewater (1.9, 1.2, 0.5, 3.2 mg kg^{-1}), 1:1 combined (1.6, 0.9, 0.4, 2.7 mg kg^{-1}), and normal water (1.5, 0.7, 0.4, 2.6 mg kg^{-1}) and *E. sativa* has the lowest contents. The plant responses reflect that the irrigation water type may be one of the strongest controls on heavy metal concentrations in each of the four, test plants. Another reason that water type is one of the strongest controls is that plants/roots absorb water from soil following the rise in negative water potential in plants. The higher the concentration of a metal in water, the higher could be the corresponding accumulation in plants. The reason is the mineral metals are uptaken by roots and mobilized to and accumulated in the stem and leaves, e.g., [41] with sap-flow and via the mass flow of water [42,43]. Though, the metal extraction (e.g., Ni, Pb), root accumulation, and subsequent transport to and accumulation in stem and leaves may be suppressed by the competitive concentrations of contemporary metals and/or nutrients in soil/water, for example, [44,45]. Differential uptake and accumulations may also be the results of exclusion or inclusion mechanisms, for example, [46]. However, few plants, for example, *B. napus* have evolved as a hyperaccumulator of Cd and Pb [47], which supports our finding of plant accumulations of the same metals in addition to Zn, Cu, Fe, and Cr we found accumulated in root > stem > leaf. The difference in plant metabolisms may also be one of the reasons different water types produced different metal content responses in plants [34]. But studying the plant metabolism, physiology or phenology was beyond the scope of this experiment. Other drivers of plant metal concentration, such as soil texture, are already published by authors [29]. Moreover, in this experiment, the soil texture may not be a contributor to the difference in plant response to different water types because we used a single homogenous soil/texture under controlled conditions (pot experiment in a greenhouse). As some of the experimental materials (soil texture, ECiw, pHiw) were the same for all treatments, therefore we agree with (Indoriaetal.(2013) [48]) who reported that plant genes involved in tolerance to heavy metals in *Brassica* plants are different from *Eruca* plants while the genes for tolerance to other biotic or abiotic stresses may not be different. However, the genes for plant tolerance to salinity and excessive concentrations of Pb and Ni may modulate each other's functions in some plants [49,50].

The increases in metal concentrations of root > stem > leaf were higher in response to wastewater compared to canal water. The influences of 1:1 mixed water on the metal contents of plant parts were diverse. Mechanisms involved in the movement of heavy metals along the soil–root–stem–leaf continuum comprise the uptake by roots and transportation into the root cells via transmembrane carriers for nutritional ions—the metals are further diffused into the xylem vessels and unloaded into the xylem sap, thereby reaching the aboveground parts of plants [40,51,52]. That is why most of the BAC and MIRS ratios were also found to be ≥ 1, except for Cd, which was similar to the previously reported data [45]. The exceptional response of plants to Cd uptake, transport, and accumulation might be the result of dominance by the contemporary nutrient or other competitive toxic metal concentrations [44,45]. In the absence of competition, the Cd is reported to have high water solubility and mobility through plant parts [44]. Cd concentrations in plants/parts (in response to different waters) we present are aligned with earlier research findings, e.g., [18,23,45]. Non-significance of BTC, MISR, MIRS, and MISL for Cd in this study compares with previous research findings (although involving different crop plants) reported by Baker [19].

4.3. Metal Extraction or Accumulation by Plants

Extraction of heavy metals was significantly different among plants: the highest by *B. napus* followed by *B. juncea* whereas, *E. sativa* extracted the lowest amounts of metals (not different from *B. rapa*), though root accumulated the highest amounts of metals followed by stem and the lowest by leaf in all crop plants (Figure 1). More specifically, *B. napus* had significantly higher metal extraction or accumulation compared to *E. sativa* for Zn (71%), Cu (69%), Fe (78%), Mn (79%), Cd (101%), Cr (57%), Ni (92%), and Pb (49%). The hyperaccumulation function of *B. napus* has been reported by Rattan, Datta [45], and is comparable with the findings of this study. Some exceptions were also recorded, for example, *B. napus* was not different from *E. sativa* for the accumulation of Ni, and *B. Juncea* for Mn. Likewise, *B. juncea* and *B. rapa* were similar in Cr concentrations. Overall, phytoextraction capacities were in the order of: *B. napus* > *B. juncea* > *B. rapa* > *E. sativa* in plants and root > stem > leaf in plant parts. The BAC, BTC, MISR, MIRS, and MISL values of all metals were also highest for *B. napus* except for Mn, which was the highest in *B. juncea*; however, the lowest Mn values were found in *B. rapa* which findings agree with those reported by Broadley et al. (2001; [53]) but disagree with Neilson and Rajakaruna (2012; [54]) who concluded that the rhizosphere processes, soil management, and the role of chelators may be more important than the physiological mechanisms (given in above paragraph) for the extraction and accumulation of metals. Contrary to their report, many accept that most crops accumulate heavy metals because of mass flow, exclusion, and/or inclusion mechanisms along the soil–root–stem–leaf continuum [47].

4.4. Metal Transfer or Accumulation along the Root-Stem-Leaf Continuum

Different water types produced different transfer or accumulation responses (BAC and BTC) in plants. Wastewater produced the highest values followed by the 1:1 combined water and canal water in that order (Table 3). Further, the metal transports or accumulatios (MISR, MIRS, MISL) increased as the irrigation water heavy metal concentrations increased—potentially, owing to the consistently higher negative water potential in the leaves compared to the potentials in stem and roots [55]. Additionally, dynamic evapotranspiration from leaves may result in in situ metal accumulations followed by lower accumulations in the stem and roots [55] in that order.

The four test plants had variably $\geq$1 BAC, BTC, MISR, MIRS, and MISL values which reflect that some crops were more efficient than others for the extractions, transports, or accumulations of some studied metals in response to the different water types. 1:1 BAC-BTC relationships showed that *B. napus* had the significant extraction, transport, or accumulation capacity for the largest number of test metals, including Zn, Cu, Fe, Mn, and Cd contents (Figure 2). The next greatest number of metal transports or accumulations (Mn, Cr) were shown by *B. juncea* and only Fe accumulation was found in *B. rapa*. Besides, these plant- and metal-specific, BAC-BTC significant regression models, the Pearson correlation coefficients shown by different crops were mostly very strong with a range of 0.76–0.99. Contrastingly, *B. rapa* did not show transfer–accumulation relationships for Cr and Cu, *B. napus* had no such relationship for Cr, and similarly, *E. sativa* did not reflect the relationship for Cd and Pb—reasons could be diverse based on previous reports, for example, the direct leaching nature of Ni, Cd, and Pb from coarse soils [40,56]. More specifically, the transitional nature and 10 valence electrons of Ni prefer long-term chelation with root exudates and/or organic acids secreted or present in the rhizosphere. Therefore, Ni or other metals leach or percolate with the mass flow of water or adhere to the root surface (due to the proton pump) until they are mineralized with free electrons required to move through the root membrane of the plant [56,57].

While most MIRS–MISL correlations (R^2) for the studied metals and crops were positive and significant, we found some relationships were negative but meaningful, for example, Mn MIRS–MISL relationship in *E. sativa* and Ni MIRS–MISL relationship in *B. juncea*. Based on most ratios $\geq$1 which reflect effective movement and/or accumulation from soil to root or root to stem or stem to leaf. The higher the coefficient value or ratio, the greater would be the phytoremediation or bioaccumulation efficiency of a crop [19,26–28], and the greater would be the risk to the health of humans consuming these vegetables. Further, there were significant correlations between most MIRS–MISL which reveal that movements or accumulations of most metals were maintained along the root-stem-leaf continuum, except the movements or accumulations of Mn in *E. sativa* and Ni in *B. juncea* were in the reverse direction—from leaves to roots. Mehes-Smith and Nkongolo [58] justified the exception of more Ni or other metal accumulations in roots than aboveground parts. Exposure to toxic metal irrigation water disturbs the plant cytological stability and causes significant mitotic disruption which leads to a heavy accumulation of toxic metals in roots. Dierssen, Herault [59] added that long-term exposure of roots to excessive metal concentrations may cause a variety of severe phenotypic syndromes. While the metal accumulation we debate, the plant/parts senescence or maturity could be one of the reasons for lower accumulation in aboveground plant parts [60,61]. While this research spanned only one growing season, repeated measures (seasons) research with the addition of soil texture and salinity variables may show consistency between the regression significance and Pearson correlation for most metal transports or accumulations along the root-stem-leaves route to generalize the results of this work. Further research is strongly recommended to investigate the edibility or valorization of the food crops irrigated with wastewater containing toxic metal concentrations.

5. Conclusions

The multivariate, toxic metal phytoextraction or accumulation we studied with the predictors of water and crop has important implications for (1) urban and suburban farmers who use wastewater for crops and/or planning to remediate their polluted soils, and (2) industrial ecologists who seek to remediate toxic heavy metal contaminated lands or treat contaminated waters using treatment wetlands in industrial settings. Irrigation with industrial wastewater contributes to an overall increase (potentially toxic levels) in heavy metal concentrations in plant roots > stem > leaf of all study crops such that *B. napus* > *B. juncea* > *B. rapa* > *E. sativa*. Overall, the BAC, BTC, MISR, MIRS, and MISL of the four plants or their parts were also affected by the wastewater irrigation in the same order. Therefore, these crops may not be used for human or animal consumption when grown with industrial wastewater of heavy metal concentrations $\geq$ permissible limits. Rather, these plants may be used for effective remediation of wastewater irrigated, heavy metal-contaminated soil. While this research spanned a single growing season, more repeated measures (seasons) research works with the addition of soil texture and salinity variables may show consistency between the BAC-BTC regression significance and Pearson correlation for most metal transports or accumulations along the root–stem–leaves route to support the results of this work. Further research is recommended to investigate the edibleness and/or valorization of the food crops irrigated with wastewater.

Author Contributions: Conceptualization, I.A., S.S. and T.M.M.; methodology, I.A. and S.A.M.; software, T.M.M.; validation, I.A. and S.S.; formal analysis, T.M.M.; investigation, I.A., S.S. and A.-u.R.; resources, S.S. and S.A.M.; data curation, I.A.; writing—original draft preparation, I.A.; writing—review and editing, T.M.M.; visualization, I.A. and A.-u.R.; supervision, S.S. and S.A.M.; project administration, S.S. and T.M.M.; funding acquisition, T.M.M. and S.S. All authors have read and agreed to the published version of the manuscript.

Funding: This research received funding from the Bahauddin Zakariya University, Multan, and Muhammad Nawaz Sharif University of Agriculture, Multan.

Institutional Review Board Statement: Not applicable.

Informed Consent Statement: Not applicable.

Data Availability Statement: The raw data supporting the findings of this study are available from the corresponding author upon reasonable request.

Conflicts of Interest: The authors declare no conflict of interest.

References

1. Qadir, M.; Wichelns, D.; Raschid-Sally, L.; McCornick, P.G.; Drechsel, P.; Bahri, A.; Minhas, P. The challenges of wastewater irrigation in developing countries. *Agric. Water Manag.* **2010**, *97*, 561–568. [CrossRef]
2. Syed, A.; Raza, T.; Bhatti, T.T.; Eash, N.S. Climate Impacts on the agricultural sector of Pakistan: Risks and solutions. *Environ. Chall.* **2022**, *6*, 100433. [CrossRef]
3. Gerland, P.; Raftery, A.E.; Ševčíková, H.; Li, N.; Gu, D.; Spoorenberg, T.; Alkema, L.; Fosdick, B.K.; Chunn, J.; Lalic, N.; et al. World population stabilization unlikely this century. *Science* **2014**, *346*, 234–237. [CrossRef]
4. IPCC. Climate Change: Impacts, Adaptation, and Vulnerability. Part B: Regional Aspects. In *Contribution of Working Group II to the Fifth Assessment Report of the Intergovernmental Panel on Climate Change*; Cambridge University Press: Cambridge, UK; New York, NY, USA, 2014.
5. Munir, T.M.; Perkins, M.; Kaing, E.; Strack, M. Carbon dioxide flux and net primary production of a boreal treed bog: Responses to warming and water-table-lowering simulations of climate change. *Biogeosciences* **2015**, *12*, 1091–1111. [CrossRef]
6. Cooper, R.C. Public health concerns in wastewater reuse. *Water Sci. Technol.* **1991**, *24*, 55–65. [CrossRef]
7. Lone, M.; Saleem, S.; Mahmood, T.; Saifullah, K.; Hussain, G. Heavy metal contents of vegetables irrigated by sewage/tubewell water in Hassanabdal area, Pakistan. *Pak. J. Arid Agric.* **2003**, *5*, 533–535.
8. DalCorso, G. Heavy Metal Toxicity in Plants. In *Plants and Heavy Metals*; Furini, A., Ed.; Springer: Dordrecht, The Netherlands, 2012; pp. 1–25.
9. Denton-Thompson, S.M.; Sayer, E.J. Micronutrients in Food Production: What Can We Learn from Natural Ecosystems? *Soil Syst.* **2022**, *6*, 8. [CrossRef]
10. Saab, M.T.A.; Daou, C.; Bashour, I.; Maacaron, A.; Fahed, S.; Romanos, D.; Khairallah, Y.; Lebbous, N.; Hajjar, C.; Saad, R.A.; et al. Treated municipal wastewater reuse for eggplant irrigation. *Aust. J. Crop Sci.* **2021**, *15*, 1095–1101. [CrossRef]
11. Martijn, E.-J.; Redwood, M. Wastewater irrigation in developing countries—Limitations for farmers to adopt appropriate practices. *Irrig. Drain.* **2005**, *54*, 563–570. [CrossRef]
12. Weng, L.; Temminghoff, E.J.; Van Riemsdijk, W.H. Contribution of individual sorbents to the control of heavy metal activity in sandy soil. *Environ. Sci. Technol.* **2001**, *35*, 4436–4443. [CrossRef] [PubMed]
13. Overesch, M.; Rinklebe, J.; Broll, G.; Neue, H.-U. Metals and arsenic in soils and corresponding vegetation at Central Elbe river floodplains (Germany). *Environ. Pollut.* **2007**, *145*, 800–812. [CrossRef] [PubMed]

14. Bradl, H.B. Adsorption of heavy metal ions on soils and soils constituents. *J. Colloid Interface Sci.* **2004**, *277*, 1–18. [CrossRef] [PubMed]
15. Arrobas, M.; de Almeida, S.F.; Raimundo, S.; da Silva Domingues, L.; Rodrigues, M.Â. Leonardites Rich in Humic and Fulvic Acids Had Little Effect on Tissue Elemental Composition and Dry Matter Yield in Pot-Grown Olive Cuttings. *Soil Syst.* **2022**, *6*, 7. [CrossRef]
16. Raskin, I.; Kumar, P.N.; Dushenkov, S.; Salt, D.E. Bioconcentration of heavy metals by plants. *Curr. Opin. Biotechnol.* **1994**, *5*, 285–290. [CrossRef]
17. Arduini, I.; Godbold, D.L.; Onnis, A. Influence of copper on root growth and morphology of *Pinus pinea* L. and *Pinus pinaster* Ait. seedlings. *Tree Physiol.* **1995**, *15*, 411–415. [CrossRef]
18. Sharma, R.; Agrawal, M.; Marshall, F. Heavy metal contamination in vegetables grown in wastewater irrigated areas of Varanasi, India. *Bull. Environ. Contam. Toxicol.* **2006**, *77*, 312–318. [CrossRef] [PubMed]
19. Baker, A.J. Accumulators and excluders-strategies in the response of plants to heavy metals. *J. Plant Nutr.* **1981**, *3*, 643–654. [CrossRef]
20. Cobbett, C.; Goldsbrough, P. Phytochelatins and metallothioneins: Roles in heavy metal detoxification and homeostasis. *Annu. Rev. Plant Biol.* **2002**, *53*, 159–182. [CrossRef]
21. Antoniadis, V.; Levizou, E.; Shaheen, S.M.; SikOk, Y.; Sebastian, A.; Baum, C.; Prasad, M.N.V.; Wenzel, W.W.; Rinklebe, J. Trace elements in the soil-plant interface: Phytoavailability, translocation, and phytoremediation—A review. *Earth Sci. Rev.* **2017**, *171*, 621–645. [CrossRef]
22. Marchiol, L.; Assolari, S.; Sacco, P.; Zerbi, G. Phytoextraction of heavy metals by canola (*Brassica napus*) and radish (*Raphanus sativus*) grown on multicontaminated soil. *Environ. Pollut.* **2004**, *132*, 21–27. [CrossRef] [PubMed]
23. Salt, D.E.; Blaylock, M.; Kumar, N.P.; Dushenkov, V.; Ensley, B.D.; Chet, I.; Raskin, I. Phytoremediation: A novel strategy for the removal of toxic metals from the environment using plants. *Nat. Biotechnol.* **1995**, *13*, 468–474. [CrossRef] [PubMed]
24. Ebbs, S.; Lasat, M.; Brady, D.; Cornish, J.; Gordon, R.; Kochian, L. Phytoextraction of cadmium and zinc from a contaminated soil. *J. Environ. Qual.* **1997**, *26*, 1424–1430. [CrossRef]
25. Kumar, K.; Gupta, S.; Baidoo, S.; Chander, Y.; Rosen, C. Antibiotic uptake by plants from soil fertilized with animal manure. *J. Environ. Qual.* **2005**, *34*, 2082–2085. [CrossRef] [PubMed]
26. Chinmayee, M.D.; Mahesh, B.; Pradesh, S.; Mini, I.; Swapna, T. The assessment of phytoremediation potential of invasive weed *Amaranthus spinosus* L. *Appl. Biochem. Biotechnol.* **2012**, *167*, 1550–1559. [CrossRef] [PubMed]
27. Galal, T.M.; Shehata, H.S. Bioaccumulation and translocation of heavy metals by *Plantago major* L. grown in contaminated soils under the effect of traffic pollution. *Ecol. Indic.* **2015**, *48*, 244–251. [CrossRef]
28. Ali, H.; Khan, E.; Sajad, M.A. Phytoremediation of heavy metals—Concepts and applications. *Chemosphere* **2013**, *91*, 869–881. [CrossRef] [PubMed]
29. Ahmad, I.; Malik, S.A.; Saeed, S.; Rehman, A.-u.; Munir, T.M. Phytoextraction of Heavy Metals by Various Vegetable Crops Cultivated on Different Textured Soils Irrigated with City Wastewater. *Soil Syst.* **2021**, *5*, 35. [CrossRef]
30. Khan, Z.M.; Asif Kanwar, R.M.; Farid, H.U.; Sultan, M.; Arsalan, M.; Ahmad, M.; Shakoor, A.; Ahson Aslam, M.M. Wastewater Evaluation for Multan, Pakistan: Characterization and Agricultural Reuse. *Pol. J. Environ. Stud.* **2019**, *28*, 2159–2174. [CrossRef]
31. Soncini, A.; Bocchiola, D.; Rosso, R.; Meucci, S.; Pala, F.; Valé, G. Water and Sanitation in Multan, Pakistan. In *Sustainable Social, Economic and Environmental Revitalization in Multan City: A Multidisciplinary Italian–Pakistani Project*; Del Bo, A., Bignami, D.F., Eds.; Springer International Publishing: Cham, Switzerland, 2014; pp. 149–162.
32. Strack, M.; Munir, T.M.; Khadka, B. Shrub abundance contributes to shifts in dissolved organic carbon concentration and chemistry in a continental bog exposed to drainage and warming. *Ecohydrology* **2019**, *12*, e2100. [CrossRef]
33. Khadka, B.; Munir, T.M.; Strack, M. Effect of environmental factors on production and bioavailability of dissolved organic carbon from substrates available in a constructed and reference fens in the Athabasca oil sands development region. *Ecol. Eng.* **2015**, *84*, 596–606. [CrossRef]
34. Assaad, R.D.A.; Saab, M.T.A.; Sellami, M.H.; Nemer, N. Kale (*Brassica oleracea* L. var acephala) production in soilless systems in the Mediterranean region. *J. Appl. Hort.* **2022**, *23*, 167–173. [CrossRef]
35. Richards, L. Diagnosis and improving of saline and alkaline soils. In *Agricultural Handbook*; US, Salinity Laboratory Staff: Washington, DC, USA, 1954.
36. Zia, M.H.; Meers, E.; Ghafoor, A.; Murtaza, G.; Sabir, M.; Zia-ur-Rehman, M.; Tack, F. Chemically enhanced phytoextraction of Pb by wheat in texturally different soils. *Chemosphere* **2010**, *79*, 652–658. [CrossRef]
37. Rice, E.W.; Baird, R.B.; Eaton, A.D. *Standard Methods for the Examination of Water and Wastewater*; American Public Health Association: Washington, DC, USA; American Water Works Association: Denver, CO, USA; Water Environment Federation: Alexandria, VA, USA, 2017.
38. Ahmad, J.U.; Goni, M.A. Heavy metal contamination in water, soil, and vegetables of the industrial areas in Dhaka, Bangladesh. *Environ. Monit. Assess.* **2010**, *166*, 347–357. [CrossRef]
39. Pescod, M. *Wastewater Treatment and Use in Agriculture-FAO Irrigation and Drainage—Paper 47*; Food and Agriculture Organization of the United Nations: Rome, Italy, 1992.

40. Babu, S.M.O.F.; Hossain, M.B.; Rahman, M.S.; Rahman, M.; Ahmed, A.S.S.; Hasan, M.M.; Rakib, A.; Emran, T.B.; Xiao, J.; Simal-Gandara, J. Phytoremediation of Toxic Metals: A Sustainable Green Solution for Clean Environment. *Appl. Sci.* **2021**, *11*, 10348. [CrossRef]
41. Kotrba, P.; Najmanova, J.; Maceka, T.; Ruml, T.; Mackova, M. Genetically modified plants in phytoremediation of heavy metal and metalloid soil and sediment pollution. *Biotechnol. Adv.* **2009**, *27*, 799–810. [CrossRef]
42. Khan, I.; Awan, S.A.; Rizwan, M.; Ali, S.; Hassan, M.J.; Brestic, M.; Zhang, X.; Huang, L. Effects of silicon on heavy metal uptake at the soil-plant interphase: A review. *Ecotoxicol. Environ. Saf.* **2021**, *222*, 112510. [CrossRef]
43. Ingwersen, J.; Streck, T. A regional-scale study on the crop uptake of cadmium from sandy soils: Measurement and modeling. *J. Environ. Qual.* **2005**, *34*, 1026–1035. [CrossRef]
44. Ismael, M.A.; Elyamine, A.M.; Moussa, M.G.; Cai, M.; Zhao, X.; Hu, C. Cadmium in plants: Uptake, toxicity, and its interactions with selenium fertilizers. *Metallomics* **2018**, *11*, 255–277. [CrossRef]
45. Rattan, R.; Datta, S.; Chhonkar, P.; Suribabu, K.; Singh, A. Long-term impact of irrigation with sewage effluents on heavy metal content in soils, crops and groundwater—A case study. *Agric. Ecosyst. Environ.* **2005**, *109*, 310–322. [CrossRef]
46. Peijnenburg, W.; Baerselman, R.; De Groot, A.; Jager, T.; Leenders, D.; Posthuma, L.; Van Veen, R. Quantification of metal bioavailability for lettuce (*Lactuca sativa* L.) in field soils. *Arch. Environ. Contam. Toxicol.* **2000**, *39*, 420–430. [CrossRef]
47. Delil, A.D.; Köleli, N.; Dağhan, H.; Bahçeci, G. Recovery of heavy metals from canola (Brassica napus) and soybean (Glycine max) biomasses using electrochemical process. *Environ. Technol. Innov.* **2020**, *17*, 100559. [CrossRef]
48. Indoria, A.K.; Poonia, S.R.; Sharma, K.L. Phytoextractability of Cd from Soil by Some Oilseed Species as Affected by Sewage Sludge and Farmyard Manure. *Commun. Soil Sci. Plant Anal.* **2013**, *44*, 3444–3455. [CrossRef]
49. Pulvento, C.; Sellami, M.H.; De Mastro, G.; Calandrelli, D.; Lavini, A. Quinoa Vikinga Response to Salt and Drought Stress under Field Conditions in Italy. *Environ. Sci. Proc.* **2022**, *16*, 5. [CrossRef]
50. Iftikhar, A.; Abbas, G.; Saqib, M.; Shabbir, A.; Amjad, M.; Shahid, M.; Ahmad, I.; Iqbal, S.; Qaisrani, S.A. Salinity modulates lead (Pb) tolerance and phytoremediation potential of quinoa: A multivariate comparison of physiological and biochemical attributes. *Environ. Geochem. Health* **2022**, *44*, 257–272. [CrossRef]
51. Lasat, M.M. Phytoextraction of Heavy Metals from Contaminated Soil: A Review of plant/Soil/Metal Interaction and Assessment of Pertinent Agronomic Issues. *J. Hazard. Subst. Res.* **1999**, *2*, 5. [CrossRef]
52. Luo, Z.-B.; He, J.; Polle, A.; Rennenberg, H. Heavy metal accumulation and signal transduction in herbaceous and woody plants: Paving the way for enhancing phytoremediation efficiency. *Biotechnol. Adv.* **2016**, *34*, 1131–1148. [CrossRef]
53. Broadley, M.R.; Willey, N.J.; Wilkins, J.C.; Baker, A.J.M.; Mead, A.; White, P.J. Phylogenetic variation in heavy metal accumulation in angiosperms. *New Phytol.* **2001**, *152*, 9–27. [CrossRef]
54. Neilson, S.; Rajakaruna, N. Roles of Rhizospheric Processes and Plant Physiology in Applied Phytoremediation of Contaminated Soils Using Brassica Oilseeds. In *The Plant Family Brassicaceae: Contribution towards Phytoremediation*; Anjum, N., Ahmad, I., Pereira, M., Duarte, A., Umar, S., Khan, N., Eds.; Springer: Dordrecht, The Netherlands, 2012; pp. 313–330.
55. Awa, S.H.; Hadibarata, T. Removal of heavy metals in contaminated soil by phytoremediation mechanism: A review. *Water Air Soil Pollut.* **2020**, *231*, 47. [CrossRef]
56. Page, V.; Feller, U. Selective Transport of Zinc, Manganese, Nickel, Cobalt and Cadmium in the Root System and Transfer to the Leaves in Young Wheat Plants. *Ann. Bot.* **2005**, *96*, 425–434. [CrossRef]
57. Du, J.; Guo, Z.; Li, R.; Ali, A.; Guo, D.; Lahori, A.H.; Wang, P.; Liu, X.; Wang, X.; Zhang, Z. Screening of Chinese mustard (*Brassica juncea* L.) cultivars for the phytoremediation of Cd and Zn based on the plant physiological mechanisms. *Environ. Pollut.* **2020**, *261*, 114213. [CrossRef]
58. Mehes-Smith, M.; Nkongolo, K.K. Physiological and Cytological Responses of Deschampsia cespitosa and Populus tremuloides to Soil Metal Contamination. *Water Air Soil Pollut.* **2015**, *226*, 125. [CrossRef]
59. Dierssen, M.; Herault, Y.; Estivill, X. Aneuploidy: From a physiological mechanism of variance to Down syndrome. *Physiol. Rev.* **2009**, *89*, 887–920. [CrossRef] [PubMed]
60. Liang, C.; Wang, Y.; Zhu, Y.; Tang, J.; Hu, B.; Liu, L.; Ou, S.; Wu, H.; Sun, X.; Chu, J. OsNAP connects abscisic acid and leaf senescence by fine-tuning abscisic acid biosynthesis and directly targeting senescence-associated genes in rice. *Proc. Natl. Acad. Sci. USA* **2014**, *111*, 10013–10018. [CrossRef] [PubMed]
61. Munir, T.M.; Westbrook, C.J. Comparison of Soil Nutrient Supply Patterns among Full and Drained Beaver Ponds and Undisturbed Peat in a Rocky Mountain Fen. *Wetlands* **2022**, *42*, 25. [CrossRef]

soil systems

MDPI

Article

Organic C Fractions in Topsoil under Different Management Systems in Northeastern Brazil

Adriano Venicius Santana Gualberto [1], Henrique Antunes de Souza [2], Edvaldo Sagrilo [2], Ademir Sergio Ferreira Araujo [3,*], Lucas William Mendes [4], Erika Valente de Medeiros [5], Arthur Prudêncio de Araujo Pereira [6], Diogo Paes da Costa [5], Renato Falconeres Vogado [7], João Rodrigues da Cunha [8], Marcos Lopes Teixeira [2] and Luiz Fernando Carvalho Leite [2]

1 Department of Agronomy, Campus de Bom Jesus, Universidade Federal do Piauí, Bom Jesus 64900-000, PI, Brazil
2 Embrapa Meio Norte, Teresina 64000-030, PI, Brazil
3 Campus da Socopo, Universidade Federal do Piauí, Teresina 64049-550, PI, Brazil
4 Centro de Energia Nuclear na Agricultura, Piracicaba 13400-340, SP, Brazil
5 Campus da Agronomia, Universidade Federal do Agreste Pernambucano, Garanhuns 55290-000, PE, Brazil
6 Cempus do Pici, Universidade Federal do Ceara, Fortaleza 60020-903, CE, Brazil
7 Campus de Areia, Universidade Federal da Paraiba, Areia 58397-000, PB, Brazil
8 CODEVASF, Teresina 64000-018, PI, Brazil
* Correspondence: ademir@ufpi.edu.br; Tel.: +55-86-998645449

Citation: Gualberto, A.V.S.; de Souza, H.A.; Sagrilo, E.; Araujo, A.S.F.; Mendes, L.W.; de Medeiros, E.V.; Pereira, A.P.d.A.; da Costa, D.P.; Vogado, R.F.; da Cunha, J.R.; et al. Organic C Fractions in Topsoil under Different Management Systems in Northeastern Brazil. *Soil Syst.* **2023**, *7*, 11. https://doi.org/10.3390/soilsystems7010011

Academic Editors: Antonella Lavini and Mohamed Houssemeddine Sellami

Received: 12 December 2022
Revised: 2 February 2023
Accepted: 3 February 2023
Published: 5 February 2023

Abstract: The conversion from native forest to other land-use systems can decline the soil organic carbon (SOC) in tropical soils. However, conservationist management could mitigate SOC losses, promoting the functioning and stability of agricultural soils. This study aimed to address the influence of conversion from native forest to different land-use systems on SOC fractions in Northeastern Brazil. Topsoil soil samples were collected in areas under pasture (PAS), no-tillage (NT1 and NT2), eucalyptus (EUC), and native forests of Cerrado in Northeastern, Brazil. Total organic C, microbial biomass (MBC), particulate (POC), and mineral-occluded organic C (MOC), as well as fulvic acids (C-FA), humic acids (C-HA), and humin (C-HUM) fractions were accessed. The results showed that land conversion maintained similar levels of humic fractions and total organic carbon (TOC) stocks in the PAS, NT1, NT2, and EUC as compared to native Cerrado. Soils with the input of permanent and diverse fresh organic material, such as NT2, PAS, and EUC, presented high levels of MBC and POC, and the lowest C-FA:TOC and C-HA:TOC ratios. The land conversion to agricultural systems that include cropping rotations associated with pasture species such as Mombasa grass and eucalyptus prevents topsoil losses of active C compartments in the Cerrado of the Brazilian Northeast. It suggests that sustainable and conservationist management should be emphasized to maintain and improve the status of soil organic C.

Keywords: Cerrado; no-tillage; soil quality; oxisols

1. Introduction

The conversion of native forests to agricultural land-use and management systems has promoted changes in soil properties and functions [1,2] mainly leading to a decline in soil organic C (SOC) [3]. This is particularly important to tropical soils since they present naturally low SOC content and significant C losses after the adoption of intensive land use [4]. Indeed, previous studies have shown that the conversion of tropical forests to intensive land use decreased by about 25% to 32% of the SOC content [5–7]. On the other hand, conservationist land uses can decrease the C losses and bring positive effects on SOC content [8,9].

SOC is an essential component for the suitable functioning and stability of soils, influencing their chemical, physical, and biological properties [10,11]. This component presents different fractions, being characterized as humic and non-humic, which have different recycling times and forms of protection [12]. The humic fractions (HF) represent the greatest fraction of SOC, which is highly persistent and stable [13], and contains three main compartments known as fulvic acids (C-FA), humic acids (C-HA), and humins (C-HUM) [14]. On the other hand, non-humic fractions (NHF) consist of a wide range of soluble substances and present more sensitivity to changes due to their rapid turnover [15]. Thus, the distinct characteristics presented by these SOC fractions may confer higher or lower stability, mainly against the effects of the conversion of native forests to different land-use and management systems.

Interestingly, it is well-known that land-use change affects the SOC content, especially in the topsoil layers. For example, a meta-analysis performed by Angers et al., and Luo et al. [16,17], demonstrated that the conservationist no-tillage (NT) system increased SOC concentration in the upper topsoil layers and the effects of NT on SOC stocks were particularly significant at a depth of 0.10 m. Similarly, Haddaway et al. [18] found that significant differences in SOC stocks across different soil management systems were noticeable in the upper soil layer and the effects disappeared when considering the full profile up to 150 cm depth. Increased SOC content in the topsoil is a relevant aspect of soil quality, considering that the soil surface is a vital interface associated with mechanisms that affect soil productivity and its environmental quality [19,20]. However, little information is available about the effects of the conversion from native forest to different land-use systems on SOC fractions in the uppermost layer of tropical soils. This is important since each fraction distinctly influences the soil functioning [21]. In addition, the assessment of SOC fractions will provide knowledge about how different land-use or management systems affect the potential losses, accumulation, mineralization, and humification processes of SOC [22,23].

In Northeastern Brazil, different land-use systems are adopted, such as no-tillage, silviculture, and pastures, which can distinctly influence the SOC status. For instance, previous studies in Brazil found that no-tillage practices increased the SOC content as compared to conventional farming [24], while the pasture system increased the SOC content as compared to a native forest [25]. In silvicultural systems, Araujo et al. [26] observed that the cultivation of eucalyptus did not reduce the SOC content as compared to native forests.

Although studies have reported positive or no significant changes in no-tillage, silviculture, and pastures on SOC content [27], little is known about the effects of these systems on SOC fractions and stocks in the topsoil of a Cerrado from the Northeast of Brazil. Here, we hypothesize that SOC and C fractions in the topsoil of areas converted from native Cerrado vegetation in the Brazilian Northeast are differently influenced by distinct agricultural systems. We also hypothesize that topsoil C status in well-managed agricultural systems can be maintained at similar values to those found in native vegetation. Our objective was to study the influence of no-tillage, eucalyptus, and pasture on SOC fractions and stock in soils from Northeastern Brazil.

2. Materials and Methods

2.1. Study Site

This study was carried out at Farm New Zealand, Uruçuí, PI, Brazil (07°33′08′′ S and 44°36′45′′ W; 378 m above sea level). The climate is Aw (Köppen) and the average temperature and rainfall are 27 °C and 817 mm, respectively [28]. The rainy season extends from November to May and the dry season, from June to October. The soil is classified as Oxisol (Yellow Latosol) [26]. In this study, four management systems were evaluated as follows: pasture (PAS); no-tillage with soybean, under maize straw, in a soybean/maize succession (NT1); no-tillage with maize, under grass straw, in a soybean/maize/Mombasa rotation (NT2); and eucalyptus (EUC) (Figure 1; Table S1). As a reference, we evaluated a native Cerrado forest (NF).

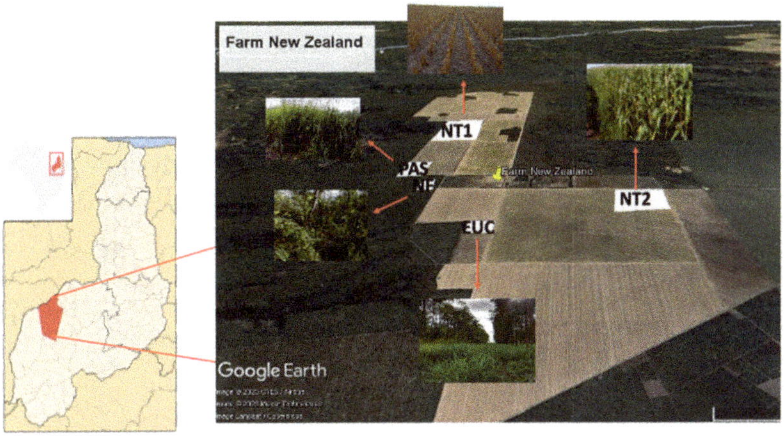

Figure 1. Map of Farm New Zealand showing the evaluated areas. PAS: Pasture species (Mombasa); NT1 and NT2: No-tillage system; EUC: Eucalyptus; NF: Native Cerrado forest.

The management systems PAS, NT1, and NT2 started in the 2004/2005 cropping season after the conversion of a native Cerrado to agricultural land use. After native forest removal, all areas received the application of lime (6 Mg ha^{-1}), which was incorporated at a depth of 0.22 m, using a 28 in disc plow. The area under EUC was implanted in the 2005/2006 cropping season using clones of *Eucalyptus urograndis* planted in the spacing of 3 m × 1.5 m (totalizing ~2200 plants per h), with each row spaced 14 m between them. Between the rows, rice followed by cowpea was sown and fertilized with 300 kg ha^{-1} of NPK 10-30-10. In the 2006/2007 cropping season, soybeans were sown using fertilization of 350 kg ha^{-1} of NPK 5-20-20. After soybean harvesting, Mombasa grass was sown for grazing. Fertilization during the implantation of the pasture was done using 150 kg ha^{-1} of NPK 06-21-06, containing 12% Ca, 3% S, and 0.3% Zn (w/w). Topdressing fertilization in the Mombasa grass between EUC rows was performed in March/2006, January/2007, and January/2008, with 200 kg ha^{-1} of NPK 20-00-20. The Mombasa grass was used for grazing at a stocking rate of 2.4 cow units ha^{-1} until 2009 and remained fallow from 2010 to 2017.

The NT1 area was cultivated with soybeans for two years following the conversion from native Cerrado. In 2007, maize was sown, followed by soybean (2008), millet (2009), maize, for two years (2009 and 2010), and cotton (2011). From this point, a maize/soybean succession was established until 2017, when soybeans were cultivated. In NT2, after native Cerrado conversion and in the same area, soybeans were sown followed by Mombasa grass used for grazing at a density of 2.4 animal unit ha^{-1}. This was repeated from 2006 to 2009. After 2010, a sequence of soybean/millet, cotton, maize, and soybean was used, and in 2016 Mombasa grass was sown followed by maize. The PAS area followed the same crop sequence as in NT1 until 2015. Mombasa grass was implanted in 2016 in the area for grazing at a density of 2.4 animal units ha^{-1}. In 2017, forage grass was desiccated with glyphosate and maize was cultivated.

The areas NT1, NT2, and PAS were fertilized according to the requirements of each plant species cultivated [29]. In 2011, dolomitic limestone was applied at 2.5 t ha^{-1}. In 2015, gypsum was applied at 1 t ha^{-1}, along with mono-ammonium phosphate at 160 kg ha^{-1}, potassium chloride at 120 kg ha^{-1}, ammonium sulfate at 250 kg ha^{-1}, and urea at 230 kg ha^{-1}.

2.2. Soil Sampling and Chemical Analysis

Soil sampling was performed in May 2017, at the end of the rainy season. The area within each management system was divided into five transects 20 m spaced from each other. In EUC, as a strategy to cover the complete system in a more representative way, each transect included samplings from the eucalyptus rows, the eucalyptus canopy projection outside the rows, and between the eucalyptus rows (Figure S1). Within each transect, five soil subsamples were taken (0–0.10 m depth) at points 20 m spaced apart. The five subsamples collected in each transect were pooled together to form a composite soil sample. In total, five replications (each one representing the composite sample from transects) were considered in each treatment (consisting of a total of 25 soil samples) covering an area of approximately 1 ha. Soil samples were sieved (2 mm) and homogenized for soil analysis. Soil pH was evaluated in a 1:2.5 soil/water extract; Ca^{2+}, Mg^{2+}, and Al^{3+} were extracted with KCl 1 mol L^{-1} – Ca^{2+} and Mg^{2+} were determined by atomic absorption spectrometry and Al^{3+} by titration; potential acidity (H+Al) was determined via extraction with 0.5 mol L^{-1} of calcium acetate and quantified by titration; and K^+ and available P were extracted with Mehlich-1 (H_2SO_4 0.0125 mol L^{-1} and of HCl 0.050 mol L^{-1})—the determination of K concentration was made through flame photometry and P was determined by colorimetry [30]. The values of the chemical properties are shown in Table S2.

2.3. Analysis of SOC Fractions

The soil microbial biomass carbon and nitrogen (MBC) were analyzed by the irradiation–extraction method [31,32] and the soil basal respiration was determined by quantifying CO_2 released after 7 days of incubation, under aerobic conditions in soil samples, with moisture content adjusted to 60% of field capacity [30]. The metabolic quotient (qCO_2) was obtained by the relationship between the soil basal respiration and MBC, according to the methodology described by Alef [33].

Total organic carbon (TOC) contents were determined by the 990.03 combustion method [34], employing an auto-analyzer, Leco CN628 (Leco Corp., St. Joseph. MI, USA). Carbon stock (C-stock) was obtained by the method of soil mass correction, using the soil bulk density (Ds) measured in the areas of each treatment and the native Cerrado forest as a reference [35]. This approach eliminates the influence of different soil bulk densities in over- or under-estimating the total C-stock across soil management systems. C-stock was calculated using the expression: C-stock= (TOC $\times$ Ds $\times$ ts), where ts represents the thickness of the soil layer considered.

Humic substance fractioning was performed according to the differential solubility technic, using the concepts of humic fractions established by the International Humic Substances Society, developed by [36], by obtaining the values of fulvic acids (C-FA), humic acids (C-HA), and humins (C-HUM). Physical fractioning of soil organic matter was performed according to [37], by obtaining the values of particulate organic carbon (POC) and mineral organic carbon (MOC).

2.4. Statistical Analysis

The soil was very homogenous across the different treatments. The data referring to soil biological attributes, carbon stocks, and chemical and physical fractioning of soil organic matter were checked for normality and homogeneity of variances and submitted for a one-way ANOVA, according to a completely randomized design. When significant, data were compared using the Tukey test (5% of probability). Additionally, a multivariate analysis was performed to compare the structure of SOC fractions among treatments using the principal components analysis (PCA) on log-transformed data. To explore the relationship between soil C fractions and microbial attributes, Spearman's rank correlation coefficients were applied, and the correction was made using Benjamini–Hochberg false discovery rate (FDR) method. Heatmaps were generated to further check for correlations. Significant ($p < 0.05$) positive and negative correlations are represented in blue and red, respectively. All statistical analyses were performed using the R software [38].

3. Results

The results showed no significant differences in humic fractions (C-HA, C-FA, and C-HUM) between the evaluated areas (Figure 2). In contrast, the microbial fraction (MBC) varied between sites. The topsoil in pasture and eucalyptus presented higher MBC values (124.7 and 117.3 mg kg^{-1}, respectively) than NT1 (54.6 mg kg^{-1}) and the native forest (69.9 mg kg^{-1}), while NT2 (97.7 mg kg^{-1}) had similar MBC values than the other sites.

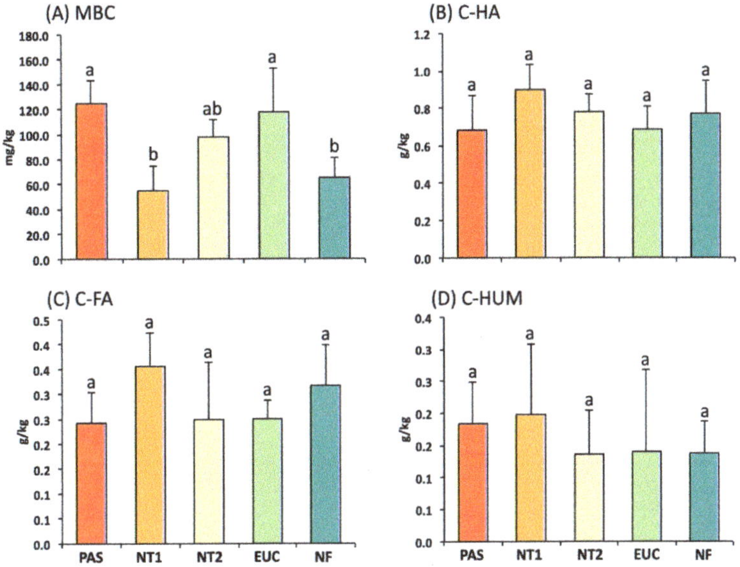

Figure 2. Microbial biomass carbon (MBC) (**A**), humic acid (C-HA) (**B**), fulvic acid (C-FA) (**C**), and humin (C-HUM) (**D**) fractions in the 0–0.10 m depth. Vertical lines on the bars indicate standard errors of means (n = 5). Treatments with different letters on the bars are significantly different ($p \leq 0.05$) by the Tukey test. PAS: Pasture species (Mombasa); NT1 and NT2: No-tillage system; EUC: Eucalyptus; and NF: Native Cerrado forest.

The values of MOC, TOC, and C-stock in topsoil did not vary between the areas under different management systems and the native forest. However, POC values in the topsoil were higher under native forest (0.047 g kg^{-1}) than under NT1 (0.022 g kg^{-1}) (Figure 3). The POC values under NT2 (0.026 g kg^{-1}), pasture (0.027 g kg^{-1}), and eucalyptus (0.031 g kg^{-1}) did not differ from each other, and native forest and NT1.

The ratios between C-fractions and TOC content showed variations between sites, except for the C-HUM:TOC ratio (Figure 4). The topsoil values of MBC:TOC were higher under pasture (1.15), NT2 (0.86), and eucalyptus (1.09) than under NT1 (0.41) and NF (0.48). In contrast, the highest topsoil values of C-HA:TOC and C-FA:TOC were observed under NT1 (8.32 and 3.31, respectively) as compared to pastures (6.13 and 2.16, respectively), and eucalyptus (4.85 and 1.77, respectively). Topsoil C-HA:TOC values in NT1 were also higher than those observed in the native forest (5.57), while C-FA:TOC in NT1 was higher than those observed in NT2 (2.10). The topsoil values regarding C-HUM:TOC did not vary among sites.

The PCA analysis explained 99.1% of the total variation in the first two axes of the graph and clustered the samples into two main groups (Figure 5A). Group 1 comprised pasture and eucalyptus that were correlated with MBC and MBC:TOC. Group 2 consisted of NT1 and native forests that were correlated with humic fractions, POC, and MOC. The heatmap showed the correlations among SOC fractions (Figure 5B). MBC correlated positively with MBC:TOC ratio and negatively with C-FA:TOC and C-HA:TOC ratios. The

C-FA and C-HA fractions correlated negatively with the MBC:TOC ratio and both fractions correlated positively with MOC. The C-FA:TOC ratio exhibited a positive correlation with C-HA:TOC.

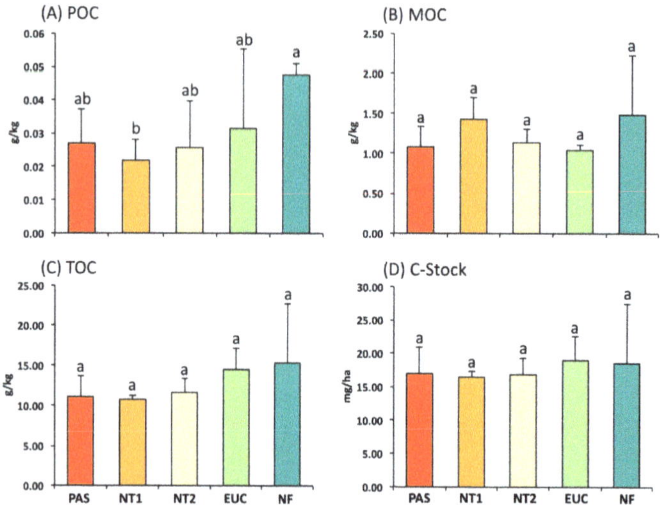

Figure 3. Particulate organic carbon (POC) (**A**), mineral organic carbon (MOC) (**B**), total organic carbon (TOC) (**C**), and carbon stock (C-Stock) (**D**) in the 0-0.10 m depth. Vertical lines on the bars indicate standard errors of means (n = 5). Treatments with different letters on the bars are significantly different ($p \leq 0.05$) by the Tukey test. PAS: Pasture species (Mombasa); NT1 and NT2: No-tillage system; EUC: Eucalyptus; and NF: Native Cerrado forest.

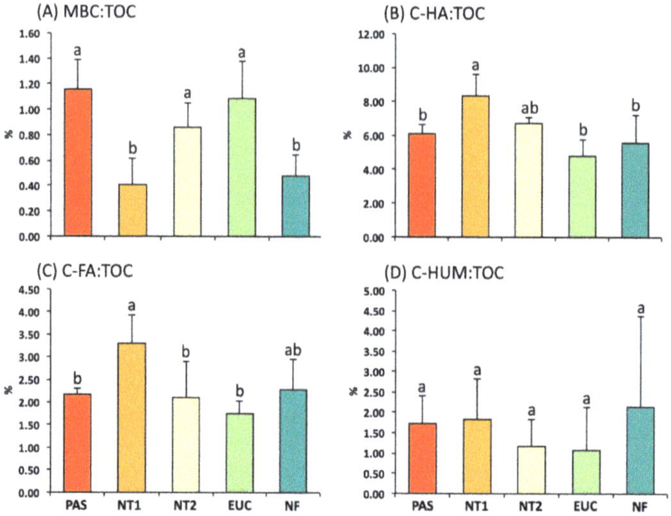

Figure 4. Microbial biomass carbon to total organic carbon (MBC:TOC) (**A**), humic acid to total organic carbon, (C-FA:TOC) (**B**), fulvic acid to total organic carbon (C-HA:TOC) (**C**), and humin to total organic carbon (C-HUM:TOC) (**D**) ratios in the 0–0.10 m depth. Vertical lines on the bars indicate standard errors of means (n = 5). Treatments with different letters on the bars are significantly different ($p \leq 0.05$) by the Tukey test. PAS: Pasture species (Mombasa); NT1 and NT2: No-tillage system; EUC: Eucalyptus; and NF: Native Cerrado forest.

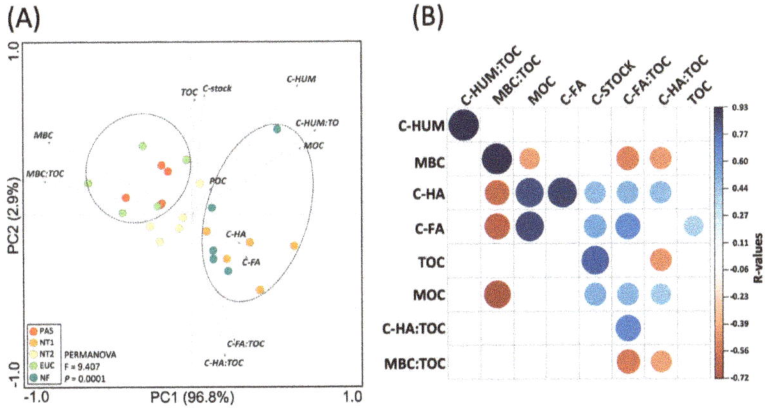

Figure 5. (**A**) Principal component analysis biplot comparing the structure of SOC fractions among treatments. The dashed lines in the graph indicate significant clusters (PERMANOVA, $p < 0.05$). (**B**) Heatmap showing the Spearman's rank correlation coefficients and statistical significance between SOC fractions. Blue and red colors indicate significant positive and negative correlations, respectively ($p < 0.05$).

4. Discussion

In general, land-use changes did not reduce TOC stocks under the conditions of this study, although previous studies have reported changes in TOC stock following land conversion from native forests to croplands in tropical soils [39–43]. Despite these previous studies, TOC stocks and their humic fractions remained at similar levels in PAS, NT1, NT2, and EUC as compared to the native Cerrado. The results showed a high standard deviation which probably occurred due to some variation along the transect [44], which contributed to the increase of the standard deviation. However, the statistical analysis was robust to show significant differences. Although no differences were observed in TOC stocks, MBC and POC fractions showed distinct responses according to the different agricultural systems ($p < 0.05$). These results are partly in line with the hypothesis that SOC and C fractions are differently influenced by distinct agricultural systems.

Changes in land use, vegetation cover, and soil management practices can increase or decrease TOC status in the soil, which depends on several factors. Plant biomass is the primary C source of the soil and, therefore, its quality and abundance strongly drive the dynamic of soil organic matter (SOM) [45]. Thus, an increase in SOM is observed when the rates of organic C inputs and incorporation are greater than the decomposition [46]. The SOM turnover occurs through the action of microbial-enzyme accessibility to the substrate, and the physical protection of soil C in aggregates plays an important role in controlling this process [40,47,48]. In this study, areas under land-use change presented diversification and abundance of plant biomass. The soil under NT1, NT2, and PAS were cultivated under crop rotation including legume (soybean) and grasses (maize and millet in NT1, while maize, millet, and pasture in NT2). Moreover, Mombasa grass was cultivated in PAS in 2016 and 2017. The management practices used in these areas ensured that a sufficient volume of organic residues in the soil environment was protected against fast decomposition due to the lack of soil disturbance. Particularly, the root system from grasses may have played an important role in TOC accumulation, as root biomass is considered the main source of C in the soil [48,49]. In EUC, the absence of a significant effect on TOC compared to NF could also be explained by the presence of Mombasa between rows and the low decomposition rate of eucalyptus leaves, as reported by Pinheiro et al. [39]. The management systems adopted following land conversion from the native forest in our study were based on sustainable practices such as crop rotation involving legumes and grasses or the adoption of perennial crops such as eucalyptus and Mombasa. The results obtained bring evidence that these management practices ensured the maintenance of soil C status after the land-use conversion, which is an essential condition for agricultural sustainability and terrestrial environmental stability [50], especially in highly weathered tropical soils [51]. Moreover, the land-use types studied, especially PAS, NT2, and EUC are some of the agricultural components of the crop-livestock-forest integrated systems, which is a technology increasingly used in intensive grain production systems in the Brazilian Cerrado. Our results therefore reinforce the agronomic and environmental feasibility of these systems in a climate change scenario.

Higher topsoil values of MBC in PAS and NT2 compared to NT1 and NF were probably a result of the inclusion of Mombasa grass in these systems, which ensured a high input of organic sources to microbes due to the exudation from the roots [52]. Moreover, grass leaves contain a greater portion of labile C fractions, while forest residues contain more recalcitrant fractions [53]. Soil microbes preferentially utilize grass-derived C as food since it is easily decomposed, allowing a further increase in SOC cycling and converting more external C into SOC [54]. The input of highly degradable substrates (plant litter with high proportions of labile fractions) on the soil surface possibly boosted microbial growth, leading to high MBC values (50). The soil under EUC also showed high values of MBC (119 mg/kg), which is possibly related to the high input of plant litter and the favorable environment provided by the trees, specifically reducing soil temperature, and maintaining high soil moisture [55]. Besides, eucalyptus rows were surrounded by Mombasa grass that was kept under fallow for eight years, which contributed with input of fresh organic material into the soil.

Land-use conversion to agricultural systems that involved crop rotation and pasture species such as Mombasa grass (NT2, PAS, and EUC), maintained POC values similar to those found in native forests. Similar trends were observed by de Moraes Sá et al. [41] in areas converted to pasture, where POC stocks did not change over time. Particulate organic carbon is a labile SOM fraction originating from newly decomposed litter biomass, greater root systems, and root exudates [56]. Although representing a small proportion of the TOC, this labile C fraction is an important component of active C pools [57]. Besides, it is considered the most sensitive indicator for land-use change because POC consists of fresh organic materials or materials with early stages of decomposition [55,58,59].

Pasture, NT2, and EUC were more efficient in incorporating organic C in the soil as microbial biomass than NT1 and NF, as indicated by the highest MBC:TOC ratios on these land-use systems. Conversely, amongst land areas converted from native forest, NT1 showed the highest proportions of humic substances in relation to TOC, notably for H-FA and H-HA. The lowest MBC:TOC in NT1 is consistent with the limited availability of permanent labile organic C for microbial activity, as also demonstrated by the lowest POC values in this management system and reinforced by the negative correlation of MBC:TOC with H-FA and H-HA. Soil MBC is a very important component in tropical soils because it represents an active pool of available nutrients for plant uptake [60], and was strongly correlated with MBC:TOC. Thus, changes in MBC:TOC are indicative of the organic matter input to the soil, microbial incorporation efficiency, soil carbon loss, and stabilization of SOC by mineral fractions [61]. Our data suggest that the high and continuous inputs of fresh material, especially by the root system of Mombasa grass (in NT2, PAS, and EUC) and eucalyptus plant residues (in EUC) were responsible for the more pronounced dynamic of C incorporation into microbial biomass. The importance of plant root systems in ensuring high MBC:TOC values is corroborated by [62], who stated that the MBC:TOC ratio is highly dependent on C inputs from the rhizosphere.

While high C inputs to the soil by root systems of pasture species favored an increased MBC:TOC ratio, the high C-FA:TOC and C-HA:TOC values observed in NT1 were conditioned by an opposite pattern, i.e., a lower efficiency in providing continuous input of fresh organic matter, especially from the root systems [63]. Such a pattern resulted in an exhaustion of the labile C fractions (e.g., POC and MBC) in NT1, leading to a more pronounced remaining proportion of humified C fractions compared to TOC. Humic substances represent a significant portion of total SOC and play an important agronomic role significantly influencing the quality and productivity of agricultural soils [21]. Despite that, the data from NT1 suggest that the higher proportion of humic substances in relation to TOC in this area compared to other land-use systems is a result of a lesser capacity to produce fresh material or the production of lower quality plant residues in NT1.

An important remark regarding NT1 is that with a few exceptions, a soybean-maize crop succession prevailed in this system. This sequence of crops is mentioned in the literature as a combination that ensures improved environmental conditions and increased profitability [64]. Nonetheless, our data suggest that for the conditions of extensive production systems in the Cerrado of Brazilian Northeast, the exclusive monocultures of soybean and maize in succession are not effective in ensuring a topsoil pool of active C fractions. This statement is reinforced by the data from Luo et al. [17], who found that increasing cropping frequency is a more efficient strategy to increase C input agroecosystems.

The changes in soil C status found in our study may partly be a consequence of the thin topsoil layer considered. We showed that intensive agricultural systems combining crop rotation and the use of pasture grasses and eucalyptus increases the contents of active C pools (MBC and POC) in the topsoil. A broad set of studies converge with our findings, showing that major changes in SOC status in no-tillage systems occur close to the soil surface, in the 0.5–0.10 m [16,17,65] or the 0.15 m soil layer [18]. The continuous input of fresh organic material in the uppermost soil layers promotes high biological activity [18] leading to more immediate changes in the C status. On the other hand, in deeper soil layers, there is a lower fresh material input and higher recalcitrance of soil C forms compared to topsoil, making C forms unavailable to microbial communities [66]. These limitations at deeper soil layers make changes in C status slower.

On one hand, our results showed that management-dependent changes in topsoil C status might be useful to allow the comprehension of the dynamics of soil C sequestration in the short span. This is because changes in labile C fractions can also promote changes in TOC contents and the uppermost soil layers can contain approximately 47–50% of the total SOC stock found in the 0–100 cm soil layer [67,68]. On the other hand, changes promoted by the management system can also point towards long-term trends in a soil profile deeper than the 0.10–0.15 m soil layer both in tropical [69,70] and temperate regions [65]. Therefore, given the restricted thickness of the soil layer considered in our study, the results cannot be extrapolated to the whole soil profile and cannot be used as a model to estimate C sequestration under the land-use systems evaluated.

Taken together, our findings show the importance of soil quality, the crop rotation with more complex crop arrangements that include forage grasses, and the combination of these forage grasses with eucalyptus trees. Our results also call for the incorporation of these practices in intensive systems in converted areas in the Brazilian Northeast, to ensure more intense topsoil C dynamics, with implications for productive sustainability [19,71]. Despite that, future research efforts should be directed towards a detailed survey of C sequestration and fractions in a soil profile of 1 m or more [70]. This is necessary for a better comprehension of the mid- or long-term effects of intensive land-use systems in the soil C status under the conditions of the Brazilian Northeast and for a complete inventory of C stocks in areas converted from native Cerrado to agricultural use.

5. Conclusions

This study showed that native Cerrado forest to agricultural land uses does not significantly influence the topsoil C-stocks and the fractions of soil humic substances but reduces the topsoil POC fraction, partially confirming our hypotheses. The adoption of agricultural systems that involve complex cropping rotations including pasture species, such as Mombasa grass and eucalyptus, is decisive to ensure a permanent input of diverse plant residues, preventing the loss of topsoil active C compartments in the Cerrado of the Brazilian Northeast. Although this study did not show significant shifts in soil organic

C fractions, it suggests that sustainable and conservationist management should be emphasized to maintain and improve the status of soil organic C. In addition, further studies should be done to monitor the pattern of soil organic C fractions in the long term.

Supplementary Materials: The following supporting information can be downloaded at: https://www.mdpi.com/article/10.3390/soilsystems7010011/s1, Figure S1: Sampling design. In NT1, NT2, PAS and NF, five 100-m transects distant approximately 20 m between them were sampled (A); In EUC, five 20-m spaced transects considered the transition from eucalyptus rows and spaces between rows (B). Five samples were taken per transect (red circles) and pooled to form a composite sample; Table S1: Historic of the crop systems applied in each area; Table S2: chemical properties (0-0.10 m depth) of the soils.

Author Contributions: Conceptualization, H.A.d.S., M.L.T. and L.F.C.L.; methodology, H.A.d.S., A.S.F.A. and E.S.; formal analysis, A.V.S.G. and J.R.d.C.; investigation, A.V.S.G. and J.R.d.C.; data curation, L.W.M., A.P.d.A.P. and D.P.d.C.; writing-original draft preparation—H.A.d.S., A.S.F.A., E.S., E.V.d.M. and R.F.V.; writing—review and editing, A.S.F.A., E.S., E.V.d.M. and L.F.C.L. All authors have read and agreed to the published version of the manuscript.

Funding: This research was funded by Empresa Brasileira de Pesquisa Agropecuária (Embrapa), grant number 22.13.11.004.00.03.

Informed Consent Statement: Not applicable.

Data Availability Statement: Not applicable.

Acknowledgments: The authors wish to thank Nova Zelândia Farm for logistical support and information.

Conflicts of Interest: The authors declare no conflict of interest.

References

1. Coyle, C.; Creamer, R.E.; Schulte, R.P.O.; O'Sullivan, L.; Jordan, P.A. Functional Land Management conceptual framework under soil drainage and land use scenarios. *Environ. Sci. Pol.* **2016**, *56*, 39–48. [CrossRef]
2. Parsapour, M.K.; Kooch, Y.; Hosseini, S.M.; Alavi, S.J. Litter and topsoil in *Alnus subcordata* plantation on former degraded natural forest land: A synthesis of age-sequence. *Soil Tillage Res.* **2018**, *179*, 1–10. [CrossRef]
3. Olorunfemi, I.E.; Fasinmirin, J.T.; Olufayo, A.A.; Komolafe, A.A. Total carbon and nitrogen stocks under different land use/land cover types in the Southwestern region of Nigeria. *Geoderma Reg.* **2020**, *22*, e00320. [CrossRef]
4. Guareschi, R.F.; Pereira, M.G.; Perin, A. Densimetric fractionation of organic matter in an agricultural chronosequence in no-till areas in the Cerrado region, Brazil. *Semin. Cienc. Agr.* **2016**, *37*, 595–610. [CrossRef]
5. Sanderman, J.; Hengl, T.; Fiske, G.J. Soil carbon debt of 12,000 years of human land use. *Proc. Natl. Acad. Sci. USA* **2017**, *114*, 9575–9580. [CrossRef]
6. Zalles, V.; Hansen, M.C.; Potapov, P.V.; Stehman, S.V.; Tyukavina, A.; Pickens, A.; Song, X.-P.; Adusei, B.; Okpa, C.; Aguilar, R.; et al. Near doubling of Brazil's intensive row crop area since 2000. *Proc. Natl. Acad. Sci. USA* **2019**, *8*, 428–435. [CrossRef]
7. Wuaden, C.R.; Nicoloso, R.S.; Barros, E.C.; Grave, R.A. Early adoption of no-till mitigates soil organic carbon and nitrogen losses due to land use change. *Soil Tillage Res.* **2020**, *204*, 104728. [CrossRef]
8. Page, K.L.; Dang, Y.P.; Dalal, R.C. The Ability of Conservation Agriculture to Conserve Soil Organic Carbon and the Subsequent Impact on Soil Physical, Chemical, and Biological Properties and Yield. *Front. Sust. Food Syst.* **2020**, *18*, 31. [CrossRef]
9. Bayer, C.; Martin-Neto, L.; Mielniczuk, J.; Pavinato, A.; Dieckow, J. Carbon sequestration in two Brazilian Cerrado soils under no-till. *Soil Tillage Res.* **2006**, *86*, 237–245. [CrossRef]
10. Bongiorno, G.; Bünemann, E.K.; Oguejiofor, C.U.; Meier, J.; Gort, G.; Comans, R.; Mäder, P.; Brussaard, L.; de Goede, R. Sensitivity of labile carbon fractions to tillage and organic matter management and their potential as comprehensive soil quality indicators across pedoclimatic conditions in Europe. *Ecol. Indic.* **2019**, *99*, 38–50. [CrossRef]
11. Kooch, Y.; Ehsani, S.; Akbarinia, M. Stratification of soil organic matter and biota dynamics in natural and anthropogenic ecosystems. *Soil Tillage Res.* **2020**, *200*, 104621. [CrossRef]
12. Ukalska-Jaruga, A.; Debaene, G.; Smreczak, B. Particle and structure characterization of fulvic acids from agricultural soils. *J. Soils Sedim.* **2018**, *18*, 2833–2843. Available online: http://link.springer.com/10.1007/s11368-018-2008-1 (accessed on 1 November 2022). [CrossRef]
13. Liu, H. Relationship between organic matter humification and bioavailability of sludge-borne copper and cadmium during long-term sludge amendment to soil. *Sci. Total Environ.* **2016**, *566–567*, 8–14. [CrossRef] [PubMed]
14. Baldock, J.; Nelson, P. Soil organic matter. In *Handbook of Soil Science*; Taylor and Francis Group: London, UK; New York, NY, USA; CRC Press: Boca Raton, FL, USA, 2000; pp. B25–B84.
15. Kaiser, K.; Kalbitz, K. Cycling downwards—Dissolved organic matter in soils. *Soil Biol. Biochem.* **2012**, *52*, 29–32. [CrossRef]

16. Angers, D.A.; Eriksen-Hamel, N.S. Full-Inversion Tillage and Organic Carbon Distribution in Soil Profiles: A Meta-Analysis. *Soil Sci. Soc. Am. J.* **2008**, *72*, 1370–1374. [CrossRef]
17. Luo, Z.; Wang, E.; Sun, O.J. Can no-tillage stimulate carbon sequestration in agricultural soils? A meta-analysis of paired experiments. *Agric. Ecosyst. Environ.* **2010**, *139*, 224–231. [CrossRef]
18. Haddaway, N.R.; Hedlund, K.; Jackson, L.E.; Kätterer, T.; Lugato, E.; Thomsen, I.K.; Jørgensen, H.B.; Isberg, P.-E. How does tillage intensity affect soil organic carbon? A systematic review. *Environ. Evid.* **2017**, *18*, 30. Available online: https://environmentalevidencejournal.biomedcentral.com/articles/10.1186/s13750-017-0108-9 (accessed on 1 November 2022). [CrossRef]
19. Franzluebbers, A. Soil organic matter stratification ratio as an indicator of soil quality. *Soil Tillage Res.* **2002**, *66*, 95–106. Available online: https://linkinghub.elsevier.com/retrieve/pii/S0167198702000181 (accessed on 1 November 2022). [CrossRef]
20. Babur, E.; Kara, O.; Fathi, R.A.; Susam, Y.E.; Riaz, M.; Arif, M.; Akhtar, K. Wattle fencing improved soil aggregate stability, organic carbon stocks and biochemical quality by restoring highly eroded mountain region soil. *J. Environ. Manag.* **2021**, *288*, 112489. [CrossRef]
21. Ukalska-Jaruga, A.; Klimkowicz-Pawlas, A.; Smreczak, B. Characterization of organic matter fractions in the top layer of soils under different land uses in Central-Eastern Europe. *Soil Use Manag.* **2019**, *35*, 595–606. [CrossRef]
22. Wang, Y.; Fu, B.; Lü, Y.; Chen, L. Effects of vegetation restoration on soil organic carbon sequestration at multiple scales in semi-arid Loess Plateau, China. *Catena* **2011**, *85*, 58–66. [CrossRef]
23. Ozlu, E.; Arriaga, F.J.; Bilen, S.; Gozukara, G.; Babur, E. Carbon Footprint Management by Agricultural Practices. *Biology* **2022**, *11*, 1453. [CrossRef] [PubMed]
24. Leite, L.F.C.; Galvão, S.R.S.; Holanda Neto, M.R.; Araújo, F.S.; Iwata, B.F. Atributos químicos e estoques de carbono em Latossolo sob plantio direto no cerrado do Piauí. *Rev. Bras. Eng. Agríc. Amb.* **2010**, *14*, 1273–1280. Available online: http://www.scielo.br/scielo.php?script=sci_arttext&pid=S1415-43662010001200004&lng=pt&tlng=pt (accessed on 1 November 2022). [CrossRef]
25. Fujisaki, K.; Perrin, A.-S.; Desjardins, T.; Bernoux, M.; Balbino, L.C.; Brossard, M. From forest to cropland and pasture systems: A critical review of soil organic carbon stocks changes in Amazonia. *Glob. Change Biol.* **2015**, *21*, 2773–2786. [CrossRef] [PubMed]
26. Araújo, A.S.F.; Silva, E.F.L.; Nunes, L.A.P.L.; Carneiro, R.F.V. The effect of converting tropical native savanna to Eucalyptus grandis forest on soil microbial biomass. *Land Degrad. Develop.* **2010**, *21*, 540–545. Available online: https://onlinelibrary.wiley.com/doi/10.1002/ldr.993 (accessed on 1 November 2022). [CrossRef]
27. Battaglia, M.L.; Thomason, W.E.; Fike, J.H.; Evanylo, G.K.; Stewart, R.D.; Gross, C.D.; Seleiman, M.F.; Babur, E.; Sadeghpour, A.; Harrison, M.T. Corn and wheat residue management effects on greenhouse gas emissions in the mid-Atlantic USA. *Land* **2022**, *11*, 846. [CrossRef]
28. INMET. Instituto Nacional de Meteorologia. 2022. Available online: http://www.inmet.gov.br/portal/ (accessed on 1 November 2022).
29. Sousa, D.M.G.; Lobato, E. *Cerrado: Correção do Solo e Adubação*; Embrapa Cerrados: Planaltina, Brasil, 2004; 416p.
30. Silva, F.C.; Abreu, M.F.; Pérez, D.V.; Eira, P.A.; Abreu, C.A.; Raij, B.V.; Gianello, C.; Coelho, A.M.; Quaggio, J.A.; Tedesco, M.J.; et al. Métodos de análises químicas para avaliação da fertilidade do solo. In *Manual de Análises Químicas de Solos, Plantas e Fertilizantes*; Embrapa Informação Tecnológica: Brasilia, Brasil, 2009; pp. 107–190.
31. Islam, K.R.; Weil, R.R. A rapid microwave digestion method for colorimetric measurement of soil organic carbon. *Commun. Soil Sci. Plant Anal.* **1998**, *29*, 2269–2284. [CrossRef]
32. Ferreira, A.S.; Camargo, F.A.O.; Vidor, C. Utilização de microondas na avaliação da biomassa microbiana do solo. *Rev. Bras. Cienc. Solo* **1999**, *23*, 991–996. Available online: http://www.scielo.br/scielo.php?script=sci_arttext&pid=S0100-06831999000400026&lng=pt&tlng=pt (accessed on 1 November 2022). [CrossRef]
33. Alef, K. Estimation of the hydrolysis of fluorescein diacetate. In *Methods in Applied Soil Microbiology and Biochemistry*; Academic Press: London, UK, 1995; pp. 232–238.
34. Mendonça, E.S.; Matos, E.S. *Matéria Orgânica do Solo: Métodos de Análises*; UFV: Viçosa, Brasil, 2005.
35. Ellert, B.H.; Bettany, J.R. Calculation of organic matter and nutrients stored in soils under contrasting management regimes. *Can. J. Soil Sci.* **1995**, *75*, 529–538. Available online: http://www.nrcresearchpress.com/doi/10.4141/cjss95-075 (accessed on 1 November 2022). [CrossRef]
36. Swift, R.S. Organic matter characterization. In *Methods of Soil Analysis*; SSSA: Madison, WI, USA, 1996; pp. 1011–1069.
37. Diekow, J.; Mielniczuk, J.; Knicker, H.; Bayer, C.; Dick, D.P.; Kögel-Knabner, I. Carbon and nitrogen stocks in physical fractions of a subtropical Acrisol as influenced by long-term no-till cropping systems and N fertilisation. *Plant Soil* **2005**, *268*, 319–328. [CrossRef]
38. R Core Team. R Core Team. R: A Language and Environment for Statistical Computing. 2020. Available online: https://www.r-project.org/ (accessed on 1 November 2022).
39. Pinheiro, F.M.; Nair, P.K.R.; Nair, V.D.; Tonucci, R.G.; Venturin, R.P. Soil carbon stock and stability under Eucalyptus-based silvopasture and other land-use systems in the Cerrado biodiversity hotspot. *J. Environ. Manag.* **2021**, *299*, 113676. [CrossRef] [PubMed]
40. Gmach, M.-R.; Dias, B.O.; Silva, C.A.; Nóbrega, J.C.A.; Lustosa-Filho, J.F.; Siqueira-Neto, M. Soil organic matter dynamics and land-use change on Oxisols in the Cerrado, Brazil. *Geoderma Reg.* **2018**, *14*, e00178. [CrossRef]

41. Sá, J.C.S.; Gonçalves, D.R.P.; Ferreira, L.A.; Mishra, U.; Inagaki, T.M.; Furlan, F.J.F.; Moro, R.S.; Floriani, N.; Briedis, C.; Ferreira, A.D.O. Soil carbon fractions and biological activity based indices can be used to study the impact of land management and ecological successions. *Ecol. Indic.* **2018**, *84*, 96–105. [CrossRef]
42. Almeida, L.L.S.; Frazão, L.A.; Lessa, T.A.M.; Fernandes, L.A.; Veloso, Á.L.C.; Lana, A.M.Q.; de Souza, I.A.; Pegoraro, R.F.; Ferreira, E.A. Soil carbon and nitrogen stocks and the quality of soil organic matter under silvopastoral systems in the Brazilian Cerrado. *Soil Tillage Res.* **2021**, *205*, 104785. [CrossRef]
43. Monroe, P.H.M.; Berreto-Garcia, P.A.B.; Barros, W.T.; Oliveira, F.G.R.B.; Pereira, M.G. Physical protection of soil organic carbon through aggregates in different land use systems in the semi-arid region of Brazil. *J. Arid Environ.* **2021**, *186*, 104427. [CrossRef]
44. Lark, R.M. Changes in the variance of a soil property along a transect, a comparison of a non-stationary linear mixed model and a wavelet transform. *Geoderma* **2016**, *266*, 84–97. [CrossRef]
45. Dalal, R.C.; Thornton, C.M.; Allen, D.E.; Owens, J.S.; Kopittke, P.M. Long-term land use change in Australia from native forest decreases all fractions of soil organic carbon, including resistant organic carbon, for cropping but not sown pasture. *Agric. Ecosyst. Environ.* **2021**, *311*, 107326. [CrossRef]
46. Torn, M.S.; Swanston, C.W.; Castanha, C.; Trumbore, S.E. Storage and turnover of organic matter in soil. In *Biophysico-Chemical Processes Involving Natural Nonliving Organic Matter in Environmental Systems*; Wiley: Hoboken, NJ, USA, 2009; pp. 219–272.
47. Liu, X.; Wu, X.; Liang, G.; Zheng, F.; Zhang, M.; Li, S. A global meta-analysis of the impacts of no-tillage on soil aggregation and aggregate-associated organic carbon. *Land Degrad. Develop.* **2021**, *32*, 5292–5305. [CrossRef]
48. Rasse, D.P.; Rumpel, C.; Dignac, M.-F. Is soil carbon mostly root carbon? Mechanisms for a specific stabilisation. *Plant Soil* **2005**, *269*, 341–356. [CrossRef]
49. Koga, N.; Shimoda, S.; Shirato, Y.; Kusaba, T.; Shima, T.; Niimi, H.; Yamane, T.; Wakabayashi, K.; Niwa, K.; Kohyama, K.; et al. Assessing changes in soil carbon stocks after land use conversion from forest land to agricultural land in Japan. *Geoderma* **2020**, *377*, 114487. [CrossRef]
50. Chenu, C.; Angers, D.A.; Barré, P.; Derrien, D.; Arrouays, D.; Balesdent, J. Increasing organic stocks in agricultural soils: Knowledge gaps and potential innovations. *Soil Tillage Res.* **2019**, *188*, 41–52. [CrossRef]
51. Ramos, F.T.; Dores, E.F.C.; Weber, O.L.S.; Beber, D.C.; Campelo, J.H.; Maia, J.C.S. Soil organic matter doubles the cation exchange capacity of tropical soil under no-till farming in Brazil. *J. Sci. Food Agric.* **2018**, *98*, 3595–3602. [CrossRef] [PubMed]
52. Pausch, J.; Kuzyakov, Y. Carbon input by roots into the soil: Quantification of rhizodeposition from root to ecosystem scale. *Glob. Change Biol.* **2018**, *24*, 1–12. [CrossRef] [PubMed]
53. Bahadori, M.; Chen, C.; Lewis, S.; Boyd, S.; Rashti, M.R.; Esfandbod, M.; Garzon-Garcia, A.; Van Zwieten, L.; Kuzyakov, Y. Soil organic matter formation is controlled by the chemistry and bioavailability of organic carbon inputs across different land uses. *Sci. Total Environ.* **2021**, *770*, 145307. [CrossRef] [PubMed]
54. You, M.; Zhu-Barker, X.; Hao, X.-X.; Li, L.-J. Profile distribution of soil organic carbon and its isotopic value following long term land-use changes. *Catena* **2021**, *207*, 105623. [CrossRef]
55. Thangavel, R.; Kanchikerimath, M.; Sudharsanam, A.; Ayyanadar, A.; Karunanithi, R.; Deshmukh, N.A.; Vanao, N.S. Evaluating organic carbon fractions, temperature sensitivity and artificial neural network modeling of CO_2 efflux in soils: Impact of land use change in subtropical India (Meghalaya). *Ecol. Indic.* **2018**, *193*, 129–141. [CrossRef]
56. Cardoso, J.A.F.; Lima, A.M.N.; Cunha, T.J.F.; Rodrigues, M.S.; Hernani, L.C.; Amaral, A.J.; Neto, M.B.D.O. Organic matter fractions in a quartzipsamment under cultivation of irrigated mango in the lower são francisco valley region, Brazil. *Rev. Bras. Cienc. Solo* **2015**, *39*, 1068–1078. Available online: http://www.scielo.br/scielo.php?script=sci_arttext&pid=S0100-06832015000401068&lng=en&tlng=en (accessed on 1 November 2022). [CrossRef]
57. Guimarães, D.V.; Gonzaga, M.I.S.; Melo Neto, J.O. Manejo da matéria orgânica do solo e estoques de carbono em cultivos de frutas tropicais. *Rev. Bras. Eng. Agríc. Amb.* **2014**, *18*, 301–306. Available online: http://www.scielo.br/scielo.php?script=sci_arttext&pid=S1415-43662014000300009&lng=pt&tlng=pt (accessed on 1 November 2022). [CrossRef]
58. Iqbal, J.; Hu, R.; Lin, S.; Ahamadou, B.; Feng, M. Carbon dioxide emissions from Ultisol under different land uses in mid–subtropical China. *Geoderma* **2009**, *152*, 63–73. [CrossRef]
59. Iqbal, J.; Ronggui, H.; Lijun, D.; Lan, L.; Shan, L.; Tao, C.; Leilei, R. Differences in soil CO_2 flux between different land use types in mid-subtropical China. *Soil Biol. Biochem.* **2008**, *40*, 2324–2333. Available online: https://linkinghub.elsevier.com/retrieve/pii/S003807170800165X (accessed on 1 November 2022). [CrossRef]
60. Santos, V.B.; Araújo, A.S.; Leite, L.F.; Nunes, L.A.; Melom, W.J. Soil microbial biomass and organic matter fractions during transition from conventional to organic farming systems. *Geoderma* **2012**, *170*, 227–231. [CrossRef]
61. Sparling, G. Ratio of microbial biomass carbon to soil organic carbon as a sensitive indicator of changes in soil organic matter. *Soil Res.* **1992**, *30*, 195. Available online: http://www.publish.csiro.au/?paper=SR9920195 (accessed on 1 November 2022). [CrossRef]
62. Haynes, R.J. Labile Organic Matter Fractions as Central Components of the Quality of Agricultural Soils: An Overview. *Adv. Agron.* **2005**, *85*, 221–268. Available online: https://linkinghub.elsevier.com/retrieve/pii/S0065211304850053 (accessed on 1 November 2022).
63. Araújo, A.S.F.; Cesarz, S.; Leite, L.F.C.; Borges, C.D.; Tsai, S.M.; Eisenhauer, N. Soil microbial properties and temporal stability in degraded and restored lands of Northeast Brazil. *Soil Biol. Biochem.* **2013**, *66*, 175–181. [CrossRef]
64. Costantini, M.; Bacenetti, J. Soybean and maize cultivation in South America: Environmental comparison of different cropping systems. *Clean. Environ. Syst.* **2021**, *2*, 100017. [CrossRef]

65. Krauss, M.; Wiesmeier, M.; Don, A.; Cuperus, F.; Gattinger, A.; Gruber, S.; Haagsma, W.; Peigné, J.; Palazzoli, M.C.; Schulz, F.; et al. Reduced tillage in organic farming affects soil organic carbon stocks in temperate Europe. *Soil Tillage Res.* **2022**, *216*, 105262. [CrossRef]
66. Lyu, M.; Noormets, A.; Ukonmaanaho, L.; Li, Y.; Yang, Y.; Xie, J. Stability of soil organic carbon during forest conversion is more sensitive in deep soil than in topsoil in subtropical forests. *Pedobiologia* **2021**, *84*, 150706. [CrossRef]
67. Balesdent, J.; Basile-Doelsch, I.; Chadoeuf, J.; Cornu, S.; Derrien, D.; Fekiacova, Z.; Hatté, C. Atmosphere–soil carbon transfer as a function of soil depth. *Nature* **2018**, *559*, 599–602. [CrossRef]
68. Basile-Doelsch, I.; Balesdent, J.; Pellerin, S. Reviews and syntheses: The mechanisms underlying carbon storage in soil. *Biogeosciences* **2020**, *17*, 5223–5242. Available online: https://bg.copernicus.org/articles/17/5223/2020/ (accessed on 1 November 2022). [CrossRef]
69. Medeiros, A.S.; Santos, T.C.; Maia, S.M.F. Effect of long-term and soil depth on soil organic carbon stocks after conversion from native vegetation to conventional tillage systems in Brazil. *Soil Tillage Res.* **2022**, *219*, 105336. [CrossRef]
70. Diekow, J.; Mielniczuk, J.; Knicker, H.; Bayer, C.; Dick, D.P.; Kögel-Knabner, I. Soil C and N stocks as affected by cropping systems and nitrogen fertilisation in a southern Brazil Acrisol managed under no-tillage for 17 years. *Soil Tillage Res.* **2005**, *81*, 87–95. Available online: https://linkinghub.elsevier.com/retrieve/pii/S0167198704001357 (accessed on 1 November 2022). [CrossRef]
71. Anghinoni, G.; Anghinoni, F.B.G.; Tormena, C.A.; Braccini, A.L.; Mendes, I.C.; Zancanaro, L.; Lal, R. Conservation agriculture strengthen sustainability of Brazilian grain production and food security. *Land Use Pol.* **2021**, *108*, 105591. Available online: https://linkinghub.elsevier.com/retrieve/pii/S0264837721003148 (accessed on 1 November 2022). [CrossRef]

soil systems

Article

Natural and Anthropogenic Sources of Cadmium in Cacao Crop Soils of Santander, Colombia

Valentina Joya-Barrero [1], Carme Huguet [1,*] and Jillian Pearse [2]

[1] Department of Geosciences, Los Andes University, Bogotá 111711, Colombia
[2] Department of Geological Sciences, California State University, Long Beach, CA 90840, USA
* Correspondence: m.huguet@uniandes.edu.co

Abstract: Elevated cadmium (Cd) levels in cacao products have been detected in a major cacao-producing region of Colombia, with concentrations well above those permitted for export and posing a potential threat to human health. Geochemical and petrographic analyses of fertilizer, soil and rocks from three farms were used to determine the origin of Cd. Parent rocks were the main source of the Cd in soils, while organic fertilizer may have further contributed to elevated metal content in one farm. High Cd levels in the organic fertilizer were most likely due to bioaccumulation, since it was sourced from animals in the same area. Even though the soil pH range, elevated OM content and the presence of Mn and K diminish bioavailability, the extremely high Cd content in soils results nonetheless in significant uptake by the plants and subsequent accumulation in cocoa beans. Traditional methods to reduce Cd adsorption, such as the addition of calcium, will not be effective in this case. Instead, the selection of cacao species that are naturally low accumulators and amendment with soil microorganisms with mineralization and biotransformation capabilities, as well as testing of fertilizers before application, could all be cost-effective solutions to reduce Cd in the final product.

Keywords: cadmium; cacao; soil analysis; autochthonous origin of Cd; allochthonous origin of Cd

Citation: Joya-Barrero, V.; Huguet, C.; Pearse, J. Natural and Anthropogenic Sources of Cadmium in Cacao Crop Soils of Santander, Colombia. *Soil Syst.* **2023**, *7*, 12. https://doi.org/10.3390/soilsystems7010012

Academic Editors: Antonella Lavini and Mohamed Houssemeddine Sellami

Received: 28 November 2022
Revised: 10 January 2023
Accepted: 11 January 2023
Published: 6 February 2023

1. Introduction

Cadmium (Cd) is one of the elements that occur naturally in the crust, with reported concentrations of 0.1–0.5 ppm [1]. Because of its mobility in soils, it is readily absorbed by plants despite not having a metabolic function [2,3]. As a non-essential heavy metal, it can cause toxic effects in plants, animals and humans, whether due to acute or chronic exposure and even at low concentrations, e.g., [1–3]. Cadmium is considered to be one of the most toxic metals for humans as it exhibits adverse effects on all biological processes, acting as a precursor of various cancers, oxidative stress and kidney malfunction, e.g., [1–4]. It is classified as a Group 1 carcinogen as it has been associated with lung and pancreatic cancer [5]. The metal can accumulate in organisms, particularly in bone and kidney tissue, leading to notorious health impacts through dietary consumption [5]. As the metal does not biodegrade, it also bioaccumulates through the food web, eventually posing a health hazard for humans, e.g., [2–4]. Chocolate and cacao powder with elevated Cd levels can be an important source of human exposure to the metal, impacting the health of consumers, e.g., [6].

High levels of Cd in soils of cacao farms pose one of the greatest challenges for producing safe cocoa products in South and Central American countries [7–10]. Latin America and the Caribbean show high Cd content in cacao beans with mean values higher than those found in Africa and Asia and above the acceptable health limits proposed internationally (Figure 1; [7–10]). The study area displays abnormally high Cd concentrations in beans, well above average for the Latin America and Caribbean (LAC) region and the acceptable level in beans (Figure 1). Aside from the potential health impacts, starting in January 2020 the European Union enforced a new regulation for the maximum values for Cd concentrations in cocoa-based products, which has put a strain on South and Central American cacao producers [11].

Since Colombia is one of the world's major cacao producers, contributing 1.5% of the global market, the high Cd concentration will have a sizeable impact on the country's economy, e.g., [12]. Santander is a well-known cacao region in Colombia, historically recognized for having vast areas of cacao crops (62,500 ha) of excellent quality and contributing the biggest portion of cacao for export, e.g., [13]. Recently, cacao has been highlighted in the context of post-conflict Colombia as the main substitution crop for coca plantations, with 25,000 ha having already been transformed from illegal crops into cacao [13]. It is thus essential to understand the possible factors that may increase Cd in soils so that new cacao farms are not established in unsuitable areas.

Cadmium can originate from bedrock, erosional-depositional and recycling processes, as well as from anthropogenic sources [14–16]. Cd has high mobility through sediment flows and erosion processes by water and wind, and material translocation, which results in accumulation in sedimentary environments [17,18]. Moreover, concentrations of Cd are higher in sedimentary rocks since this metal can also be easily adsorbed into fine particles and porosity sites [16,19]. The Cd content tends to be higher in fine-grained acidic sedimentary rocks [16,20,21]. The relative accumulation of Cd in sedimentary environments may also be due to the degassing of the Earth and mantle processes, in which excess volatile elements such as Cd are liberated and accumulated within empty spaces [22].

The anthropogenic sources of Cd are mainly related to the addition of both organic and inorganic fertilizers, and the potential contamination from mining or construction sites, e.g., [20]. Cd can also be reinserted into the soil through the plant's leaves or branches, as farmers leave plant debris as a fertilizer, thus recycling the metal into the ground, e.g., [6,9].

Even though other forms of Cd co-exist with Cd^{2+}, contributing to the total content of the metal in the bedrock and soil, only the ion is available for plant uptake [6,8,15]. The parent rock and the demineralization and weathering processes during pedogenesis will determine the amount of bioavailable Cd^{2+} in soils [23,24]. Soils inherit many of the bedrocks characteristics and retain a large portion of its elements; for example, carbonate rocks with high Cd content have been shown to produce soils enriched in the metal after pedogenesis [24]. Several soil properties can regulate Cd bioavailability; while high electrical conductivity and salinity, as well as loamy and clayey soil textures, result in an increase, near-neutral pH range, medium-high organic matter content or the presence of certain elements (e.g., Mn, K) reduce its availability [4,25].

The country's production is year-round and consists mainly of Criollo and Trinitario varieties, which are known for their fine chocolate flavor but relatively low yield [26]. The *Theobroma cacao* plant bioaccumulates Cd, which is easily absorbed by its roots from soil and water in its available Cd^{2+} form along with the other nutrients the plant needs, accumulating then within the structures of the plant, e.g., [6,8,15]. The accumulation occurs preferentially in cacao beans, followed by fruit shells, and the smallest quantity accumulates in leaves [6,7]. It has been found that in some cases the proportion of Cd content in soil and beans is about 1:4, but that even if the proportion varies, there is always substantial accumulation in beans compared to soil concentration [6,7].

Previous studies show that the concentration of Cd in plant structures can vary depending on the farm location and the soil characteristics, e.g., [9,21]. In a recent study of Colombian cacao soils, a mean level of Cd of 1.43 mg/kg for a total of 1837 soils was reported, well above the natural concentrations found in soils worldwide [27]. Santander shows the second highest Cd soil concentration, with a mean of 1.90 mg/kg and a maximum value of 27 mg/kg Cd—far beyond those of other regions in the country [27]. Perhaps not surprisingly, average Cd concentrations as high as 4.3 mg/kg for beans have been reported in the area, far exceeding the threshold of 0.60 mg/kg applied for cacao bean exports to the EU and the maximum level of 1.3 mg/kg for cocoa powder proposed by the Codex Committee on Contaminants in Food (CCCF) (Figure 1; [14,27]).

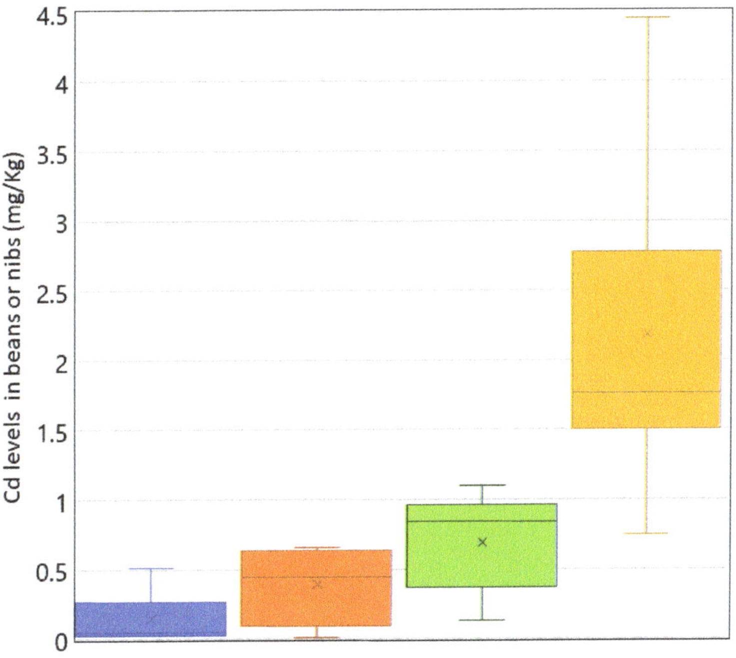

Figure 1. Cadmium levels in cacao beans and nibs from Africa, Asia, Latin America and the Caribbean (LAC) and San Vicente de Chucurí (SVC) (data from [10,14]). Overlaid: the ideal maximum level of 0.5 mg/kg (IML, dashed grey line); the acceptable limit in beans of 0.8 mg/kg (ALB, dashed black line); and the maximum proposed limit of 1.3 mg/kg Cd for cocoa powder (100% total cocoa solids on a dry matter basis) set by the Codex Committee on Contaminants in Food (MPL, dashed red line) [14,28].

Our aim was to improve the understanding of the possible natural and anthropogenic sources of the metal in the study area. We also analyzed soil parameters that may regulate the bioavailability of the metal to gain a better insight into the problem and be able to propose better solutions. Providing a sound baseline of Cd levels and sources is a first step towards better management practices and, when needed, remediation strategies in the area. Beyond improving the situation in the study area, it is crucial to establish the conditions for low Cd in soils to aid with the planned cacao farm expansion to replace illicit crops.

2. Materials and Methods

2.1. Study Area

San Vicente de Chucurí is located in the Northeastern region of the department of Santander, Colombia, and it is known as the "cocoa capital of Colombia" (Figure 2). The annual mean temperature is 23.7 °C and average annual rainfall is 1820 mm [14] with a tropical rainforest climate according to the Köppen-Geiger classification. The main soil types in the area are Humic Cambisols (CMu) in both low (0–600 m.a.s.l.) and mid-altitude (600–900 m.a.s.l.) terrains and Umbric Leptosols (LPu) in high altitudes (900–1200 m.a.s.l.) [14]. All our study sites fall in the CMu soils category, with some soil differentiation and the presence of a humus-rich horizon, considered ideal for cacao cultivation.

The three farms used in the study are over the geological unit b6b6-Sm within the "Simití" formation, which is made up mainly of laminated black claystones and carbonaceous and locally calcareous fine-grained rocks, with a significant presence of calcareous concretions (Figure 2; [29,30]).

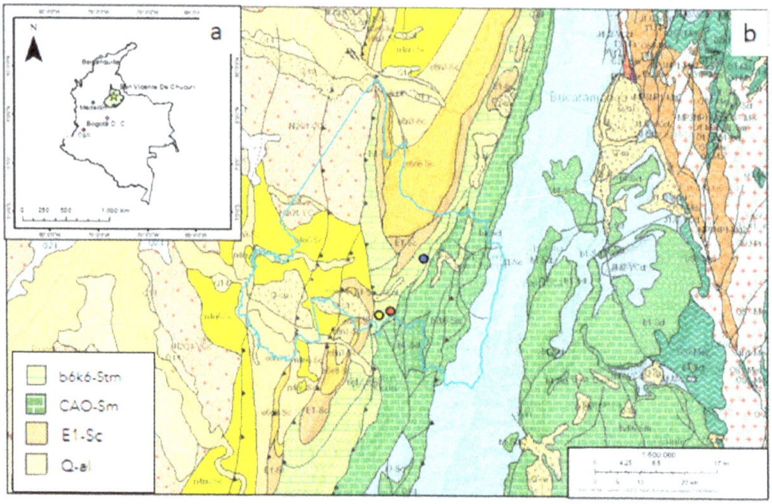

Figure 2. Maps of the study area. (**a**) Map of Colombia with the country's main cities shown. The department of Santander is highlighted in green with a star indicating the location of San Vicente de Chucurí. (**b**) Geological map of San Vicente de Chucurí [29,30] with the municipality limits indicated by the light blue line. Sample collection sites Farm 1 (red dot), Farm 2 (yellow dot) and Farm 3 (blue dot) are indicated.

2.2. Sample Collection

A total of 37 samples (23 soils, 12 source rocks and 2 fertilizers) were collected and analyzed at three different farms in the study area (Figure 2). All soil samples were taken at a depth of 30 cm and each weighed approximately 1000 g. Samples from fertilizers in Farms 1 and 2 were also collected. An organic fertilizer was used regularly (subjective dose applied 1 to 2 times a month depending on rainfall) in Farm 1 and an inorganic one in Farm 2 (the last application was a year before sampling). The organic fertilizer was composed of cocoa and banana tree leaf litter, chicken and pig manure (animals were fed with food residues and organic matter from crops grown in the farm soils) and topsoil, all collected within the farm and managed by the farmer using a rustic composting system to transform the residues. We did not measure the nutrient content of the organic fertilizer as we focused solely on its Cd content; however, as its main component was chicken manure, the nutrient content could be similar to that measured in other uses of chicken manure: 27% raw protein, 18% mineral matter, 4% Calcium (Ca) and 2% phosphorous (P).

The inorganic fertilizer was Diammonium Phosphate (DAP) which is commercially sold. This fertilizer is used as a source of phosphorus (P) and nitrogen (N) for plant nutrition. It is highly soluble and has an alkaline pH that develops around the dissolving granules. Its composition is 18% total nitrogen (N), 18% ammoniacal nitrogen (NH_4) and 46% water-soluble phosphorus (P_2O_5). This fertilizer was preferred as it was less aggressive with seedlings and was recommended by the farmers' technical assistants.

In Farms 1 and 2, which are plantations of less than 1 ha, the sampling was carried out in transects with a distance of around 8 m between each point. In the first farm we collected a rock sample at each soil sampling point, while in the second plantation only one rock sample was collected for each transect. In Farm 3, which is more extensive (~15 ha), but also more homogeneous, five soil samples were taken using randomized sampling with a distance of around 12 m between each point. No rock samples were collected in this case since soil was homogenized in the first 2 m and rock fragments were not present. As the third farm is located over the same geological unit, we expect the parental rock material to be of the same composition as both the other farms.

2.3. Petrographic Analysis

Petrographic analysis was performed on 3 rock samples (from Farms 1 and 2) and 3 soils (one from each farm), selected for having the highest Cd concentrations. The samples were processed into 6 polished thin-sections (of 30 μm thickness), and sent for analysis at Alicante University, Spain. Samples were studied using a ZEISS Assioskop microscope and pictures were taken with a Photometrics CoolSNAPcf digital camera and the RS ImageTM v.1.8.6 software (Seattle, WA, USA). Mineral chemical composition was established using a scanning electron microscope (SEM). Backscattered Electro (BSE) and X-ray spectroscopy (EDS) images were obtained using a Hitachi microscope, model S3000N at an accelerating voltage of 20 kV.

The XMET-7000 manual specifies all the Limits of Detection (LOD) in parts per million (ppm), but there is no given LOD for Cd measurements [30]. The Limit of Detection relates to repeatability, but does not indicate the instrument accuracy. LODs are dependent on matrix interferences, overlapping elements, level of statistical confidence and testing time [31]. To reduce potential effects on instrument precision, we used longer than recommended testing times (5 min) and verified that the statistical confidence level was high.

2.4. Soil Analysis

A LAQUAact-PC110 probe was used to measure the pH in soil samples. To do so, 1 g of humid soil was placed in 9 mL of distilled water and the solution was mixed for 2 min in a vortex and left to rest for 30 min. The probe was placed in the supernatant and pH was measured 3 times.

In order to quantify humidity, 100 g of each soil were placed in the oven at 45 °C for three days. The soil was re-weighed and the difference in mass was assumed to be water content, e.g., [31].

To measure the organic matter (OM) content, the same dry soil samples were then placed in the oven at a temperature of 250 °C for 24 h. Samples were then weighted and the difference in mass was assumed to be combusted OM, e.g., [31].

For carbonate content, the dry-inorganic soil samples were placed in a muffle at a temperature of 450 °C for 24 h. Samples were then weighed, and the mass value of carbonate was calculated from the mass difference, e.g., [31].

All instruments within the laboratories of the Universidad de los Andes used for the sample analysis above are calibrated on a biweekly or monthly basis by the department technicians.

2.5. X-ray Fluorescence (XRF) Measurements

Element composition was measured in the soil, fertilizer and rock samples. For soils and fertilizers, 250 g of fresh sample were dried in an oven at 45 °C for three days in order to remove moisture, as the presence of water may influence the signal intensity and increase instrumental error [32]. Dry samples were homogenized using a mortar and pestle and divided in four; then, two opposite quadrants were mixed and rearranged again as a circle. This process was repeated as many times as needed until reaching ~2 g of sample for analysis. This was performed in order to minimize measurement variation due to sample heterogeneity. The rocks were washed and measured directly.

The chemical elements from Beryl (Z = 4) to Uranium (Z = 92) were measured by XRF (X-Ray Fluorescence) using an Oxford XRF 7500 probe. All samples were measured with an XRF gun (XMET-7000) with the preprogrammed soil setting for soils and the mine setting for rock samples. Each sample was measured a minimum of three times and reported values in the present study represent the mean measurement with error bars shown. The procedure with the organic and inorganic fertilizers was the same as for the soil samples, only changing the preprogrammed setting in the XRF gun to the standard element set. Soil samples show smaller errors, most likely due to the removal of humidity and the grinding process which homogenized the samples and enabled a more uniform XRF reading, e.g., [33,34]. For the rock samples the error is generally larger, most likely caused by measuring directly on solid rock. Nevertheless, these measurements on direct rock are reliable as specified by the manufacturer and still fall within an acceptable standard error 3 [32–34].

2.6. Statistical Analysis

A Principal Component Analysis (PCA) was performed on the normalized soil parameters studied to provide a general view of the variability within the studied soils and the main factors determining Cd concentrations. The data were normalized using the corresponding mean and standard deviation as follows:

$$\text{Normalized data} = (\text{Data value} - \text{Data mean})/\text{Stdev} \quad (1)$$

This allows the comparison of data that have different units.

A paired-group (UPGMA) Euclidean distance cluster analysis was performed to understand data grouping of the individual samples. All analyses were performed with PAST 4.07 [35].

3. Results

3.1. Cadmium Concentrations

Measured Cd values ranged from 8 mg/kg ± 7 for soil in Farm 2 to 90 mg/kg ± 7 in a rock sample of Farm 1 (Figure 3). The highest value (90 mg/kg ± 7) was found in a rock fragment collected in Farm 1, which was at double the average Cd concentration of all other rock samples analyzed (50 mg/kg ± 3.8; Figure 3). While all farms display a range of Cd concentrations and thus values are not significantly different between farms, generally the highest concentrations are detected in Farm 1 (42 ± 19.3 mg/kg), intermediate ones are found in Farm 2 (32 ± 13.3 mg/kg) and lower ones in Farm 3 (25 ± 11.7 mg/kg) (Figure 3; Table S1).

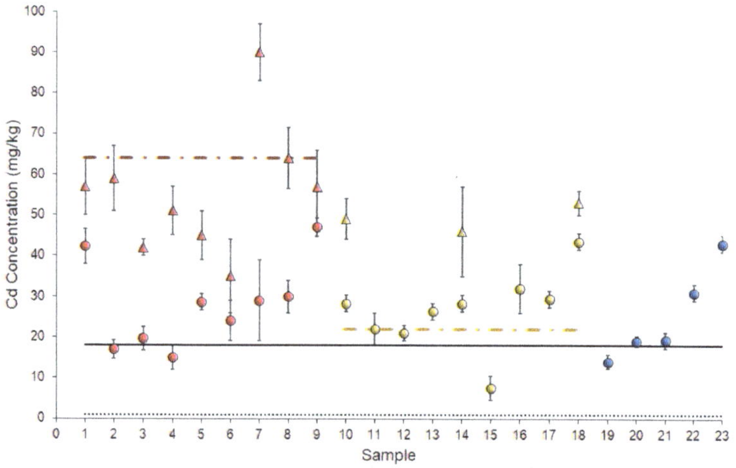

Figure 3. Cadmium concentration in soil (circles), rock samples (triangles), concentrations of organic (red dashed line) and chemical fertilizers (yellow dashed line). Red is for samples from Farm 1, yellow Farm 2 and blue Farm 3. The Cd critical threshold for human health risks (18 mg/kg; continuous line [36,37]) and the widely used threshold values for Cd concentrations in agricultural soils (1 mg/kg, dotted line; [9,27] and references therein) are indicated.

Based on the farms being in the same geological setting, if we instead separate matrices we find that the rocks have a significantly higher Cd average (54 mg/kg ± 13.9) than soils

(27 mg/kg ± 10.1) (Figure 3; Table S1). Since the sampling of rocks was uneven between farms, and rocks have the highest Cd concentrations, when averaging only soils we find no significant difference between the studied Farms (Figure 3; Table S1): Cd levels for Farm 1 were found to be 28 ± 10.8 mg/kg, Farm 2 slightly lower at 26 ± 9.6 mg/kg and in Farm 3 Cd concentrations were at 25 ± 11.7 mg/kg.

In Farm 1, an organic fertilizer was used and it had the second highest Cd concentration in the sample set (64 ± 2 mg/kg; Figure 3). In the case of Farm 2, the inorganic fertilizer applied one year prior to the study had Cd values of 22 ± 2 mg/kg (Figure 3; Table S1).

All soils displayed values above the reported "natural" Cd levels in agricultural soils of 1 mg/kg ([27] and references therein). The critical threshold for human health risks (18 mg/kg; [36,37]) was also surpassed by most soils, with the exception of soil 4 in Farm 1, soil 15 in Farm 2 and soil 19 in Farm 3 (Figure 3; Table S1).

3.2. Petrographic Analysis

Based on the petrographic analysis the bedrock types in the area were classified as limestone, marl and shale, and in all cases, rocks were found to be carbonate-rich (Figure 4a). This is mostly related to them being fossil-rich; abundant fragments of sea urchins and minor bivalves, in both cases built from calcium carbonate, can be observed in thin sections (Figure 4b).

The Backscattered electron images (BSE) of rock sample thin sections indicate an abundance of calcite (Cc), minor quartz (Qz) and some hematite (Fe_2O_3) and barite ($BaSO_4$) (Figure 4c). The rocks analyzed also have fine-grained quartz and accessory minerals are zircons and frankolite (carbonate-fluorapatite) (not shown).

The soil thin sections revealed an abundance of organic matter (OM) as well as unconsolidated carbonates (Figure 4d). We observe microaggregates of OM to also contain fragments of carbonate (Figure 4d).

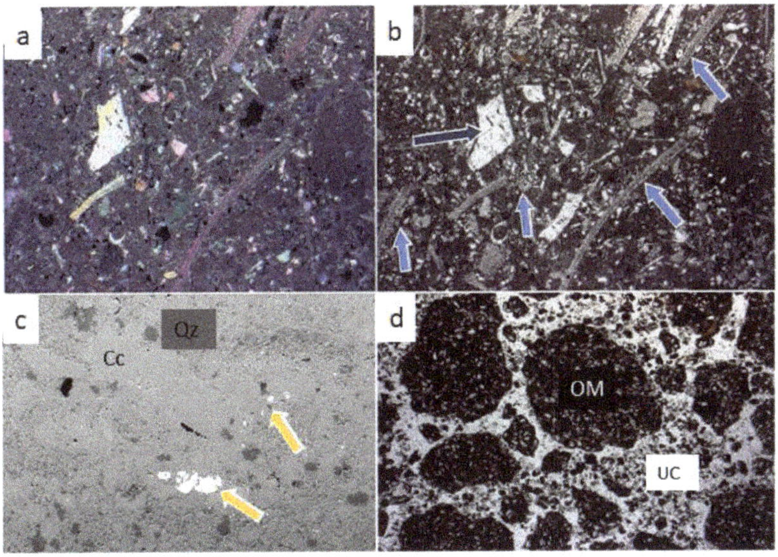

Figure 4. Petrographic analysis images of (**a**) cross polarized light (XPL) image of rock sample thin-section; (**b**) plane polarized light (PPL) of rock sample thin-section with abundant fragments of sea urchins (light blue arrows) and minor bivalves (dark blue arrow); (**c**) Backscattered electron image (BSE) of rock sample thin-section indicate abundance in calcite (Cc), minor quartz (Qz) and some hematite (Fe_2O_3) and barite ($BaSO_4$) grains (yellow arrows); (**d**) soil sample thin-section showing mostly unconsolidated carbonate (UC) with organic matter (OM).

4. Discussion

Santander is a well-known region for the cultivation of cacao and has been historically recognized for having cacao crops of great extension and quality, representing 40 to 45% of Colombian production. Addressing elevated Cd content is crucial not only as a public health concern, but also for the potentially negative impact it will have on cacao exports, a key economic sector for the region [26,28]. Moreover, since there are plans to expand cacao farming in the country as a substitute for illicit crops, it is paramount to establish the conditions that determine the presence and bioavailability of Cd in soils. Understanding the origin and fluxes of Cd can give insight into the best management strategies for the soil and translate into practical solutions for farmers in the region. Cadmium can be of autochthonous and/or allochthonous origin and sources include bedrock, erosional-depositional and recycling processes, as well as anthropogenic input; moreover, several soil properties regulate its bioavailability [4,6,15,25].

4.1. Cadmium Levels in the Study Area

All the samples analyzed show elevated Cd values compared to both international standards and national averages (Figures 1 and 3; [27]). In fact, the entire west flank of the Colombian Eastern cordillera shows significantly higher Cd values in sediments and soils (ranging from 1.8 to 74 mg/kg) than any other region in the country, resulting in a potential health hazard (Figure 5). Even though the study farms are within an area with no reported geochemical Cd values, we can extrapolate high concentrations from neighboring regions, which have concentrations falling in the ranges of 1.8 to 4.6 mg/kg and 4.6 to 74 mg/kg (in sediments; Figure 5). Additionally, values reported in other farms of San Vicente show average Cd concentrations of 3.3 mg/kg, with farms on the east side of the municipality having values around 2.1 to 4.3 mg/kg of total Cd and those on the west side having a maximum of around 2.5 mg/kg total and 0.1 mg/kg of available Cd [14].

The parent rock in the study area presents unusually high Cd concentrations with values of up to 3 times what was previously reported in Central Colombia [38]. Previously reported anomalously high Cd levels in rocks range between 8.15 mg/kg and 21.4 mg/kg worldwide ([21] and references therein), which is consistent with our findings of extremely elevated Cd concentrations in the study area rocks. Rock Cd values in the studied farms, even though always unusually elevated, vary from 30 to 50 mg/kg, suggesting heterogeneity of the parent material's metal content (Figure 3). Thus, the adsorption of Cd into fine particles and porosity sites varied within the Simití formation, probably because of the different types of sedimentary rocks and micro-meso lithification environments (Figure 1; [16,19]).

Total Cd content in cacao soils has been found to be mainly present as residual and oxidable fractions, associated with the weathering of the bedrock and other pedogenesis processes, so it follows that when the parent rock presents with extremely high Cd content, the resulting soils will be contaminated, e.g., [16,21,23]. In our samples total Cd content was on average 50% lower in soils than in rocks, which is a typical range of decrease after weathering of the parent material, e.g., [16,21,38].

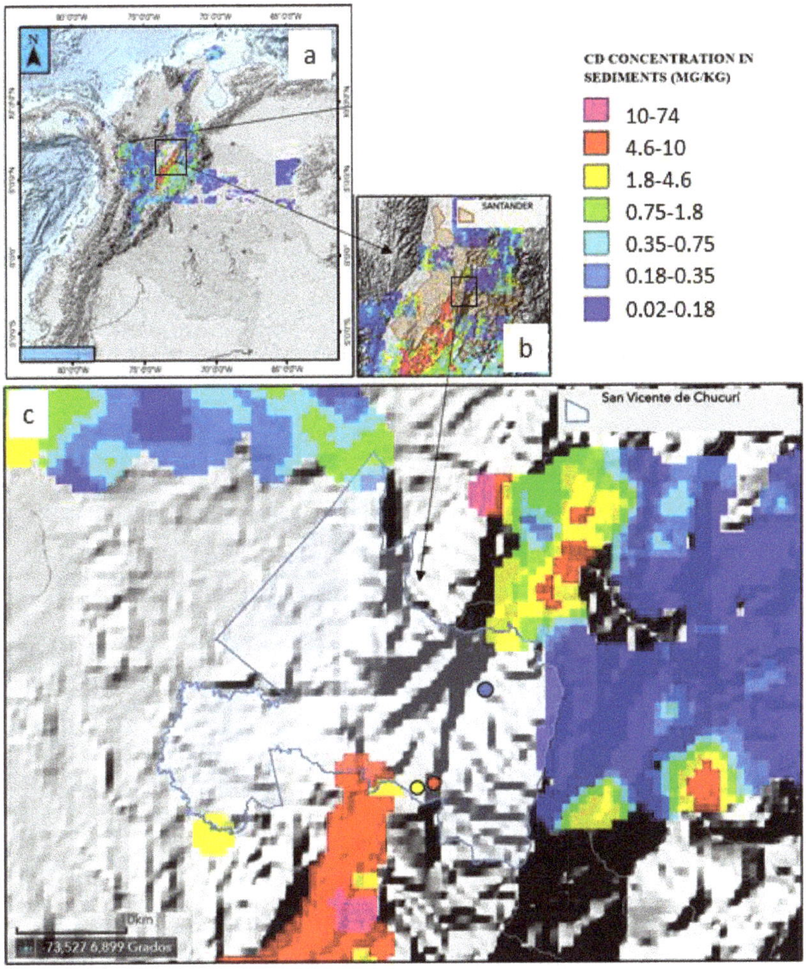

Figure 5. Map of cadmium content in sediments (mg/kg) in Colombia [39]. (**a**) Geochemical map of Cd for Colombia. (**b**) Geochemical map of Cd for Santander. (**c**) Geochemical map of Cd for San Vicente de Chucurí, Farm 1 (red dot), Farm 2 (yellow dot) and Farm 3 (blue dot) are indicated.

The soil samples studied have much higher average Cd concentrations (27 mg/kg ± 10.1) than those reported in other farms in San Vicente, which show an average of 3.3 mg/kg total Cd in soils, as well as mean values reported for soils in Santander and generally in Colombia [14,27]. Soil Cd concentrations in San Vicente de Chucurí are also much higher than international average Cd concentrations in cacao soils and well above the 1 mg/kg level usually found in soils (Figure 3; [9,27]). In fact, the measured values between 10 and 47 mg/kg in the soils studied fall within the range of typical polluted and metal-rich soils worldwide ([21] and references therein). While elevated Cd concentrations are not necessarily available for uptake and incorporation by the cacao plant, a correlation between geological substrate and Cd bioavailability and/or Cd content in cacao tissues has been previously reported in Colombia, e.g., [10]. Regardless of the specific values, it is clear that Cd in soils in the region is very elevated and represents a threat to cacao cultivation and commercialization, which is the basis of the local economy.

Of even more concern is that the vast majority of our samples do not comply with international regulations for tolerable values for mammals' exposure, ecosystems or critical thresholds for human health risks (Figure 3; [36,37]). Soils in the area also exceed the Ecological Soil Screening Levels set by the EPA (Eco-SSL) (0.36 mg/kg dry weight), which if surpassed is considered damaging to mammals and ecosystems [37]. This poses a serious health risk for people exposed to the soils and consuming crops in this particular area, as Cd accumulates in organisms and is potentially carcinogenic, as well as being linked to oxidative stress and kidney malfunction [1,21]. The values reported imply possible health issues, as Cd enters the human body through foods produced in soils that are contaminated with the metal, and after accumulation can cause poisoning or organ malfunction [20,21]. The farmers in the region also grow subsistence crops as well as raise animals for consumption, thus further increasing their exposure to the metal through bioaccumulation from consuming crops and animal products directly linked to the contaminated soils.

It is important to note that a high variability in concentrations of the metal at several scales has been reported [27] and is in fact observed within our study area (Figures 3 and 5; [14,27]). Total Cd concentrations in cacao soils from Santander range from 0.01 to 27 mg/kg, thus small-scale heterogeneity and/or additional Cd sources in our studied farms may be a variation factor [27]. Heterogeneity was previously reported, with farms in the east flank of the municipality having higher Cd averages than those in the west flank (Figure 5; [14]). Metal concentrations vary significantly within Farms 1 and 2, both smaller than 1 ha, suggesting that small-scale heterogeneity plays a significant role in Cd distribution, most likely coupled to the heterogeneous metal content in the parent material reported in the present study. Furthermore, additional sources of Cd may play a role, and will be discussed in the following sections.

It is crucial to consider that only about 4% of the total metal is in Cd^{2+} form, and hence bioavailable for plant uptake [6,8,14,15]. Therefore, we would expect to have about 1 mg/kg of available Cd in the study area (1.16 mg/kg in Farm 1; 1.04 mg/kg in Farm 2; and 1 mg/kg in Farm 3). This would be a reasonably standard level of Cd in soils were it not for the fact that the element bioaccumulates. Moreover, the high levels of Cd reported in nibs at the study area indicate a strong uptake and accumulation by the cacao plants (Figure 1).

4.2. Natural Cadmium Sources

Autochthonous sources are those intrinsic sources from which a component or element might be introduced into an ecosystem, area, or in this case, plantation. Cd has high mobility through sediment flows and erosion processes by water and wind, and material translocation, resulting in accumulation in sedimentary plains and rivers, e.g., [17]. High dependence of 'total' soil Cd has been linked to geological substrates, with the highest median concentration being found in alluvial sediments and soils developed on sedimentary rocks [20,25].

Our sampling sites are above the geological unit b6b6-Sm within the "Simití" formation that is made of laminated black claystones, carbonaceous and calcareous fine-grained rocks with calcareous concretions (Figure 2; [29,30]). This was confirmed by our petrographic analyses that indicated the presence of limestone, marl and shale in all our samples (Figure 4). All rocks analyzed are carbonaceous, thus they can hold substantial quantities of Cd by adhesion and absorption into pores (Figure 4; [40,41]). Calcite will bind Cd through cation exchanges due to the similar sizes and charges of Ca and Cd [7,8,40,41]. This is supported by the significant Pearson correlation between Cd and Ca in rocks (Pearson correlation = 0.79, $p < 0.05$), which points to the calcium-rich rocks as the main source of Cd in the present study, as has been seen in other studies [8,9,24].

Carbonate rocks with high Cd content, through the pedogenesis process, tend to produce soils enriched in the metal [24]. According to the petrographic analysis, the humic cambisols studied are mainly composed of unconsolidated carbonates and are rich in humus (Figure 4d). The abundance of carbonates is further confirmed by the ignition analysis of soils, which shows carbonates to range from 2.2 to 6.5 g (±0.1), consistent with soil forming from the underlying carbonate rocks. While the petrographic analysis confirms that soils in the area were formed by weathering of the carbonate-rich parental material rather than by transport of allochthonous material, no significant correlation is found between the Cd contents of rocks and soils. Such a strong association between parental material and Cd content in soils has been previously reported in other areas [20,21]. The lack of correlation for our dataset may be due to the heterogeneity of Cd concentrations in rocks as well as in weathering and pedogenesis processes.

Even though the carbonate rocks can have high Cd concentrations due to their porosity, it is strongly bound in their structure, which makes it less available for plant absorption. In fact, Ca addition has been used as a remediation strategy for high Cd concentrations in arable land as it binds the metal, e.g., [42,43].

The analyzed rocks also show the presence of Barite ($BaSO_4$), which has been found to be enriched with Cd, with levels of up to 9.9 mg/kg (Figure 4c; [41]). The Barite in parental rocks may further add Cd to the resulting soil after chemical weathering associated with pedogenesis.

4.3. Anthropogenic Cadmium Input

Fertilization, mining, construction sites and industrial activities that introduce extrinsic elements into the farm soils, as well as natural processes such as mass movements, water flows or aeolian deposition coming from Cd-rich areas, all constitute possible allochthonous sources [14,20,30]. Even though the region is known for hosting mining, there is no reported activity near the sampling area (confirmed by the farmers), which means that it can be ruled out as a significant source of Cd for the studied soils.

As we established that soils in the area are formed from the weathering of parental rock, mass movements, water and eolian transport can also be ruled out as significant Cd sources.

Another potential allochthonous source of the metal is fertilizer of sedimentary and phosphorous origin [44]. Approximately 85% of phosphate used in inorganic fertilizers is sourced from sedimentary deposits with high Cd levels [20,44,45]. Mineral fertilizers have been identified as the main source of Cd in some agricultural soils and organic waste-derived fertilizers have been shown to contribute metal content in some cases [44].

In the present study an inorganic fertilizer was used in Farm 2, but had only a marginally higher level of average concentrations compared to Farm 3 (where no fertilizer was used), which indicates a negligible effect on Cd soil content. In contrast, the organic fertilizer added in Farm 1 seems to slightly increase the average Cd soil concentration (Figure 3). Therefore, the organic fertilizer might be an additional source of Cd in the studied soils, which is a rare case since usually those types of fertilizers are less contaminating than chemical ones [44]. This was confirmed by the clustering of all the samples in Farm 1 when PCA was performed (Figure 6). We see that fertilizer had the strongest loading successfully separating Farm 1, where the organic fertilizer was applied, from the other two (Figure 6). This confirms the much stronger impact of the organic fertilization versus the inorganic one, which was already noted when comparing soil Cd concentrations (Figures 3 and 6).

The organic fertilizer applied to Farm 1 contained 64 ± 2 mg/kg Cd, well above the values for soils and most rock fragments measured in the area, and exceeding acceptable Cd levels for fertilizers in Europe (40 mg/kg; Figure 3; [45]). The organic fertilizer was sourced from chickens and pigs which were fed food residues and organic matter from crops grown in the same soils, clearly resulting in the biomagnification of Cd, and thus extremely high levels of the metal (Figure 3; [2,4]). This case is particularly worrisome since the Cd introduced with the fertilizer will be easily absorbed by the plants and result in further metal accumulation in cocoa beans.

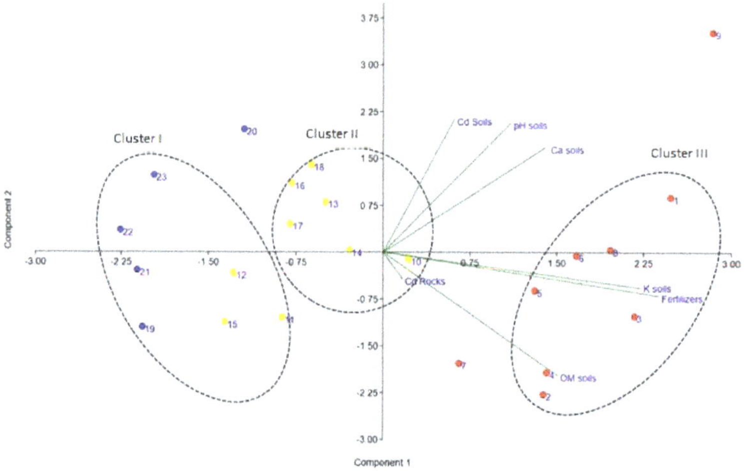

Figure 6. Principal component analysis including the main soil factors measured in the study soils. Clusters resulting from the paired-group analysis are overlaid. Farm 1 is indicated by red dots, Farm 2 by yellow dots and Farm 3 by blue dots; analysis carried out with PAST 4.07 [35].

In Farm 2, the inorganic fertilizer was used one year prior, but no addition was carried out close to the time of sampling. This could mean that part of the Cd added with the fertilizer may have already been removed from the soil, and that is why the average metal concentrations are the same as in the case of soils without fertilizer. Moreover, Cd concentrations measured in the inorganic fertilizer are below those measured in soil samples; thus, its addition would not be expected to raise metal concentrations significantly (Figure 3). We also see no clustering of Farms 2 and 3 related to the fertilizer variable (Figure 6). Despite the fact that Cd content in the inorganic fertilizer does not exceed the international legal limits, concentrations are still higher than the upper critical threshold; thus, it should not be used indiscriminately as it may cause Cd enrichment over time (Figure 3; [44]). Additionally, there is no guarantee that another batch of the same fertilizer might not have a much higher concentration of Cd, as lots tend to be heterogeneous due to production changes [44].

The metal can also be reintroduced into the soil through falling plant leaves or branches, and in some cases, farmers will leave plant debris as a fertilizer [8,9]. In all three plantations most of the organic matter falling from the cacao tree was left to degrade and decompose in situ; thus, the higher Cd concentration in top-soils (first 30 cm sampled) may in part be due to the accumulation of the metal over the years from leaves and husks. In farms in Central Colombia, cacao leaf litter has been found to have higher Cd content than both cacao beans and green leaves, with an average of 85.5 mg/kg, which implies a high level of Cadmium cycling [25,38]. This is supported by a study that coupled Cd concentrations and isotopes in cocoa soils indicating that the use of tree litter may indeed be an additional source of the element [46]. The use of vegetation waste originating within the same plantation as the fertilizer will recirculate bioavailable heavy metals, though the relative importance of this recycling process as a contributor to Cd accumulation in cacao beans is not yet fully understood. This may make the problem worse by re-introducing easily absorbable Cd^{2+} into the system, resulting in an enrichment of the element in the root area and enhanced uptake. This could be easily avoided by shifting to cultivation practices that recover waste away from the crops.

4.4. Soil Factors Affecting Cadmium Bioavailability

While soil Cd content is directly correlated with the bioavailability and cycling of the element in soils, not all the Cd present in soils will be available for uptake and incorporation. If the metal is either not bioavailable or irreversibly bound to the soil matrix, no transfer will take place [4]. In addition to the actual content of Cd in soils, several soil properties affect the Cd bioavailability and consequent root uptake [4]. High electrical conductivity and salinity concentrations, as well as loamy and clayey soil textures, result in higher Cd availability; fortunately, none of those are found in the soils of the study area. Indeed, most of the soil parameters including pH range, organic matter content or chemical composition result in reduced Cd availability [4,25].

Soils in humid tropical climates have been associated with the migration of Cd by leaching from the topsoil layer, which reduces concentrations in the cacao root area, e.g., [2,3]. However, we do not observe Cd leaching, which could be explained by the high slope of the cacao plantations we examined, which causes elevated run-off versus percolation, despite the humid climate. The lack of washing of surface soils may also be due to the pH range, carbonates and high organic matter, resulting in strong Cd binding [4,25].

The pH values of our soils ranged between 6 and 7.8 ($\pm$0.1), with little variation between farms and within the ideal range for retention of Cd within the soil matrix [4]. At pH higher than 5.5 the metal converts into insoluble carbonate and phosphate forms, making it unavailable [2,3]. Even when the soil pH is in the ideal range, Cd remobilization, absorption and accumulation in plant tissues may still occur as plants exude acids from their roots to improve the solubility of nutrients and ions, creating small acid pH zones where the metal can become available, e.g., [47]. For instance, Cd has been found to be a significant problem in cacao grown in near-neutral pH soils in the north of Peru and Honduras [9,48]. While microheterogeneity may partly explain Cd absorption in the present study, the high concentrations in beans are most likely linked to the extremely high levels of Cd in soils. Nonetheless, we observe that pH is one of the stronger loading variables in the PCA and is associated with the Cd in soils even though no significant linear correlation could be found between the two variables (Figure 6). In a study made in cacao systems in Ecuador, they reported that for pH values below 6.3 the beans can accumulate up to 8 times the Cd concentration found in soils, which would only be the case for two samples in our dataset (8.7%; [46]). Most of the data fall in the intermediate pH values (57%) with a predicted enrichment of 4-fold in the beans and the remaining 34.3% present a pH higher than 7 with only a 3.2 enrichment [46]. This would mean an averaged 4.1 enrichment factor for the beans in our farms. In fact, using the reported ratio of accumulation of 1:4 in soil to beans, and based on the highest Cd content of 9.34 mg/kg reported for beans, this would mean that only a maximum of 2.3 mg/kg of metal will be bioavailable in the soil for plant uptake [2,14,49]. Therefore, despite the extremely high reported values, the soil characteristics are significantly reducing the cacao plant uptake of Cd. Other studies in the area confirm the presence of Cd rich soils coupled with a lower proportion of the metal in the cacao tree structures [10,38].

While we find no correlation between Cd content in soils and manganese (Mn) or potassium (K), the PCA results show a strong loading of the K in the separation of cluster III (Figure 6). While the presence of both these elements may not alter the total Cd measured in soils, they have been shown to reduce plant Cd incorporation, probably due to ion competition [9,25]. As we were not able to measure the Cd content of beans, we cannot directly test this hypothesis. However, based on a ratio of 47.5 Mn to 1 Cd, well above the necessary 20 to 1, we assume that the presence of Mn in the studied soils will likely result in reduced Cd intake by plants [9,25]. We also find high K concentrations that work in a similar way, increasing ion competition and therefore decreasing Cd adsorption.

Organic soils have high sorption affinity for Cd: up to 30 times higher than mineral soils [2]. Soil OM will efficiently bind Cd^{2+}, especially if associated with pH levels between 5.5 and 7.5 [4]. Organic matter is known to have a significant surface area and micropores that can serve as sorption sites to retain humidity and nutrients, but also the positively charged Cd^{2+} [4]. The sampled soils had from 1.7 to 7 (±0.1) grams/kg of OM, with Farm 1 having the highest values (Figure 7). While data generally show the tendency, only Farm 1 showed a significant negative correlation with Cd content in soils (Figure 7). This may be because soils in Farm 1 have more bioavailable Cd, which is the one preferentially sequestered in OM. It may also be that due to factors such as salinity or more clay-rich textures, which were not considered in the present study, that will affect Cd presence and availability.

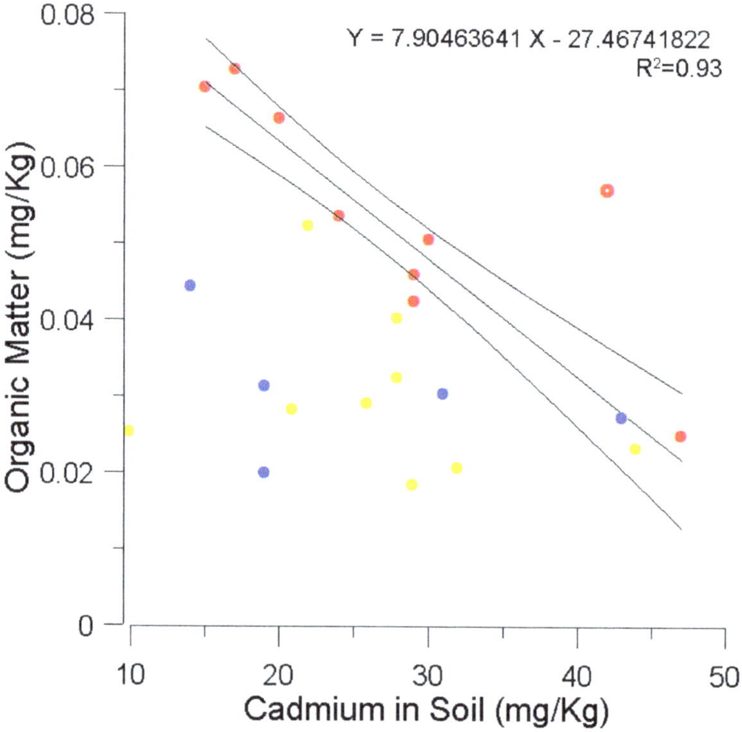

Figure 7. Correlation between cadmium in soils and organic matter content. Farm 1 is indicated in red, Farm 2 is indicated in yellow and Farm 3 is shown in blue. A linear fit has been added only for Farm 1 and the 95% confidence interval is shown; the void red dot has been excluded as an outlier.

Since factors such as the pH, organic matter content and the presence of carbonates mitigate the plant's uptake of the metal as they buffer and absorb the element, we conclude that the Cd availability is considerably below what it potentially could be, given the metal concentrations found in the area's soils. Without these mitigating soil characteristics, the exposure and availability of the metal would likely pose a much higher health risk and could mean even higher accumulation in cacao beans [1,4,21]. This would ultimately make them toxic and not suitable for consumption or commercialization.

The adsorption of Cd has also been reported for hematite and has been shown to be pH dependent with an exponential increase above pH values of 7 [50]. Our soils display an average pH of 6.9 and thus some of the Cd could be adsorbed into the hematite present (Figure 4c). However, this hypothesis could not be corroborated with the correlation between Fe and Cd, most likely because there are other iron sources besides hematite.

4.5. Suggested Mitigation Strategies

Solutions should be cost-effective and be a true aid for cacao farmers so that they can produce cacao beans with no or lower Cd values to sell to manufacturers, and potentially export. A Cd mitigation hierarchy approach should be implemented, by considering actions from farm to final product that are adapted to the specific conditions of the cacao value chain in question [14,48,51].

Eradicating the addition of organic fertilizers to minimize the amount of bioavailable Cd accumulation might be an easy and effective strategy [44] but, of course, a better approach would be to test fertilizers both chemical and organic in nature to guarantee that they have low metal values and can be applied as needed.

While the addition of soil amendments that alter pH, calcium or soil organic matter content to reduce the bioavailability of Cd for the cacao plants is widely used, it would not be useful in this particular case [8,52]. Amendments would not have an impact in the present study area since the soil parameters are already optimal for Cd sequestration.

Another potential solution is leaching, which can remove fertilizer and contaminant components over time [9]. The leaching or washing would move the Cd lower into the soil profile where the roots of the trees cannot uptake it (deeper than 100 cm) [53,54], stopping the accumulation in their structures. However, as we have seen, the chemical characteristics of the soils will strongly bind Cd to the matrix and prevent effective leaching of the element.

The selection of cacao species that are naturally low accumulators of Cd or with low Cd transfer from vegetative parts into the beans has high potential to keep Cd accumulation in cacao beans at levels that are safe for consumption [6]. However, this strategy can only be applied for new producers or for existing farms when they renovate their trees, and thus would have a limited impact in the study area.

Demineralization processes in cacao soils, linked to both biological and physical routes, are under consideration as solutions that would reduce the availability of the metal by its mineralization or biotransformation, e.g., [49–52]. Cadmium-tolerant bacteria (CdtB) and other microorganism existing in these Cd enriched cacao soils have been identified, with about 26 phylogenetically diverse bacteria (Actinobacteria, Alphaproteobacteria, Bacilli, Betaproteobacteria, Gammaproteobacteria) already described and under study for their biotransformation capabilities [8,50,52].

5. Conclusions

In all soil, rock and fertilizer samples in our study area, high levels of Cd were found, with the majority of them by far exceeding the limits set by international regulations.

The concentration of Cd in soils is determined by the geological substrate as an autochthonous source. The petrological analysis indicates the presence of Cd-bearing minerals and sedimentary rocks with high porosity that can hold Cd.

Fertilizers also showed a positive correlation with the Cd content in soils as the main allochthonous source, which can be managed with better practices. Testing fertilizers for heavy metals before their application should be a standard practice.

Even though the pH range, OM content and presence of Mn and K in the soils significantly diminish Cd bioavailability, there is still a high metal content in cocoa beans of the area.

We consider the selection of cacao species that are naturally low accumulators of Cd and the trial of microorganism with biotransformation and mineralization capabilities to be relatively fast and cost-effective solutions to the observed soil Cd enrichment.

Supplementary Materials: The following supporting information can be downloaded at: https://www.mdpi.com/article/10.3390/soilsystems7010012/s1, Table S1: Content of cadmium (Cd) for rocks and soils in mg/Kg. The error of the measurements is indicated (ε).

Author Contributions: For the present article the authors contributed as follows: Conceptualization, V.J.-B. and C.H.; methodology, V.J.-B. and C.H.; software, V.J.-B. and C.H.; validation, V.J.-B., C.H. and J.P.; formal analysis, V.J.-B., C.H. and J.P.; investigation, V.J.-B. and C.H.; resources, C.H. and J.P.;

data curation, V.J.-B. and C.H.; writing—original draft preparation, V.J.-B., C.H. and J.P.; writing—review and editing, V.J.-B., C.H. and J.P.; visualization, V.J.-B. and C.H.; supervision, C.H.; project administration, C.H.; funding acquisition, C.H. All authors have read and agreed to the published version of the manuscript.

Funding: The authors want to thank the Faculty of Science at the Universidad de los Andes for providing funding through project: INV-2019-84-1814 to CH.

Data Availability Statement: Data are available in the supplementary material.

Acknowledgments: We also want to acknowledge the countless hours spent by S. Salamanca perfecting the technique to produce soil thin sections.

Conflicts of Interest: The authors declare no conflict of interest.

References

1. Das, S.C.; Al-Naemi, H.A. Cadmium Toxicity: Oxidative Stress, Inflammation and Tissue Injury. *Occup. Dis. Environ. Med.* **2019**, *7*, 144–163. [CrossRef]
2. Kabata-Pendias, A.; Szteke, B. *Trace Elements in Abiotic and Biotic Environments*; Kabata-Pendias, A., Szteke, B., Eds.; Taylor & Francis: Boca Raton, FL, USA, 2015. [CrossRef]
3. Kabata-Pendias, A. *Trace Elements in Soils and Plants*, 4th ed.; Kabata-Pendias, A., Szteke, B., Eds.; CRC Press: Boca Raton, FL, USA, 2010. [CrossRef]
4. Mortensen, L.H.; Rønn, R.; Vestergård, M. Bioaccumulation of cadmium in soil organisms—With focus on wood ash application. *Ecotoxicol. Environ. Saf.* **2018**, *156*, 452–462. [CrossRef]
5. Nordberg, G.F. Historical perspectives on cadmium toxicology. *Toxicol. Appl. Pharmacol.* **2009**, *238*, 192–200. [CrossRef]
6. Engbersen, N.; Gramlich, A.; Lopez, M.; Schwarz, G.; Hattendorf, B.; Gutierrez, O.; Schulin, R. Cadmium accumulation and allocation in different cacao cultivars. *Sci. Total Environ.* **2019**, *678*, 660–670. [CrossRef]
7. Chavez, E.; He, Z.L.; Stoffella, P.J.; Mylavarapu, R.S.; Li, Y.C.; Moyano, B.; Baligar, V.C. Concentration of cadmium in cacao beans and its relationship with soil cadmium in southern Ecuador. *Sci. Total Environ.* **2015**, *533*, 205–214. [CrossRef]
8. Chavez, E.; He, Z.L.; Stoffella, P.J.; Mylavarapu, R.S.; Li, Y.C.; Baligar, V.C. Chemical speciation of cadmium: An approach to evaluate plant-available cadmium in Ecuadorian soils under cacao production. *Chemosphere* **2016**, *150*, 57–62. [CrossRef] [PubMed]
9. Argüello, D.; Chavez, E.; Lauryssen, F.; Vanderschueren, R.; Smolders, E.; Montalvo, D. Soil properties and agronomic factors affecting cadmium concentrations in cacao beans: A nationwide survey in Ecuador. *Sci. Total Environ.* **2019**, *649*, 120–127. [CrossRef]
10. Meter, A.; Atkinson, R.J.; Laliberte, B. *Cadmium in Cacao from Latin America and The Caribbean. A Review of Research and Potential Mitigation Solutions*; Development Bank of Latin America: Caracas, Venezuela, 2019; Available online: http://scioteca.caf.com/handle/123456789/1506 (accessed on 28 November 2022).
11. The European Commission. Commission Regulation (EU) no 488/2014 of 12 may 2014 amending regulation (EC) no 1881/2006 as regards maximum levels of cadmium in foodstuffs. *Off. J. Eur. Union* **2014**, *488*, 75–79.
12. Abbott, P.C.; Benjamin, T.J.; Burniske, G.R.; Croft, M.M.; Fenton, M.; Kelly, C.R.; Lundy, M.; Camayo, F.R.; Wilcox, M.D. An Analysis of the Supply Chain of Cacao in Colombia. Technical Report. In Purdue University International Center for Tropical Agriculture (CIAT). 2018. Available online: https://pdf.usaid.gov/pdf_docs/PA00W4KG.pdf (accessed on 27 November 2022).
13. Fernández-Niño, M.; Rodríguez-Cubillos, M.J.; Herrera-Rocha, F.; Anzola, J.M.; Cepeda-Hernández, M.L.; Mejía, J.L.A.; Chica, M.J.; Olarte, H.H.; Rodríguez-López, C.; Calderón, D.; et al. Dissecting industrial fermentations of fine flavour cocoa through metagenomic analysis. *Sci. Rep.* **2021**, *11*, 8638. [CrossRef] [PubMed]
14. de Walque, B.; Boeckx, P.; Dewettinck, K. Biophysical Control on Cocoa Quality in Santander, Colombia, 2017–2018. Ph.D. Thesis, Univeristy of Ghent, Ghent, Belgium, August 2018. Available online: https://lib.ugent.be/fulltxt/RUG01/002/509/472/RUG01-002509472_2018_0001_AC.pdf (accessed on 28 November 2022).
15. Smolders, E. Cadmium uptake by plants. *Int. J. Occup. Med. Environ. Health* **2001**, *14*, 177–183. [PubMed]
16. Liu, Y.; Xiao, T.; Perkins, R.B.; Zhu, J.; Zhu, Z.; Xiong, Y.; Ning, Z. Geogenic cadmium pollution and potential health risks, with emphasis on black shale. *J. Geochem. Explor.* **2017**, *176*, 42–49. [CrossRef]
17. Aflizar; Aprisal; Alarima, C.I.; Masunaga, T. Effect of soil erosion and topography on distribution of cadmium (Cd) in Sumani watershed, west Sumatra, Indonesia. *MATEC Web Conf.* **2018**, *229*, 03001. [CrossRef]
18. Page, A.L.; Chang, A.C.; El-Amamy, M. Cadmium levels in soils and crops in the United States. In *Lead, Mercury, Cadmium and Arsenic in the Environment*; John Wiley and Sons: New York, NY, USA, 1987; pp. 119–146.
19. Carrillo-González, R.; Šimůnek, J.; Sauvé, S.; Adriano, D. Mechanisms and pathways of trace element mobility in soils. *Adv. Agron.* **2006**, *91*, 111–178.
20. Smolders, E.; Mertens, J. Cadmium. In *Heavy Metals in Soils: Trace Metals and Metalloids in Soils and Their Bioavailability*; Alloway, B.J., Ed.; Springer Science & Business Media: Berlin/Heidelberg, Germany, 2013; Volume 22, pp. 283–312.

21. He, S.; He, Z.; Yang, X.; Stoffella, P.J.; Baligar, V.C. Soil biogeochemistry, plant physiology, and phytoremediation of cadmium-contaminated soils. *Adv. Agron.* **2015**, *134*, 135–225.
22. Marowsky, G.; Wedepohl, K.H. General trends in the behavior of Cd, Hg, Tl and Bi in some major rock forming processes. *Geochim. Cosmochim. Acta* **1971**, *35*, 1255–1267. [CrossRef]
23. Johnson, D.L.; Watson-Stegner, D. Evolution model of pedogenesis. *Soil Sci.* **1987**, *143*, 349–366. [CrossRef]
24. Xia, X.; Ji, J.; Yang, Z.; Han, H.; Huang, C.; Li, Y.; Zhang, W. Cadmium risk in the soil-plant system caused by weathering of carbonate bedrock. *Chemosphere* **2020**, *254*, 126799. [CrossRef]
25. Gramlich, A.; Tandy, S.; Gauggel, C.; López, M.; Perla, D.; Gonzalez, V.; Schulin, R. Soil cadmium uptake by cocoa in Honduras. *Sci. Total Environ.* **2018**, *612*, 370–378. [CrossRef] [PubMed]
26. Federación Nacional de Cacaoteros. Available online: https://www.fedecacao.com.co/ (accessed on 10 October 2022).
27. Bravo, D.; Leon-Moreno, C.; Martínez, C.A.; Varón-Ramírez, V.M.; Araujo-Carrillo, G.A.; Vargas, R.; Quiroga-Mateus, R.; Zamora, A.; Rodríguez, E.A.G. The First National Survey of Cadmium in Cacao Farm Soil in Colombia. *Agronomy* **2021**, *11*, 761. [CrossRef]
28. Vanderschueren, R.; Argüello, D.; Blommaert, H.; Montalvo, D.; Barraza, F.; Maurice, L.; Schreck, E.; Schulin, R.; Lewis, C.; Vazquez, J.L.; et al. Mitigating the level of cadmium in cacao products: Reviewing the transfer of cadmium from soil to chocolate bar. *Sci. Total Environ.* **2021**, *781*, 146779. [CrossRef]
29. Moreno, G.; Sarmiento, G. Estratigrafía Cuantitativa de las Formaciones Tablazo y Simití en las localidades de Sáchica (Boyacá) y Barichara—San Gil (Santander), Colombia. *Geología Colomb.* **2002**, *27*, 51–74.
30. Gómez, J.; Nivia, Á.; Montes, N.E.; Almanza, M.F.; Alcárcel, F.A.; Madrid, C.A. *Notas Explicativas: Mapa Geológico de Colombia*; Gómez, J., Almanza, M.F., Eds.; Compilando la geología de Colombia: Una visión a 2015. Servicio Geológico Colombiano; Publicaciones Geológicas Especiales: Bogotá, Colombia, 2015; pp. 9–33.
31. Nelson, D.W.; Sommers, L.E. Total carbon, organic carbon, and organic matter. In *Methods of Soil Analysis: Part 2 Chemical and Microbiological Properties*; Page, A.L., Ed.; Soil Science Society of America and American Society of Agronomy: Madison, WI, USA, 1983; pp. 539–579. [CrossRef]
32. Oxford Instruments Ltd. Manual de operador X-MET 7500. Industrial Analysis. 2012. Available online: https://fccid.io/ANATEL/02758-12-08305/Manual-X-MET-7500/B28832EF-BF4A-4C25-817D-C8675B600D4E (accessed on 1 October 2022).
33. Imanishi, Y.; Bando, A.; Komatani, S.; Wada, S.I.; Tsuji, K. Experimental parameters for XRF analysis of soils. *Powder Diffr.* **2010**, *53*, 248–255.
34. Bortolotti, M.; Lutterotti, L.; Pepponi, G. Combining XRD and XRF analysis in one Rietveld-like fitting. *Powder Diffr.* **2017**, *32*, S225–S230. [CrossRef]
35. Hammer, Ø.; Harper, D.A.; Ryan, P.D. PAST: Paleontological statistics software package for education and data analysis. *Palaeo Electron.* **2001**, *4*, 9.
36. Brus, D.J.; de Gruijter, J.J.; Walvoort, D.J.J.; de Vries, F.; Bronswijk, J.J.B.; Römkens, P.F.A.M.; de Vries, W. Mapping the Probability of Exceeding Critical Thresholds for Cadmium Concentrations in Soils in the Netherlands. *J. Environ. Qual.* **2002**, *31*, 1875–1884. [CrossRef]
37. United States Environmental Protection Agency. Regional Screening Levels (RSLs)—User's Guide. Available online: https://www.epa.gov/risk/regional-screening-levels-rsls-users-guide (accessed on 10 October 2022).
38. Albarrcín, H.S.R.; Contreras, A.E.D.; Henao, M.C. Spatial regression modeling of soils with high cadmium content in a cocoa producing area of Central Colombia. *Geoderma Reg.* **2019**, *16*, e00214. [CrossRef]
39. Atlas Geoquímico de Colombia. Available online: https://srvags.sgc.gov.co/JSViewer/Atlas_geoquimico_2018/ (accessed on 10 October 2022).
40. Shirvani, M.; Kalbasi, M.; Shariatmadari, H.; Nourbakhsh, F.; Najafi, B. Sorption—desorption of cadmium in aqueous palygorskite, sepiolite, and calcite suspensions: Isotherm hysteresis. *Chemosphere* **2006**, *65*, 2178–2184. [CrossRef]
41. Pivovarov, S. Adsorption of cadmium onto hematite: Temperature dependence. *J. Colloid Interface Sci.* **2001**, *234*, 1–8. [CrossRef]
42. Lu, Q.; Xu, Z.; Xu, X.; Liu, L.; Liang, L.; Chen, Z.; Dong, X.; Li, C.; Wang, Y.; Qiu, G. Cadmium contamination in a soil-rice system and the associated health risk: An addressing concern caused by barium mining. *Ecotoxicol. Environ. Saf.* **2019**, *183*, 109590. [CrossRef]
43. Marini, M.; Caro, D.; Thomsen, M. The new fertilizer regulation: A starting point for cadmium control in European arable soils? *Sci. Total Environ.* **2020**, *745*, 140876. [CrossRef] [PubMed]
44. Roberts, T.L. Cadmium and phosphorous fertilizers: The issues and the science. *Procedia Eng.* **2014**, *83*, 52–59. [CrossRef]
45. News European Parliament. Available online: https://www.europarl.europa.eu/news/en/press-room/20181119IPR19407/fertilisers-cadmium-parliament-and-council-negotiators-reach-provisional-deal (accessed on 10 October 2022).
46. Barraza, F.; Moore, R.E.T.; Rehkämper, M.; Schreck, E.; Lefeuvre, G.; Kreissig, K.; Coles, B.J.; Maurice, L. Cadmium isotope fractionation in the soil–cacao systems of Ecuador: A pilot field study. *RSC Adv.* **2019**, *9*, 34011–34022. [CrossRef]
47. Dong, J.; Mao, W.H.; Zhang, G.P.; Wu, F.B.; Cai, Y. Root excretion and plant tolerance to cadmium toxicity—A review. *Plant Soil Environ.* **2007**, *53*, 193–200. [CrossRef]
48. Zug, K.L.M.; Yupanqui, H.A.H.; Meyberg, F.; Cierjacks, J.S.; Cierjacks, A. Cadmium Accumulation in Peruvian Cacao (Theobroma cacao L.) and Opportunities for Mitigation. *Water Air Soil Pollut.* **2019**, *230*, 72. [CrossRef]
49. Khan, M.A.; Khan, S.; Khan, A.; Alam, M. Soil contamination with cadmium, consequences and remediation using organic amendments. *Sci. Total Environ.* **2017**, *601–602*, 1591–1605. [CrossRef]

50. Topcu, A.; Bulat, T. Removal of cadmium and lead from aqueous solution by Enterococcus faecium strains. *J. Food Sci.* **2010**, *75*, 13–17. [CrossRef] [PubMed]
51. Bravo, D.; Pardo-Díaz, S.; Benavides-Erazo, J.; Rengifo-Estrada, G.; Braissant, O.; Leon-Moreno, C. Cadmium and cadmium-tolerant soil bacteria in cacao crops from northeastern Colombia. *J. Appl. Microbiol.* **2018**, *124*, 1175–1194. [CrossRef] [PubMed]
52. Liu, J.; Hue, N.V. Amending subsoil acidity by surface applications of gypsum, lime, and composts. *Commun. Soil Sci. Plant Anal.* **2001**, *32*, 2117–2132. [CrossRef]
53. Dermont, G.; Bergeron, M.; Mercier, G.; Richer Laflèche, M. Soil washing for metal removal: A review of physical/chemical technologies and field applications. *J. Hazard. Mater.* **2008**, *152*, 1–31. [CrossRef]
54. Torres, L.G.; Lopez, R.B.; Beltran, M. Removal of As, Cd, Cu, Ni, Pb, and Zn from a highly contaminated industrial soil using surfactant enhanced soil washing. *Phys. Chem. Earth* **2012**, *37–39*, 30–36. [CrossRef]

soil systems

MDPI

Article

Paper Mill Biosolids and Forest-Derived Liming Materials Applied on Cropland: Residual Effects on Soil Properties and Metal Availability

Bernard Gagnon and Noura Ziadi *

Agriculture and Agri-Food Canada, Quebec Research and Development Centre, 2560 Hochelaga Boulevard, Québec, QC G1V 2J3, Canada
* Correspondence: noura.ziadi@agr.gc.ca

Abstract: Combined paper mill biosolids (PB) and forest-derived liming by-products improve soil properties, but their residual effects following several years of application have hardly been investigated. A 13-year (2009–2021) field study was initiated at Yamachiche, QC, Canada, to assess the residual effects of PB and liming materials on the properties of a loamy soil. The PB was applied during nine consecutive years (2000–2008) at 0, 30, 60, and 90 Mg wet·ha^{-1}, whereas the 30 Mg PB·ha^{-1} rate also received one of three liming materials (calcitic lime, lime mud, wood ash) at 3 Mg wet·ha^{-1}. No amendment was applied during residual years. Past liming materials continued to increase soil pH but their effect decreased over time; meanwhile, past PB applications caused a low increase in residual soil NO_3-N. Soil total C, which represented 40% of added organic C when PB applications ceased, stabilized to 15% after six years. Soil Mehlich-3-extractable contents declined over the thirteen residual years to be not significant for P, K, and Cu, while they reached half the values of the application years for Zn and Cd. Conversely, Mehlich-3 Ca was little affected by time. Therefore, land PB and liming material applications benefited soil properties several years after their cessation.

Keywords: liming; organic matter; paper mill biosolids; residual effect; trace metals

Citation: Gagnon, B.; Ziadi, N. Paper Mill Biosolids and Forest-Derived Liming Materials Applied on Cropland: Residual Effects on Soil Properties and Metal Availability. *Soil Syst.* **2023**, *7*, 40. https://doi.org/10.3390/soilsystems7020040

Academic Editors: Antonella Lavini, Mohamed Houssemeddine Sellami and Andreas Schaeffer

Received: 4 February 2023
Revised: 16 March 2023
Accepted: 3 April 2023
Published: 19 April 2023

1. Introduction

Every year in North America, large amounts of paper mill biosolids (PB) from treated effluents (7.1 Mg dry) are generated from the forest industry along with liming materials, such as wood ash (WA) and lime mud (LM) (6.6 Mg dry) [1,2]. Adding these residues to cropland improves soil organic matter content, pH, and major nutrient phyto-availability [3–5]. As long as they respect the local regulatory standards for metal concentrations [4], they can be efficiently and more ecologically used on fields rather than being disposed of in landfill [6].

The application of PB has been largely evaluated for its impact on soil organic matter and N and P contribution to crops in the first year of application and in the subsequent year [3,4,7,8]. However, the residual effects of past PB application on soil properties have been little investigated for a longer period of time, especially after several years of application. Longer-lasting soil organic C improvement has been reported following repeated applications of PB during a few years [9–12], but few studies addressed how long soil C may be improved after PB cessation. Using municipal biosolids, Cogger et al. [13] reported increases in soil total C and Bray-1 extractable P that were maintained for nine years after the end of a 10-year application. They noted, however, a significant decrease in fall NO_3-N during the residual period. In contrast, other studies on municipal biosolids reported between 25% and 40% loss in soil organic C within 10 years following the cessation of repeated applications [14–16].

The residual effect of forest-derived liming materials has also been examined, but much more so after a one-time application. Some studies reported no significant change in soil pH occurring over time with lime mud (LM) and wood ash (WA) three years following a one-time application [17,18]. Moreover, Arshad et al. [17] observed that WA provided 75% more available P than agricultural lime during all the residual years and attributed that to the rise in organic P mineralization, reduced P fixation, and increased P solubility.

The soil bioavailability of metals once material application ceases is a main concern in agriculture and has been part of many studies on municipal biosolids. It was reported that biosolids continue to decompose after cessation and may release the adsorbed metals into the soil solution, particularly in the first five years [16], although the reverse was also observed with the occlusion of metals in Fe-oxides or chemisorption [19]. Biosolid decomposition can also lead to a decrease in soil pH, making metals more available [20]. By contrast, research is yet to be carried out with PB, especially after repeated application and with liming materials. Compared with municipal biosolids, PB mostly represents a low risk for soil metal contamination [21,22] with pollutants rarely exceeding regulatory limits [23,24]. During the years of application, PB adds C and mineral components to soil, thus increasing sorption sites and binding metals in insoluble forms [25]. Despite low inherent metal concentrations, some increases in metal availability in agricultural soils were observed following a fresh addition of PB or deinking paper sludge [26,27], but to our knowledge the residual effects after many consecutive years of application of such products have never been evaluated and would deserve additional study to see any potential impact on soil metal accumulation.

The present study reports the residual effects of the application of PB and forest-derived liming materials on main soil properties and metal availability. We hypothesized that upon material decay and following the cropping season, soil main properties (organic matter, pH, major nutrients) would decrease while metal availability would increase with time. A companion paper that evaluated the effects on plant parameters, including yield and nutrient accumulation, has been previously released [28]. In this study, the liming materials were added to PB with a view to counteract possible soil acidification related to the mineralization of organic N from PB.

2. Materials and Methods

2.1. Site and Treatment History

The study was carried out at Yamachiche (lat. 46°17′ N, long. 72°48′ W, alt. 10 m), Quebec, Canada, on a flat, imperfectly drained Chaloupe loam soil (fine, mixed, frigid Typic Humaquept, according to U.S. soil classification). This study was part of a long-term experiment in which the different materials were applied in early summer at sidedress from 2000 to 2008, and residual effects were followed from 2009 to 2021. The site was under minimum tillage until 2018, including the years of material application, with spring harrowing to prepare the seedbed for annual field crops. In 2019, the site was converted into a perennial hay crop.

The different treatments, plot design, and cultural practices for the years of application have been previously detailed in [29]. Concretely, PB was applied at 0, 30, 60, and 90 wet Mg·ha^{-1}, whereas the 30 Mg PB·ha^{-1} rate received also one of the three following liming materials each at 3 wet Mg·ha^{-1}: calcitic lime (CL), LM, and WA. Typically, PBs are applied at a rate of 30 Mg wet ha^{-1} year^{-1} in Québec with some additional fertilizer, N, to meet crop requirements [30]. The 60 Mg PB ha^{-1} rate corresponded to rate without N addition, whereas the 90 Mg PB ha^{-1} simulated long-term nutrient accumulation. The experimental layout consisted of plots of 3 × 10 m replicated four times in a randomized complete block design.

Material characteristics were reported previously [29] and summarized for the last six years of application (2003–2008) in Table 1. Globally, the PB had a good fertilizing potential with a low metal concentration. For their part, WA and LM had a high total K and Ca content, respectively, but both exceeded the provincial Cd criterion for unrestricted agricultural land use (3.0 mg·kg^{-1}). However, these materials were applied at 50% of the mandated provincial limit [31].

Table 1. Main chemical characteristics of applied papermill biosolids and liming materials (mean ± standard deviation of 2003–2008 application years).

Attribute	Units	PB	LM	WA	CL
pH_{water}		5.0 ± 0.9	10.6 ± 1.9	12.3 ± 0.5	9.2 ± 0.5
Moisture	$g \cdot kg^{-1}$ FM	677 ± 51	280 ± 34	199 ± 97	11 ± 9
Total C	$g \cdot kg^{-1}$ DM	438 ± 5	-	-	-
Total N	$g \cdot kg^{-1}$ DM	21.9 ± 5.8	0.3 ± 0.3	0.3 ± 0.1	0.1 ± 0.1
Total P	$g \cdot kg^{-1}$ DM	3.5 ± 0.8	1.6 ± 0.7	5.4 ± 1.8	0.5 ± 0.2
Total K	$g \cdot kg^{-1}$ DM	0.6 ± 0.1	1.6 ± 2.2	20.4 ± 8.0	1.9 ± 0.6
Total Ca	$g \cdot kg^{-1}$ DM	7 ± 2	263 ± 41	137 ± 26	235 ± 24
Total Cu	$mg \cdot kg^{-1}$ DM	7 ± 3	20 ± 10	33 ± 8	3 ± 0
Total Zn	$mg \cdot kg^{-1}$ DM	57 ± 20	196 ± 42	364 ± 41	3 ± 1
Total Cd	$mg \cdot kg^{-1}$ DM	0.7 ± 0.2	4.1 ± 0.9	5.2 ± 1.2	2.0 ± 0.1
Total Mo	$mg \cdot kg^{-1}$ DM	1.3 ± 0.7	0.1 ± 0.1	0.0 ± 0.1	0.1 ± 0.1

PB, paper mill biosolids; LM, lime mud; WA, wood ash; CL, calcitic lime; FM, fresh matter; DM, dry matter.

2.2. Field Operations during Residual Years

In residual years (2009–2021), no amendment was applied to the plots except for the mineral N treatment. Annual field crops were grown from 2009 to 2018 and comprised, successively, dry bean (*Phaseolus vulgaris* L.), winter wheat (*Triticum aestivum* L.), grain corn (*Zea mays* L.), malting barley (*Hordeum vulgare* L.), winter wheat, soybean (*Glycine max* (L.) *Merr.*), winter rye (*Secale cereale* L.), soybean, winter wheat, and spring barley. For 2019 to 2021, a forage mixture of timothy (*Phleum pratense* L.), meadow bromegrass (Bromus biebersteinii Roem. and Schult.), and birdsfoot trefoil (*Lotus corniculatus* L.) was planted as a companion crop with spring barley. For all annual field crops, grains were harvested while straw residues were returned to the field, except for small cereals (wheat, barley, rye) where the straw was sold to other farmers. Forage was harvested three times during each growing season.

2.3. Soil Sampling and Analysis

After each crop harvest (the third cut for forage) in residual years (2009–2019), soils were sampled (composite of five cores) to 30 cm depth. Samples were kept moist at 4 °C for pH and NO_3-N determination or air-dried and sieved to <2 mm for other soil analysis. The soil pH and NO_3-N analysis was performed each year, whereas total C, Mehlich-3-extractable P, K, and Ca, and bio-available metal (Cu, Zn, Cd, Mo) concentrations were measured at a 3 years interval, namely, in 2011, 2014, and 2017. Additional soil sampling was carried out after the first forage cut at the beginning of June 2021 to determine the effect in the course of the 13th year on soil pH and total C, major nutrient (P, K, Ca), and metal (Cu, Zn, Cd) concentrations. For this sampling, Mo availability was not determined since most soil pH $\leq$ 6.0, which makes Mo barely available to plants [32].

The pH was determined in a 1:2 soil to water ratio. Results from the 10 first residual years (2009–2018) have been reported in [28], and updated here with two more samplings showing a further decrease, where the liming materials reached 0.3–0.6 units over the untreated control for the last sampling (Figure 1).

The NO_3-N concentration was extracted with 2.5 g field-moist sample and 20 mL 1 $mol \cdot L^{-1}$ KCl with mechanical agitation for 30 min before filtration [33]. The concentrations were determined through the Cd-Cu reduction method using a continuous-flow injection auto-analyzer (QuickChem 8000 FIA+ analyzer, Lachat Instruments, Loveland, CO, USA). A subsample of dried soil was finely ground to 0.20 mm and then analyzed for total C determination on a LECO TruSpec CN (Leco Inc., St. Joseph, MI, USA).

Soil P, K, Ca, and metal (Cu, Zn, Cd) concentrations were extracted with the Mehlich-3 solution [34]. Metal concentrations (except 2021 samples) were also extracted using DTPA (Cu, Zn, Cd [35]) or water (Mo [36]). Single multielement laboratory Mehlich-3 extraction is relatively simple, cost-effective, and provides generally reproducible results [37].

Concentrations of available P were determined by colorimetry (DU720, Beckman Coulter, Mississauga, ON, Canada) with reaction with the ascorbic acid–molybdate complex [38], whereas concentrations in other elements were determined by the inductively coupled plasma optical emission spectrometer procedure (Optima 4300DV, Perkin Elmer, Shelton, CT, USA).

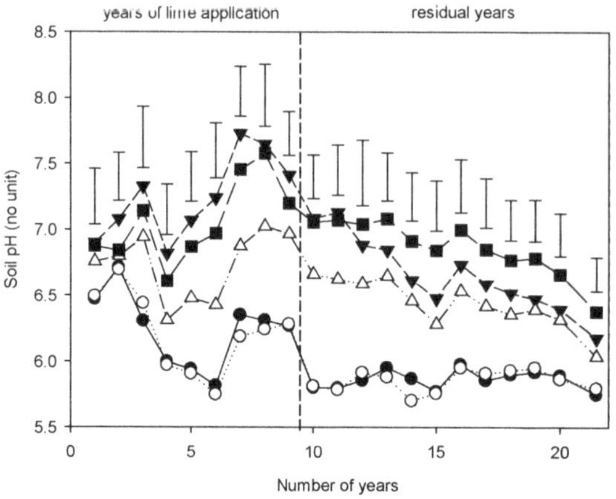

Figure 1. Direct and residual effects of liming materials applied during nine consecutive years on the soil pH in the 0 to 30 cm layer. Vertical bars represent the LSD (5%). • control (0N); ○ combined paper mill biosolids applied alone at 30 Mg wet·ha^{-1} or with 3 Mg wet·ha^{-1} of ▼ lime mud, Δ wood ash, or ■ calcitic lime.

2.4. Statistical Analysis

All data were first analyzed using the univariate procedure to see their normality and then subjected to a Bartlett's test; Zn (2011, all methods) and Mo (2014) were log-transformed to improve variance homogeneity. Data analysis was performed for separate years to see any effect due to material decomposition using the MIXED procedure of SAS (SAS Studio v.1.21). Main treatment effects were compared using orthogonal linear contrast for PB rate and single degree-of-freedom contrasts for comparison of PB 30 Mg·ha^{-1} alone vs. PB 30 Mg·ha^{-1} + liming, and for CL vs. LM + WA, and LM vs. WA within the PB 30 Mg·ha^{-1} + liming treatments. Statistical significance was set as $p < 0.05$.

3. Results and Discussion

3.1. Soil Total C

Soil total C continued to benefit from PB after the cessation of land application, but the extent decreased once the C was mineralized (Figure 2). From a respective increase of 3.7, 7.9, and 9.2 g·kg^{-1} compared to the control in fall 2008, it was 2.4, 3.9, and 5.4 g·kg^{-1} in fall 2011, and 1.4, 2.3, and 2.8 g·kg^{-1} at 13 years after the end of application (spring 2021) with the 30, 60, and 90 Mg PB·ha^{-1} rates, respectively. This corresponded roughly to an estimated ~40% of the added organic C being present at the end of continuous application, which fell to 25% after three residual years, and was around 15% thereafter. Based on regression analysis, the model showed a significant decrease in soil C in the first 6 years and then it stabilized at 9.5 years after PB cessation. This indicates that repeated PB application for several years had a long-lasting positive impact on soil total C. It also suggests that a more stable and recalcitrant form could account for the soil total C content some years after PB cessation. Mao et al. [22] estimated that resistant C added with paper waste contributes significantly to the long-term soil-stable C.

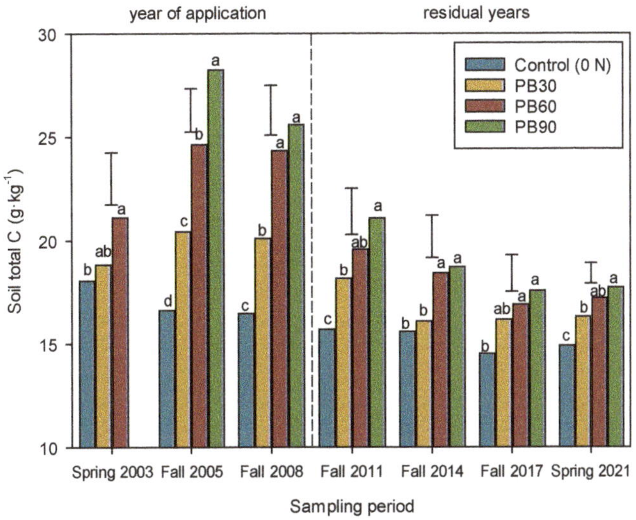

Figure 2. Direct and residual effects of paper mill biosolids (PB) application on the soil total C content in the 0 to 30 cm layer. Vertical bars represent the LSD (5%). Different letters within a sampling date represent difference at $p = 0.05$ according to an LSD test. Data for years of application come from [29].

Previous works reported that repeated annual application of PB for three or more years promoted a sustained increase in soil C which was less apparent with a single application [9–12,39]. Cogger et al. [13] applied municipal biosolids continuously for 10 years to a tall fescue (Festuca arundinacea Schreb.), obtaining an increase in soil C equivalent to 28% of material C added, but the increase in soil C was nearly the same (27% of added C) at the end of the nine-year residual period, suggesting soil C stabilization. By contrast, other studies reported between 25% and 40% of soil organic C losses within 10 years after repeated sewage sludge application for 6 to 11 years [14–16]. In this study, by using a linear function of soil total C data versus time after PB cessation at 0, 3, 6, and 9 years for each PB rate treatment (before C stabilization occurring at 9.5 years), we estimated that for the 60 Mg PB·ha^{-1} rate, which summed 70 Mg C·ha^{-1} for the total period of application, net soil C losses reached 0.57 g·kg^{-1} per year (Figure 3).

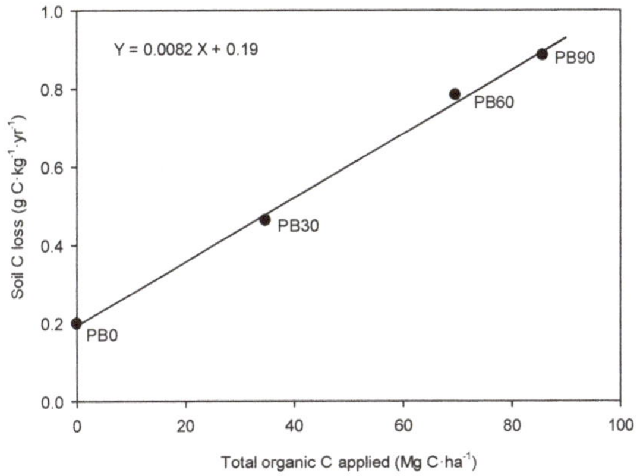

Figure 3. Soil total C losses per year from 2008 to 2017 in the paper mill biosolids (PB)-amended plots as a function of the total amount of organic C applied by each treatment in 2000–2008 which totalled, respectively, 0, 35, 70, and 86 Mg·ha^{-1} for each PB rate [29].

3.2. Soil NO_3-N

Soil NO_3-N concentrations at harvest were increased by the residual PB rate, notably when long-season crops were grown, such as grain corn in 2011 and soybean in 2014, or when grain yields were low, as in 2017 with winter wheat (Figure 4) [28]. However, the increases over the control in those years were much smaller, averaging 2.0 mg·kg^{-1} for the 90 Mg·ha^{-1} rate, which were nine times lower that the levels found in the years of PB application [29]. Considering the benefits in plant N accumulation for the first years after cessation [28], it is not expected that PB would cause large environmental NO_3-N leaching losses in residual years. Cogger et al. [13] observed a rapid decline in fall NO_3-N levels with municipal biosolids, which reached the level of zero-N treatment three years after application cessation.

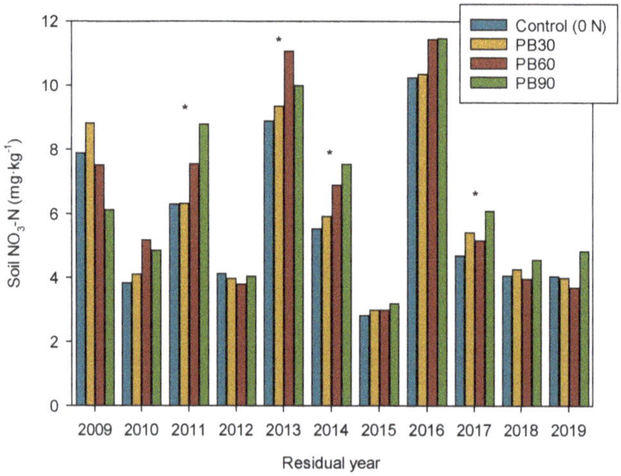

Figure 4. Residual effects of paper mill biosolids (PB) application on the soil NO_3-N concentrations in the 0 to 30 cm layer at crop harvest. Asterisks indicate significant differences at $p = 0.05$ between the 90 Mg ha^{-1} rate and the control except in 2013 (60 Mg ha^{-1} rate vs. control).

3.3. Soil Available P, K, and Ca

Soil available P, as determined by the Mehlich-3 extraction, was significantly increased by the PB rate at the third and ninth residual years but not thereafter (Table 2). This increase in soil P translated into higher crop P accumulation up to five years after PB cessation [28]. This meant that PB constituted an efficient source of P for plants and recycled P for a period surpassing the years of application, which has to be considered in fertilizer P recommendations [40]. Cogger et al. [13] found small changes in soil Bray-1 P even nine years after municipal biosolids cessation.

Past liming material application generally increased soil Mehlich-3 P in residual years except 2021 (Table 2). This was related to the increased soil pH, mobilization of P following labile organic carbon added from previous PB, and also a supply of P from WA [3,41,42]. Soil pH exerts an effect on P availability through increasing competition of OH- with PO_4^{3-} ions for binding sites, stimulating microbial activity at neutral pH, and complexing P (precipitation/sorption) with Ca, Fe, and Al ions [43,44].

Soil Mehlich-3 K was not affected by PB in residual years (data not shown) as it was in the years of application [29], indicating that this material was a poor source of K for crops. However, WA increased soil Mehlich-3 K three years after cessation (+43 mg·kg^{-1} when 30 Mg PB·ha^{-1} received WA) but not thereafter. This material comprises large concentrations of highly bioavailable K [3,41,45]. Upon crop uptake and likely leaching due to the low binding of K to soil particles, this study indicated that the soil Mehlich-3 K increase was relatively short-lived after cessation of WA [46].

The CL and LM treatments still had significant effects on soil Mehlich-3 Ca in residual years, followed by WA (Table 2). The high contribution of added Ca from CL and LM to soil Mehlich-3 Ca was reported during incubation (82% and 97%, respectively), whereas it was 68% for WA [41]. Despite PB decay and without addition of fresh amendment, our study showed a stability in Ca availability in residual years.

Table 2. Effect of residual application of paper mill biosolids and alkaline materials on the soil Mehlich-3-extractable P and Ca in the 0 to 30 cm layer.

Treatment	Mehlich-3 P ($mg \cdot kg^{-1}$)				Mehlich-3 Ca ($mg \cdot kg^{-1}$)			
	2011	2014	2017	2021	2011	2014	2017	2021
Control (0N)	69 b	75 b	66 c	65 a	1095 d	1256 c	1097 c	1123 c
PB 30 $Mg \cdot ha^{-1}$	71 b	77 b	68 c	62 a	1224 cd	1302 c	1161 c	1230 bc
PB 60 $Mg \cdot ha^{-1}$	92 ab	81 ab	75 bc	69 a	1279 cd	1430 c	1285 bc	1384 bc
PB 90 $Mg \cdot ha^{-1}$	101 a	94 ab	97 ab	78 a	1251 cd	1403 c	1275 bc	1283 bc
PB 30 Mg + 3 Mg LM ha^{-1}	102 a	113 a	107 a	89 a	1738 ab	1912 ab	1710 a	1738 a
PB 30 Mg + 3 Mg WA ha^{-1}	94 ab	86 ab	85 abc	68 a	1481 bc	1597 bc	1422 b	1393 b
PB 30 Mg + 3 Mg CL ha^{-1}	90 ab	103 ab	91 abc	83 a	1788 a	1995 a	1774 a	1813 a
LSD (5%)	26	32	26	28	273	318	230	242
	Statistical analysis (*F*-value)							
Treatment	2.1	1.5	2.6 *	0.9	6.9 ***	6.0 ***	9.2 ***	8.0 ***
Contrasts								
PB—linear	7.7 *	1.5	5.3 *	0.9	1.4	1.2	3.0	2.6
PB vs. PB + liming	6.0 *	3.6	6.3 *	2.7	17.4 ***	18.3 ***	27.8 ***	19.5 ***
CL vs. LM + WA	0.6	0.1	0.2	0.1	2.5	3.3	4.8 *	6.1 *
LM vs. WA	0.4	3.0	3.1	2.6	3.8	4.3	6.8 *	8.9 **

PB, paper mill biosolids; LM, lime mud; WA, wood ash; CL, calcitic lime. The years 2011, 2014, 2017, and 2021 correspond to 3, 6, 9, and 13 years after the end of material application. Statistical significance at 5%, 1%, and 0.1% is denoted by *, **, and ***, respectively. Means within a column followed by the same lowercase letter are not statistically significant at $p = 0.05$ according to an LSD test.

3.4. Soil Metal Availability

In the years of application, four metals (Cu, Zn, Cd, and Mo) were closely followed, considering their total concentrations in the materials [26]. Copper was the least affected metal. Although no significant effect was detected during the years of application [26], the past PB rate induced a linear effect on Mehlich-3-extractable Cu in 2011 and 2014 and on DTPA Cu in 2014 (Table 3). This meant that Cu was released into the soil in residual years with PB decomposition, which can be attributed to Cu being most preferentially sorbed to organic matter [47].

Soil Zn availability was largely increased by the PB rate in all residual years (Table 4). Soil Zn further increased three years after the last repeated application (fall 2011) and then sharply declined in the sixth year (fall 2014), with fewer variations thereafter (Figure 5A). Soil Cd availability was also increased by the PB rate in residual years (Table 5), but unlike Zn, it decreased soon after application ended and subsequently maintained a fairly constant level throughout those years (Figure 5B). At the time of PB cessation (fall 2008), mean increases in soil available Zn and Cd represented, respectively, 76 and 92% for Mehlich-3 extraction and 55 and 107% for DTPA extraction of the total amounts of each metal added during the years of application. These percentages decreased in residual years to constitute around half the values found in the final application year at the last sampling (36 and 49% for Mehlich-3 and 29 and 67% for DTPA). During the time of application, PB and soil interact to form sorbing sites that retain metals in the soil [25]. Once PB application ceases, PB organic C continues to mineralize at a slower rate, which releases adsorbed metals into the soil solution, but metals may also react with oxides (Fe, Al, Mn) in PB and soil, rendering them less soluble and then less available to plants [14,48]. Trace metals sorbed to oxide surfaces would remain sequestered for a longer period than those complexed by organic C [49]. In this study, metals such as Zn were more available to plants in the earlier years after PB cessation than during the time of application, likely due to the decomposition of PB, which was still quite significant (Figure 2) [50], and the mobilization of metals by the dissolved organic C produced [51]. Organically bound Zn fraction represented a significant pool in the PB-amended soil, contrarily to Cd [47]. Afterwards, the soil Zn availability was reduced, sorbed to inorganic and very recalcitrant organic components in the soil–residual biosolids mixture [49].

Table 3. Effect of residual application of paper mill biosolids and alkaline materials on the soil-available copper in the 0 to 30 cm layer.

Treatment	Mehlich-3 Cu (mg·kg^{-1})				DTPA Cu (mg·kg^{-1})		
	2011	**2014**	**2017**	**2021**	**2011**	**2014**	**2017**
Control (0N)	3.1 c	4.7 bc	5.2 a	4.5 a	7.1 b	4.7 bc	5.1 a
PB 30 Mg·ha^{-1}	4.1 abc	4.5 c	4.9 a	4.6 a	7.7 ab	4.4 bc	4.8 a
PB 60 Mg·ha^{-1}	5.0 a	5.1 bc	5.4 a	5.3 a	9.0 a	4.7 b	5.0 a
PB 90 Mg·ha^{-1}	4.3 ab	5.5 a	5.5 a	4.8 a	8.1 ab	5.6 a	5.1 a
PB 30 Mg + 3 Mg LM ha^{-1}	4.1 abc	5.4 ab	5.4 a	5.3 a	7.6 ab	4.4 bc	4.8 a
PB 30 Mg + 3 Mg WA ha^{-1}	4.4 ab	5.2 bc	5.4 a	4.8 a	8.1 ab	4.9 b	4.9 a
PB 30 Mg + 3 Mg CL ha^{-1}	3.3 bc	4.6 bc	4.9 a	4.2 a	6.9 b	4.0 c	4.3 a
LSD (5%)	1.2	0.8	0.9	1.2	1.7	0.7	0.8
	Statistical analysis (*F*-value)						
Treatment	2.3	2.0	0.7	0.8	1.1	4.2 **	1.0
Contrasts							
PB—linear	6.5 *	6.3 *	1.3	0.7	2.5	10.1 **	0.0
PB vs. PB + liming	0.2	3.5	1.1	0.2	0.1	0.0	0.2
CL vs. LM + WA	3.4	4.0	1.8	2.8	1.7	4.5 *	2.7
LM vs. WA	0.4	0.4	0.0	0.7	0.3	2.2	0.1

PB, paper mill biosolids; LM, lime mud; WA, wood ash; CL, calcitic lime. The years 2011, 2014, 2017, and 2021 correspond to 3, 6, 9, and 13 years after materials application ending. Statistical significance at 5% and 1% denoted by * and **, respectively. Means within a column followed by the same lowercase letter are not statistically significant at $p = 0.05$ according to an LSD test.

Table 4. Effect of residual application of paper mill biosolids and alkaline materials on the soil available zinc in the 0 to 30 cm layer.

Treatment	Mehlich-3 Zn (mg·kg^{-1})				DTPA Zn (mg·kg^{-1})		
	2011	**2014**	**2017**	**2021**	**2011**	**2014**	**2017**
Control (0N)	6.4 bc	6.3 bc	5.6 bc	5.2 bc	4.0 bc	3.4 bcd	2.8 bc
PB 30 Mg·ha^{-1}	5.9 bc	5.9 bc	5.5 bc	4.9 bc	3.9 bc	3.5 bc	2.8 bc
PB 60 Mg·ha^{-1}	8.5 ab	7.1 ab	7.0 ab	5.5 bc	5.6 ab	4.1 b	3.6 ab
PB 90 Mg·ha^{-1}	11.2 a	8.4 a	7.8 a	6.9 a	7.4 a	5.2 a	4.0 a
PB 30 Mg + 3 Mg LM ha^{-1}	6.8 bc	6.9 abc	6.5 abc	6.0 ab	3.6 bc	3.4 bcd	2.9 bc
PB 30 Mg + 3 Mg WA ha^{-1}	6.5 bc	5.8 bc	5.9 bc	4.9 bc	3.8 bc	3.0 cd	2.7 c
PB 30 Mg + 3 Mg CL ha^{-1}	5.7 c	5.5 c	4.9 c	4.7 c	3.0 c	2.5 d	2.0 c
LSD (5%)	3.2	1.5	1.6	1.3	2.8	0.9	0.8
	Statistical analysis (*F*-value)						
Treatment	3.1 *	3.5 *	3.0 *	2.9 *	3.7 **	7.1 ***	4.6 **
Contrasts							
PB—linear	12.6 **	10.5 **	11.3 **	8.4 **	11.4 **	19.6 ***	12.7 **
PB vs. PB + liming	0.2	0.1	0.3	0.3	0.5	2.0	0.8
CL vs. LM + WA	1.0	1.7	3.7	2.1	1.5	3.3	4.7 *
LM vs. WA	0.1	2.4	0.6	3.4	0.1	0.6	0.1

PB, paper mill biosolids; LM, lime mud; WA, wood ash; CL, calcitic lime. The years 2011, 2014, 2017, and 2021 correspond to 3, 6, 9, and 13 years after materials application ending. Statistical significance at 5%, 1%, and 0.1% denoted by *, **, and ***, respectively. Means within a column followed by the same lowercase letter are not statistically significant at $p = 0.05$ according to an LSD test.

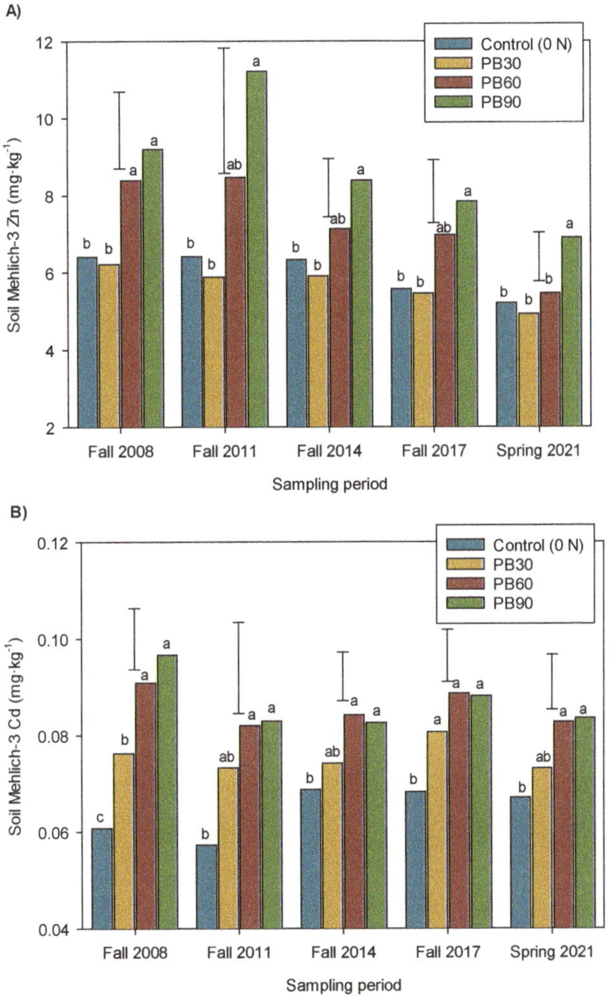

Figure 5. Direct (2008) and residual (2011, 2014, 2017, 2021) effects of paper mill biosolids application on the soil Mehlich-3 Zn (**A**) and Cd (**B**) concentrations in the 0 to 30 cm layer. Vertical bars represent the LSD (5%). Different letters within a sampling date represent difference at $p = 0.05$ according to an LSD test. Data for 2008 come from [26].

Past addition of CL decreased the soil-available Cd and, to a lesser extent, the soil-available Zn in residual years (Tables 4 and 5). Their concentrations in the amended soil were similar or slightly lower than those in the unamended control after application cessation. This was related to the long-term positive effect of CL on soil pH (Figure 1) [32], low cumulative metal supply, and lower labile extracted fractions in this treatment [47]. It was reported that the soil pH variations induced by the addition of sewage sludge and liming were a dominant factor in metal solubility [52].

Soil water-soluble Mo responded positively to liming materials in all residual years, but we observed a general decline, from 4 to 6 $\mu g \cdot kg^{-1}$ over the 30 Mg PB ha^{-1} rate applied alone at the last year of application, to 3 to 5 $\mu g \cdot kg^{-1}$ after three residual years, and 1–2 $\mu g \cdot kg^{-1}$ after nine residual years (Figure 6). Even at these low concentrations, increased soil Mo related to past liming addition induced substantial Mo concentrations in soybean grains [28]. For all the duration of the residual years, CL treatment provided the highest soil Mo concentrations. In this study, a regression analysis indicated that at a pH > 6.2, there is a steady increase in soil water-soluble Mo (Figure 7). O'Connor et al. [53] reported that Mo sorption becomes negligible above pH 6. Moreover, following addition of liming materials, the sorptivity of Mo in acidic soils is partially reversible due to the high solubility of Al-molybdate [54].

Soil water-soluble Mo did not respond to the PB rate in residual years (data not shown) as it did in application years [26]. Previous studies with municipal biosolids, which were richer in Mo than PB, indicated only a small effect on soil Mo sorption due to the presence of Fe and Al oxides but a change in soil pH that can affect Mo retention and release [55].

Table 5. Effect of residual application of paper mill biosolids and alkaline materials on the soil available cadmium in the 0 to 30 cm layer.

Treatment	Mehlich-3 Cd ($mg{\cdot}kg^{-1}$)				DTPA Cd ($mg{\cdot}kg^{-1}$)		
	2011	2014	2017	2021	2011	2014	2017
Control (0N)	0.057 bc	0.069 c	0.068 c	0.067 b	0.071 bc	0.061 bc	0.063 bc
PB 30 $Mg{\cdot}ha^{-1}$	0.073 ab	0.074 bc	0.081 ab	0.073 ab	0.082 ab	0.073 ab	0.075 ab
PB 60 $Mg{\cdot}ha^{-1}$	0.082 a	0.084 ab	0.089 a	0.083 a	0.090 a	0.076 a	0.078 a
PB 90 $Mg{\cdot}ha^{-1}$	0.083 a	0.083 ab	0.088 a	0.084 a	0.096 a	0.080 a	0.083 a
PB 30 Mg + 3 Mg LM ha^{-1}	0.069 abc	0.090 a	0.084 ab	0.082 a	0.072 bc	0.068 ab	0.071 abc
PB 30 Mg + 3 Mg WA ha^{-1}	0.070 abc	0.073 c	0.082 ab	0.078 ab	0.069 bc	0.062 bc	0.070 abc
PB 30 Mg + 3 Mg CL ha^{-1}	0.053 c	0.065 c	0.075 bc	0.076 ab	0.059 c	0.054 c	0.059 c
LSD (5%)	0.019	0.010	0.011	0.011	0.017	0.013	0.013
	Statistical analysis (F-value)						
Treatment	2.9 *	6.1 ***	3.4 *	2.3	4.6 **	4.6 **	3.1 *
Contrasts							
PB—linear	8.9 **	11.3 **	16.8 ***	11.6 **	10.3 **	10.3 **	10.2 **
PB vs. PB + liming	1.6	0.3	0.0	1.5	5.0 *	5.5 *	2.8
CL vs. LM + WA	4.5 *	14.9 ***	3.1	0.6	2.5	4.8 *	4.6 *
LM vs. WA	0.0	12.6 **	0.2	0.7	0.2	1.2	0.0

PB, paper mill biosolids; LM, lime mud; WA, wood ash; CL, calcitic lime. The years 2011, 2014, 2017, and 2021 correspond to 3, 6, 9, and 13 years after the end of material application. Statistical significance at 5%, 1%, and 0.1% is denoted by *, **, and ***, respectively. Means within a column followed by the same lowercase letter are not statistically significant at $p = 0.05$ according to an LSD test.

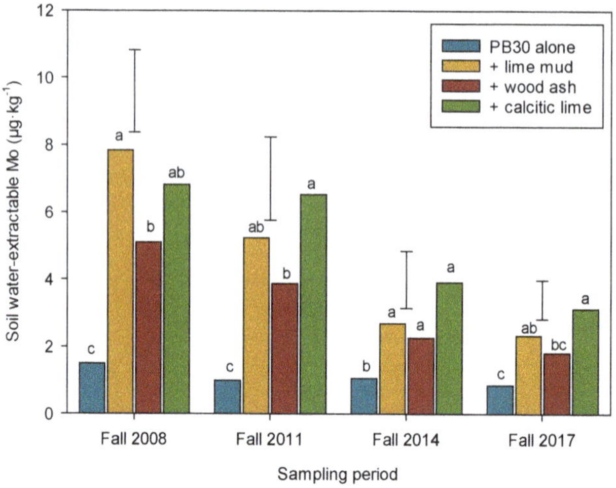

Figure 6. Direct (2008) and residual (2011, 2014, 2017) effects of liming materials application on the soil water-extractable Mo concentrations in the 0 to 30 cm layer. Vertical bars represent the LSD (5%). Different letters within a sampling date represent difference at $p = 0.05$ according to an LSD test. Data for 2008 come from [26].

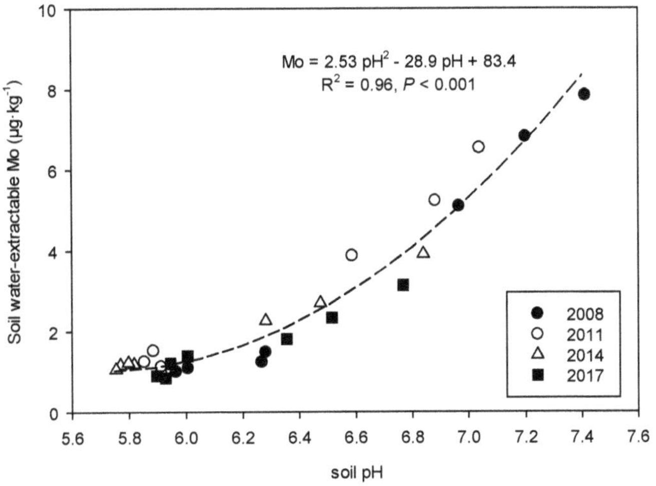

Figure 7. Relationship between soil pH and soil water-extractable Mo content for all treatments (PB rates and PB + liming materials) across time samplings.

4. Conclusions

This study aimed to determine the residual effects, over a thirteen-year period, of nine consecutive years of application of PB and liming materials on main soil properties and metal availability. During this period, soil pH was higher in past amended CL and LM plots, but gradually declined over time to represent 0.3–0.6 units over the untreated control at the end of the study. The past PB addition continued to improve soil total C, but its effect also declined to stabilize at around 15% of total added organic C after six years. Past PB rate also increased soil NO_3-N, but the contribution was negligible, indicating a low risk of leaching. Soil Mehlich-3 P and K decreased in residual years to be not significant at the end of the study, while liming materials continued to positively affect Mehlich-3 Ca. Three years after cessation, the availability of Cu and Zn increased due to PB decomposition and then decreased thereafter. Conversely, the availability of Cd diminished soon after PB cessation. By the end of the residual study, the availability of Zn and Cd reached half the values of those at the end of the material application year, while no more effects were detected for the availability of Cu. For its part, the level of soil water-soluble Mo decreased as the soil pH dropped back towards the control value. Therefore, our study indicated that land application of PB and forest liming by-products promoted soil properties up to 9 years after treatments ceased without increasing the metal availability.

Author Contributions: Conceptualization and design of the experiment: N.Z.; methodology, data acquisition, and validation: B.G.; writing—original draft preparation: B.G.; writing—review and editing: N.Z.; funding acquisition: N.Z. All authors have read and agreed to the published version of the manuscript.

Funding: This study was funded by Agriculture and Agri-Food Canada's A-base program. AAFC-1555.

Data Availability Statement: Data are available upon request.

Acknowledgments: We thank Ferme Firmin Fréchette et fils for providing the experimental site for all those years. We are also grateful to Sylvie Côté, Sylvie Michaud, Claude Lévesque, and Gabriel Lévesque for providing technical assistance.

Conflicts of Interest: The authors declare no conflict of interest.

References

1. Reid, I.D. Solid residues generation and management at Canadian pulp and paper mills in 1994 and 1995. *Pulp Pap. Can.* **1998**, *99*, T109–T112.
2. Bird, M.; Talberth, J. *Waste Stream Reduction and Re-Use in the Pulp and Paper Sector*; Center for Sustainable Economy, Washington State Department of Ecology: Santa Fe, NM, USA, 2008.
3. Cabral, F.; Ribeiro, H.M.; Hilário, L.; Machado, L.; Vasconcelos, E. Use of pulp mill inorganic wastes as alternative liming materials. *Bioresour. Technol.* **2008**, *99*, 8294–8298. [CrossRef]
4. Camberato, J.J.; Gagnon, B.; Angers, D.A.; Chantigny, M.H.; Pan, W.L. Pulp and paper mill by-products as soil amendments and plant nutrient sources. *Can. J. Soil Sci.* **2006**, *86*, 641–653, Erratum *Can. J. Soil Sci.* **2007**, *87*, 118. [CrossRef]
5. Muse, J.K.; Mitchell, C.C. Paper mill boiler ash and lime by-products as soil liming materials. *Agron. J.* **1995**, *87*, 432–438. [CrossRef]
6. Royer-Tardif, S.; Whalen, J.; Rivest, D. Can alkaline residuals from the pulp and paper industry neutralize acidity in forest soils without increasing greenhouse gas emissions? *Sci. Total Environ.* **2019**, *663*, 537–547. [CrossRef]
7. Joseph, C.-A.; Khiari, L.; Gallichand, J.; Bouslama, S. Classification and assessment models of first year byproducts nitrogen plant-availability from literature data. *Sci. Total Environ.* **2017**, *586*, 976–984. [CrossRef]
8. Khiari, L.; Joseph, C.-A.; Gallichand, J.; Beecher, N.; Bouslama, S. Classification and assessment models of first year biosolids phosphorus bioavailability. *Waste Biomass Valor.* **2020**, *11*, 1443–1452. [CrossRef]
9. Bipfubusa, M.; Angers, D.A.; N'Dayegamiye, A.; Antoun, H. Soil aggregation and biochemical properties following the application of fresh and composted organic amendments. *Soil Sci. Soc. Am. J.* **2008**, *72*, 160–166. [CrossRef]
10. N'Dayegamiye, A. Mixed paper mill sludge effects on corn yield, nitrogen efficiency, and soil properties. *Agron. J.* **2006**, *98*, 1471–1478. [CrossRef]
11. Newman, C.M.; Rotenberg, D.; Cooperband, L.R. Paper mill residuals and compost effects on particulate organic matter and related soil functions in a sandy soil. *Soil Sci.* **2005**, *170*, 788–801. [CrossRef]
12. Zibilske, L.M.; Clapham, W.M.; Rourke, R.V. Multiple applications of paper mill sludge in an agricultural system: Soil effects. *J. Environ. Qual.* **2000**, *29*, 1975–1981. [CrossRef]
13. Cogger, C.G.; Bary, A.I.; Myhre, E.A.; Fortuna, A.-M. Biosolids applications to tall fescue have long-term influence on soil nitrogen, carbon, and phosphorus. *J. Environ. Qual.* **2013**, *42*, 516–522. [CrossRef]
14. Granato, T.C.; Pietz, R.I.; Knafl, G.J.; Carlson, C.R., Jr.; Tata, P.; Lue-Hing, C. Trace element concentrations in soil, corn leaves, and grain after cessation of biosolids applications. *J. Environ. Qual.* **2004**, *33*, 2078–2089. [CrossRef]
15. Hyun, H.; Chang, A.C.; Parker, D.R.; Page, A.L. Cadmium solubility and phytoavailability in sludge-treated soil: Effects of soil organic carbon. *J. Environ. Qual.* **1998**, *27*, 329–334. [CrossRef]
16. Walter, I.; Martínez, F.; Alonso, L.; de Gracia, J.; Cuevas, G. Extractable soil heavy metals following the cessation of biosolids application to agricultural soil. *Environ. Pollut.* **2002**, *117*, 315–321. [CrossRef]
17. Arshad, M.A.; Soon, Y.K.; Azooz, R.H.; Lupwayi, N.Z.; Chang., S.X. Soil and crop response to wood ash and lime application in acidic soils. *Agron. J.* **2012**, *104*, 715–721. [CrossRef]
18. Simson, C.R.; Kelling, K.A.; Liegel, E.A. Papermill lime-sludge as an alternative liming material. *Agron. J.* **1981**, *73*, 1003–1008. [CrossRef]
19. Sukkariyah, B.F.; Evanylo, G.; Zelazny, L.; Chaney, R.L. Cadmium, copper, nickel, and zinc availability in a biosolids-amended piedmont soil years after application. *J. Environ. Qual.* **2005**, *34*, 2255–2262. [CrossRef]
20. Sharma, B.; Sarkar, A.; Singh, P.; Singh, R.P. Agricultural utilization of biosolids: A review on potential effects on soil and plant grown. *Waste Manag.* **2017**, *64*, 117–132. [CrossRef]
21. Kashem, A.; Singh, B.R. Soil amendment with city sewage sludge increases soil extractable cadmium, nickel and zinc more than tannery, pharmaceutical and paper mill wastes. *Acta Agric. Scand. Sect. B-Soil Plant Sci.* **2007**, *57*, 134–139. [CrossRef]
22. Mao, L.; Keenor, S.G.; Cai, C.; Kilham, S.; Murfitt, J.; Reid, B.J. Recycling paper to recarbonise soil. *Sci. Total Environ.* **2022**, *847*, 157473. [CrossRef]
23. Charbonneau, H.; Hébert, M.; Jaouich, A. Portrait de la valorisation agricole des matières résiduelles fertilisantes au Québec-partie 2: Contenu en éléments fertilisants et qualité environnementale. *Vecteur Environ.* **2001**, *34*, 56–60.
24. Morris, L.A.; Nutter, W.L.; Miller, W.P.; Overcash, M. Treatment and use of pulp and paper and textile industry residues in southern U. S. forests. In *The Forest Alternative: Principles and Practice of Residuals Reuse*; Henry, C.L., Harrison, R.B., Bastian, R.K., Eds.; College of Forest Resources, University of Washington: Seattle, WA, USA, 2000; pp. 208–217.
25. Battaglia, A.; Calace, N.; Nardi, E.; Petronio, B.M.; Pietroletti, M. Paper mill sludge–soil mixture: Kinetic and thermodynamic tests of cadmium and lead sorption capability. *Microchem. J.* **2003**, *75*, 97–102. [CrossRef]
26. Gagnon, B.; Ziadi, N.; Robichaud, A.; Karam, A. Metal availability following paper mill and alkaline residuals application to field crops. *J. Environ. Qual.* **2013**, *42*, 412–420. [CrossRef]
27. Price, G.W.; Voroney, R.P. Papermill biosolids effect on soil physical and chemical properties. *J. Environ. Qual.* **2007**, *36*, 1704–1714. [CrossRef]
28. Gagnon, B.; Ziadi, N. Residual effects of papermill biosolids and forest-derived alkaline materials on crop yield and plant metal accumulation. *Can. J. Soil Sci.* **2021**, *101*, 248–260. [CrossRef]

29. Gagnon, B.; Ziadi, N. Papermill biosolids and alkaline residuals affect crop yield and soil properties over nine years of continuous application. *Can. J. Soil Sci.* **2012**, *92*, 917–930. [CrossRef]
30. Hébert, M. *Bilan 2015 du Recyclage des Matières Résiduelles Fertilisantes;* Gouvernement du Québec, Ministère du Développement durable, de l'Environnement et de la Lutte contre les Changements Climatiques: Québec, QC, Canada, 2016; pp. 1–30.
31. Québec MDDEP. *Guide Sur Le Recyclage des Matières Résiduelles Fertilisantes: Critères de Référence et Normes Réglementaires;* Gouvernement du Québec, Ministère du Développement durable, de l'Environnement et des Parcs: Québec, QC, Canada, 2012; pp. 1–160.
32. Kabata-Pendias, A. *Trace Elements in Soils and Plants*, 4th ed.; CRC Press: Boca Raton, FL, USA, 2010.
33. Maynard, D.G.; Kalra, Y.P.; Crumbaugh, J.A. Nitrate and Exchangeable Ammonium Nitrogen. In *Soil Sampling and Methods of Analysis*, 2nd ed.; Carter, M.R., Gregorich, E.G., Eds.; CRC Press: Boca Raton, FL, USA, 2008; pp. 71–80.
34. Mehlich, A. Mehlich-3 soil test extractant: A modification of Mehlich-2 extractant. *Commun. Soil Sci. Plant Anal.* **1984**, *15*, 1409–1416. [CrossRef]
35. Liang, J.; Karamanos, R.E. DTPA-extractable Fe, Mn, Cu, and Zn. In *Soil Sampling and Methods of Analysis;* Carter, M.R., Ed.; Lewis Publishers: Boca Raton, FL, USA, 1993; pp. 87–90.
36. Gagnon, B.; Simard, R.R.; Goulet, M.; Robitaille, R.; Rioux, R. Soil nitrogen and moisture as influenced by composts and inorganic fertilizer rate. *Can. J. Soil Sci.* **1998**, *78*, 207–215. [CrossRef]
37. Zbíral, J. Determination of plant-available micronutrients by the Mehlich 3 soil extractant—A proposal of critical values. *Plant Soil Environ.* **2016**, *62*, 527–531. [CrossRef]
38. Murphy, J.; Riley, J.P. A modified single solution method for the determination of phosphates in natural waters. *Anal. Chem. Acta* **1962**, *27*, 31–36. [CrossRef]
39. Rasa, K.; Pennanen, T.; Peltoniemi, K.; Velmala, S.; Fritze, H.; Kaseva, J.; Joona, J.; Uusitalo, R. Pulp and paper mill sludges decrease soil erodibility. *J. Environ. Qual.* **2021**, *50*, 172–184. [CrossRef]
40. USEPA. *Process Design Manual: Land Application of Sewage Sludge and Domestic Septage;* EPA/625/R-95/001; Office of Research and Development: Cincinnati, OH, USA, 1995.
41. Gagnon, B.; Ziadi, N. Forest-derived liming by-products: Potential benefits to remediate soil acidity and increase soil fertility. *Agron. J.* **2020**, *112*, 4788–4798. [CrossRef]
42. Stutter, M.I.; Shand, C.A.; George, T.S.; Blackwell, M.S.A.; Dixon, L.; Bol, R.; MacKay, R.L.; Richardson, A.E.; Condron, L.M.; Haygarth, P.M. Land use and soil factors affecting accumulation of phosphorus species in temperate soils. *Geoderma* **2015**, *257–258*, 29–39. [CrossRef]
43. Faucon, M.-P.; Houben, D.; Reynoird, J.-P.; Mercadal-Dulaurent, A.-M.; Armand, R.; Lambers, H. Advances and perspectives to improve the phosphorus availability in cropping systems for agroecological phosphorus management. *Adv. Agron.* **2015**, *134*, 51–79.
44. Stewart, J.W.B.; Tiessen, H. Dynamics of soil organic phosphorus. *Biogeochemistry* **1987**, *4*, 41–60. [CrossRef]
45. Sharifi, M.; Cheema, M.; McVicar, K.; LeBlanc, L.; Fillmore, S. Evaluation of liming properties and potassium bioavailability of three Atlantic Canada wood ash sources. *Can. J. Plant Sci.* **2013**, *93*, 1209–1216. [CrossRef]
46. Ferreiro, A.; Merino, A.; Díaz, N.; Piñeiro, J. Improving the effectiveness of wood-ash fertilization in mixed mountain pastures. *Grass Forage Sci.* **2011**, *66*, 337–350. [CrossRef]
47. Gagnon, B.; Robichaud, A.; Ziadi, N.; Karam, A. Repeated annual paper mill and alkaline residuals application affects soil metal fractions. *J. Environ. Qual.* **2014**, *43*, 517–527. [CrossRef]
48. Hettiarachchi, G.M.; Scheckel, K.G.; Ryan, J.A.; Sutton, S.R.; Newville, M. μ-XANES and μ-XRF investigations of metal binding mechanisms in biosolids. *J. Environ. Qual.* **2006**, *35*, 342–351. [CrossRef]
49. Basta, N.T.; Ryan, J.A.; Chaney, R.L. Trace element chemistry in residual-treated soil: Key concepts and metal bioavailability. *J. Environ. Qual.* **2005**, *34*, 49–63. [CrossRef] [PubMed]
50. Stietiya, M.H.; Wang, J.J. Effect of organic matter oxidation on the fractionation of copper, zinc, lead, and arsenic in sewage sludge and amended soils. *J. Environ. Qual.* **2011**, *40*, 1162–1171. [CrossRef] [PubMed]
51. Antoniadis, V.; Alloway, B.J. The role of dissolved organic carbon in the mobility of Cd, Ni and Zn in sewage sludge-amended soils. *Environ. Pollut.* **2002**, *117*, 515–521. [CrossRef] [PubMed]
52. Kim, B.; McBride, M.B.; Richards, B.K.; Steenhuis, T.S. The long-term effect of sludge application on Cu, Zn, and Mo behavior in soils and accumulation in soybean seeds. *Plant Soil* **2007**, *299*, 227–236. [CrossRef]
53. O'Connor, G.A.; Brobst, R.B.; Chaney, R.L.; Kincaid, R.L.; McDowell, L.R.; Pierzynski, G.M.; Rubin, A.; Van Riper, G.G. A modified risk assessment to establish molybdenum standards for land application of biosolids. *J. Environ. Qual.* **2001**, *30*, 1490–1507. [CrossRef]
54. Yang, P.-T.; Wang, S.-L. Sorption and speciation of molybdate in soils: Implications for molybdenum mobility and availability. *J. Hazard. Mater.* **2021**, *408*, 124934. [CrossRef]
55. Brinton, S.R.; O'Connor, G.A. Sorption of molybdenum in soils field-equilibrated with biosolids. *Commun. Soil Sci. Plant Anal.* **2003**, *34*, 1331–1346. [CrossRef]

soil systems

Review

Symbiotic and Asymmetric Causality of the Soil Tillage System and Biochar Application on Soil Carbon Sequestration and Crop Production

Amare Assefa Bogale [1,2,*], Anteneh Agezew Melash [3] and Attila Percze [1]

1 Institute of Crop Production, Hungarian University of Agriculture and Life Sciences, 2100 Gödöllő, Hungary; percze.attila@uni-mate.hu
2 Department of Horticulture, College of Agriculture and Natural Resource, Mekdela Amba University, Tulu Awulia P.O. Box 32, Ethiopia
3 Faculty of Agricultural and Food Sciences and Environmental Management, Institute of Crop Sciences, University of Debrecen, Boszorm'enyi Str. 138, 4032 Debrecen, Hungary
* Correspondence: bogale.amare.assefa@phd.uni-mate.hu

Abstract: Agriculture faces a significant challenge in maintaining crop production to meet the calorie demand of the ever-growing population because of limited arable land and climate change. This enforces a search for alternative multifarious agricultural-based solutions to meet the calorie demand. In search of alternatives, agricultural soil management has been highlighted and is expected to contribute to climate change mitigation through soil carbon sequestration and reduce greenhouse gas emissions through effective agricultural management practices. The addition of biochar to the soil significantly improves the soil nitrogen status, soil organic carbon, and phosphorus, with greater effects under the different tillage systems. This symbiosis association could further change the bacterial structure in the deeper soil layer which thus would be important to enhancing productivity, particularly in vertisols. Biochar also has an environmental risk and negative consequences. Heavy metals could be present in the final food products if we use contaminated raw materials to prepare biochar. However, there is a need to investigate biochar application under different climatic conditions, seasons, soil tillage systems, and crop types. These indicate that the positive effect of proper biochar fertilization on the physiology, yield formation, nutrient uptake, and soil health indicators substantiate the need to include biochar in the form of nutrients in the crop production sector, especially in light of the changing climate and soil tillage systems.

Keywords: biochar; carbon sequestration; conservation tillage; conventional tillage; greenhouse gases

Citation: Bogale, A.A.; Melash, A.A.; Percze, A. Symbiotic and Asymmetric Causality of the Soil Tillage System and Biochar Application on Soil Carbon Sequestration and Crop Production. *Soil Syst.* **2023**, *7*, 48. https://doi.org/10.3390/soilsystems7020048

Academic Editors: Antonella Lavini and Mohamed Houssemeddine Sellami

Received: 14 March 2023
Revised: 6 May 2023
Accepted: 7 May 2023
Published: 10 May 2023

1. Introduction

In agriculture, crop production is a prominent sector intended to ensure food security, enhance food self-sufficiency, and supply input for different industries worldwide. However, there is a need to enhance the productivity of the crop production sector to satisfy the calorie demand of the world's ever-increasing population. Hence, to meet this demand, the producers should keep in mind the soil health of where the crops are grown and the environment in which the living things survive, since soil management is an important aspect affecting the functionality of the soil [1–3]. However, these aims have been hampered due to improper agronomic management practices and environmental degradation. It has been observed that the degradation of soil, due to the loss of soil carbon (C) and nitrogen (N) pools, is decreasing crop productivity and intimidating food security [1]. Tillage is one of the critical components of agricultural systems and practices, commonly working internationally in croplands to reduce climatic and soil restrictions, even though sustaining several ecosystem services is an issue [4]. This practice can affect several soil-mediated processes, such as soil carbon sequestration (SCS) or depletion, water pollution, and greenhouse gas (CO_2, CH_4, and N_2O) emissions [4]. Maximum and intensive tillage without proper residues and crop nutrient management practices is among the reasons for the loss of soil C and N pools and the decreased productivity and quality of the crops [1].

In addition to tillage's effect on the soil nutrient status, it can also further influence N_2O emissions by affecting the nitrification and denitrification processes, influence CH_4 emissions by affecting methanotrophs [5], and impact the quantity and composition of the soil organic matter [6]. Increments of N_2O and CH_4 are greenhouse gases with a global warming potential of approximately 300 and 30 times that of CO_2 [7,8]. However, both CH_4 and N_2O emissions could be counterbalanced through the positive benefits gained by soil carbon sequestration practices. Hence, enhancing soil organic carbon through proper implementation of agronomic crop management practices is very important to increasing organic matter inputs into the soil, reducing the decomposition of soil organic matter, and oxidizing soil organic carbon; each of these, or a combination of them, act as an alternative agronomic practice for a sustainable agriculture production system designed to enhance overall crop productivity [9,10].

A number of agronomic practices, such as biochar application, have been previously proposed to be aimed at mitigating greenhouse gas emission and simultaneously improving the yield and quality of crops. Biochar is fine-grained charcoal made up of a stable carbon-based material and applied to soils to realize the net carbon sequestration, which is encouraged principally for balancing the effect of climate change and maintaining the soil fertility status to improve the yield of the crops [1]. It can be prepared from organic materials, including paper mill sludge, crop and forestry residues, and poultry waste. The adoption of biochar is gaining growing attention as a sustainable technique that helps improve soil fertility in weathered and degraded soils [11]. It consists of a stable C compound produced via pyrolysis and gasification [12,13]. Biochar mainly subsidizes the stable SOC pool, although fresh crop residues add to the more bio-accessible fraction of the soil C reserve. This is why the decomposition of biochar proceeds naturally at a rate that is 10 to 100 times slower than uncharred biomass [12,14]. Therefore, our hypothesis is that the combination of a soil tillage system and biochar application can have a symbiotic effect on soil carbon sequestration and crop production. Specifically, the effects of each practice may be enhanced by the other, resulting in greater benefits than if they were used independently. For example, the use of biochar can mitigate the negative effects of tillage on soil structure, while the use of tillage can increase the availability of nutrients for plant growth. Hence, the objective of the review is to illustrate the independent and symbiotic effect of soil tillage systems and biochar on crop production, carbon sequestration, and other soil properties.

2. Materials and Methods

Several techniques were implemented to assure a high-quality literature review. The first step was selecting the topic or the title and placing the subtitles under the main topic. This review paper's subtitles include 'the effect of conventional tillage systems on crop production', 'the effect of conservation tillage systems on crop production', 'carbon sequestration as influenced by those tillage systems and the role of biochar for crop production', 'the effect of biochar in the environment', and 'soil carbon sequestration'. Then, based on the selected titles, the second step was collecting different scientific articles from different sources, such as Google® Scholar, PubMed, ResearchGate, Web of Science, and Scopus. Thirdly, the selected articles were organized based on their relevancy and latest publication year. Following thorough reading of the selected articles, a review was composed based on the ideas that were relevant to the topics. Only English-language articles were taken into consideration for this review.

3. Overview of Soil Tillage System and Its Role in Agriculture

Identification of an environmentally friendly and crop-yield-sustaining tillage strategy is required to remedy the growing concern for food security through enhanced soil management practices [15]. The mechanical alteration of soil for crop production, also known as tillage, has dramatically changed soil properties, including soil temperature, infiltration, and evapotranspiration. This intentional soil disturbance during tillage causes an impact on the ecosystem and the growing of crops [15,16]. It is an important component of agricultural technology [17–20] because it alters the primary soil layer's root zone, the soil is nourished with mineral or organic fertilizer and plant leftovers, and a good seedbed is made [20,21]. Soil tillage is a crop production factor contributing up to 20% [22,23] and affecting the sustainable use of soil resources through its influence on the physical and chemical properties of soil [24]. This indicates that tillage is a fundamental crop production practice that emphasizes forming a good seedbed for germination and subsequent plant growth; there are changes in the soil bulk density and resistance, improving soil aeration and providing ideal conditions for plant life. Hence, based on this information, our intentions are concerned with conservation tillage and the conventional tillage systems as follows.

3.1. *Effect of Conservation Tillage on Soil Properties*

Conservation tillage can be defined as a crop management strategy that leaves at least 30% of the crop residue on the soil's surface after planting to prevent soil erosion by water [25]. Conservation tillage systems are a part of the practice of managing agricultural leftovers on the soil's surface with minimum or no tillage. The techniques are also referred to as direct drill, stubble mulching, eco-fallow, restricted tillage, reduced tillage, minimum tillage, and no tillage. These approaches to plant residue management have three main objectives: lowering energy use, saving soil and water, and often leaving enough plant residue on the soil's surface to minimize wind and water erosion [26]. It is worth mentioning that conservation tillage also has the potential to reduce the negative effects of soil disturbance and nutrient loss, which arise due to erosion, increased rainfall infiltration, reduced subsurface compaction, and maximized soil organic carbon (SOC) accumulation, with this mitigation positively affecting many soils' physical and chemical properties [27].

The magnitude and the level of influence of conservation tillage practices on soil nutrient status have been observed as divergent, according to a given set of environmental conditions. In an environment where wind erosion is an issue, conservation tillage refers to any strategy that keeps at least 1000 pounds per acre of crop residue from small grains on the surface during the crucial erosion period [28]. It has been observed that conservational tillage practices can also attain the highest organic matter accumulation, result in the maximum root mass density (0–15 cm soil depth), and improve the physical and chemical properties of the soil. However, bulk densities were reduced due to the tillage practices, having the highest reduction of these properties and the highest increase of porosity and field capacity, particularly at zero tillage. Zero tillage with 20% residue retention, is suitable for soil health and attaining the optimum yield [23]. According to [1], conservation tillage with proper nutrient management and crop residues improves rice yield by 51.1–52.2%. This indicates that conservation tillage contributes significantly to improving grain yield compared to conventional tillage. On the other hand, the no-tillage system can also provide several benefits, including enabling the growers to manage more significant areas of cropland with less total input [29], reducing the soil erosion caused by wind and water, and enabling the accretion of soil organic carbon (SOC) and an improved soil structure [30]. Although the no-tillage system provides all of the aforementioned benefits, there are some concerns about the long-term sustainability due to the accumulation of herbicide-resistant (HR) weed populations, compacted soil, soil cracks, and varmint (Taxidea taxus) holes causing a rough surface for field operations; the stratification of pH, nutrients, and SOC in the first few cm of the soil profile are additionally problematic [31]. These conditions can lead to a decrease in the plant's ability to supply and absorb nutrients as well as an increased risk of nitrogen (N) and phosphorus (P) losses in surface runoff to the environment [32].

In comparison with conventional tillage, conservation tillage, such as no-tillage or reduced tillage measures, tends to increase the soil's moisture-holding capacity and improve water permeability. Although the soil is susceptible to wind and water erosion [4], the application of minimal tillage systems increases the organic matter content from 0.8% to

22.1% and the stable aggregate content in the water from 1.3% to 13.6%, at 0 to 30 cm deep, compared to the conventional tillage system [33]. This effect could further affect soil water purification and retention functions, particularly under no-tillage conditions [3]. Similarly, two studies [34,35] reported that conservation tillage, such as no tillage and reduced tillage, increases soil organic matter and total soil nitrogen by 18.7% [36,37] and the reverse is true in conventional soil tillage systems (Table 1); (Figures 1 and 2). This indicates that conservation tillage is very important for improving soil productivity, and ultimately, crop yield.

Table 1. Effect of soil tillage systems on soil organic matter and soil bulk density.

Tillage Techniques	0–8 cm		8–30 cm		30–40 cm	
	OM	Bd	OM	Bd	OM	Bd
DS	1.99 [a]	1.29 [f]	1.72 [bcd]	1.47 [b]	1.61 [cd]	1.52 [a]
MT	1.82 [b]	1.33 [e]	1.54 [d]	1.45 [bc]	1.24 [e]	1.51 [a]
CT	1.79 [bc]	1.39 [d]	0.76 [f]	1.41 [cd]	1.24 [e]	1.51 [a]

NB: **DS** = Direct sowing, **MT** = Minimum Tillage, **CT** = Conventional Tillage, **OM** = Organic Matter, **Bd** = Bulk density, letters a, b, c, d, e, f = the significant difference of tillage techniques on organic matter and bulk density [35].

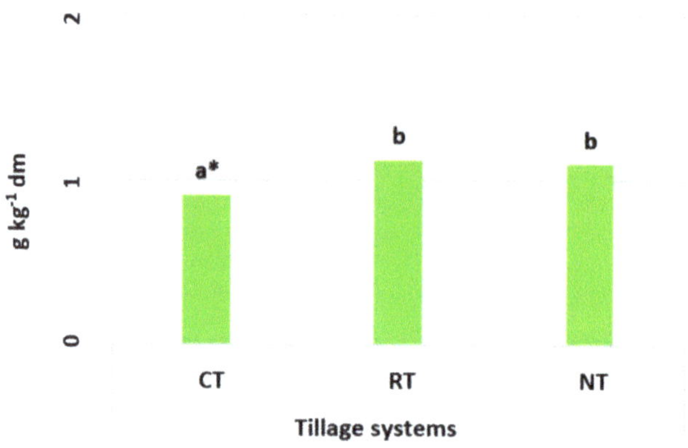

Figure 1. The total nitrogen content in the 0–25 cm soil layer, depending on the soil tillage system (average Agriculture of 2014–2016), were CT: Conventional Tillage, RT: Reduced Tillage, NT: Non-Tillage, *: level of significance difference [37].

3.2. *Effect of Conservation Tillage on Crop Production*

Conservation tillage is a farming practice that aims to minimize soil disturbance during planting, thereby reducing erosion and maintaining soil structure and fertility. One of the primary benefits of conservation tillage is the increased crop productivity. By reducing soil disturbance, conservation tillage helps to maintain the structure of the soil, which allows for better water infiltration and retention. This, in turn, leads to improved soil fertility and nutrient availability, which can result in increased crop yields. Numerous studies have examined the effect of conservation tillage on crop productivity and production. For instance, one study [38] found that conservation tillage practices, such as no till and reduced tillage, can increase crop yields by 10–20% compared to conventional tillage methods. Similarly, a meta-analysis [39] revealed that conservation tillage practices increased crop yields by 4.6% on average across different crops and regions. Moreover, a study conducted in the United States found that conservation tillage practices increased corn and soybean yields by up to 3.3% and 0.74%, respectively, compared to conventional tillage methods [40]. However, the effect of conservation tillage on crop productivity and production may vary depending on soil type, climate, crop type, and management practices. For instance,

A meta-analysis of 63 worldwide studies found that conservation tillage increased cereal crop yields but had no discernable effect on legume crop yields [41].

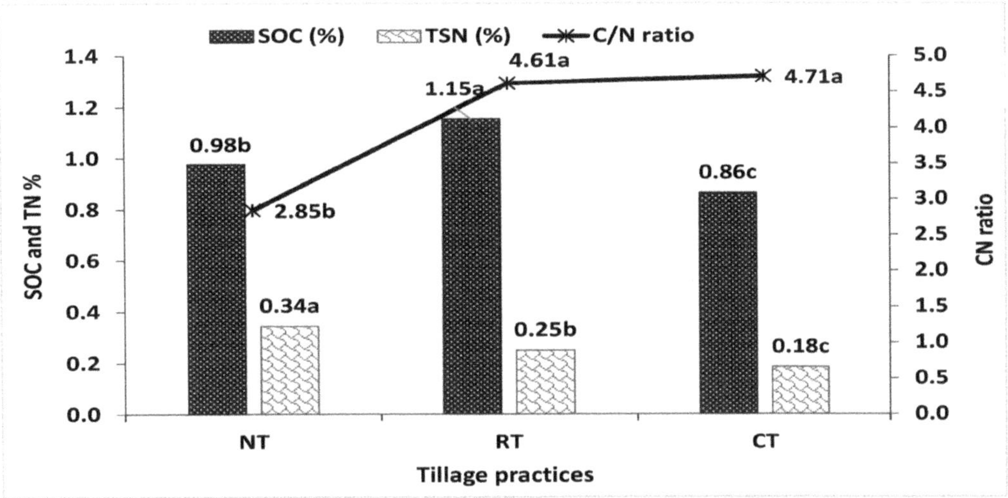

Figure 2. The effect of different soil tillage systems on soil organic Carbone (SOC), total soil nitrogen (TSN), and C: N ratio; NT = No tillage, RT = Reduced tillage, and CT = Conventional tillage [36].

In addition to increasing crop productivity, another benefit of conservation tillage is that it can reduce the need for herbicides and other chemicals. By maintaining a permanent soil cover, conservation tillage helps to suppress weed growth, which can reduce the need for herbicides. For instance, a study conducted in Australia found that conservation tillage resulted in lower herbicide use and reduced weed pressure, compared to conventional tillage methods [42].

3.3. *Effect of Conventional Tillage on Crop Production*

Conventional tillage refers to practices considered standard for a specific location and set of crops with the primary intention to bury crop residues. It refers to plowing (inverting the soil) followed by secondary tillage activities, such as tilling and harrowing [43]. This practice is usually considered the basis for determining the cost-effectiveness of erosion control measures [44]. Conventional tillage is used for turning and plowing a deep layer of soil, combining and destroying plant debris, exposing soil pests to the sun to control them, demolition, and ground leveling; this kind of tillage causes excessive soil fragmentation, collapse, and compaction.

On the contrary, conventional tillage also leads to erosion, run-off, impoverishment, and drying out of the land [35]. This could affect soil organic matter and soil bulk density at different soil depths. For instance, as presented in Table 1, the soil organic matter is significantly affected by the conventional tillage system compared to direct seeding and minimum soil tillage systems [34,35]. This implies that a conventional tillage system could affect crop yield and soil productivity, mainly through the decline of soil organic matter status. In addition, under conventional tillage practice, the soil organic carbon and total nitrogen levels were the lowest, leading to the highest C: N ratio (Figure 2).

The conventional soil tillage system affects not only the production and productivity of crop yields but also the total cost of technology implementation during crop implantation [45]. This means that the cost of fuel consumption, the cost of mechanized services, and the time taken to plow the land require maximum costs compared to the conservation soil tillage system.

3.4. Effect of Soil Tillage System on Soil Carbon Sequestration

It is internationally accepted that global climate changes result from human intervention in the biogeochemical cycles of water and materials, and soil carbon sequestration practices are considered necessary interventions intended to limit these changes [3]. Soil organic carbon (SOC) plays a significant role in agricultural ecosystems, including soil fertility, soil tilth, nutrient cycling, soil sustainability, and crop productivity, by influencing the physical, chemical, and biological aspects of the soil [46–49]. Soil organic carbon storage in soil-crop ecosystems represents the net balance between the soil carbon breakdown processes brought on by microbial oxidation and the continuous carbon (C) accumulation caused by inputs of crop biomass [49–51]. A positive imbalance results in soil carbon sequestration when carbon dioxide inputs to the soil outweigh carbon dioxide outflow [52–54]. By changing the quantity and quality of crop residues in the soil, the microbial dynamics, and the supply of nutrients, soil and crop management practices (such as tillage, crop type and variety, rotation, chemical fertilizer application, and manure application) could significantly affect the soil organic carbon of cultivated land [46,49,50,55].

Applying agricultural practices to increase organic carbon (SOC) content in the soil is part of a "climate-smart" agricultural approach that sequesters CO_2 from the atmosphere [56]. In addition to significantly impacting soil health and agricultural production, the loss of SOC exacerbates climate change. When the soil organic matter (SOM) breaks down, carbon-based greenhouse gases are released into the atmosphere. If this happens at too high a rate, the soil could be contributing to our planet's warming. On the other hand, many soils have the potential to increase their soil organic carbon (SOC) stocks, thereby mitigating climate change by reducing atmospheric CO_2 concentrations [57]. The choice of cropping system in farming has a powerful effect on soil health, crop yield, and the broader environment. Conservation tillage with non-inversion techniques will save soil carbon, reduce erosion risk, and enhance soil quality. In addition, it has been demonstrated that conservation tillage sequesters more carbon and total soil nitrogen than conventional tillage (Figure 2).

Stable, well-structured topsoil developed by long-term conservation farming leads to more energy-efficient systems by reducing the energy requirements for farming [58]. Among the significant agricultural strategies, soil organic carbon sequestration is prominent in mitigating greenhouse gas (GHG) emissions, food security enhancement, and agricultural sustainability improvement [59]. According to [12,54] one description, a key carbon sequestration strategy to enhance food security and agricultural sustainability improvement includes (i) restoration of degraded soil through conversion to agricultural land; (ii) application of recommended management practices, such as conservation tillage agriculture, organic farming using manure, and compost; and (iii) using biochar as a soil improver.

4. General Overview of Biochar in Agriculture

Biochar is a type of charcoal that is produced through the thermal decomposition of organic material in the absence of oxygen, a process known as pyrolysis [13]. It is also a multifunctional carbon substance, which is used to solve soil fertility and climate change issues [60]. It is considered a novel soil treatment and carbon sequestration pathway that has improved soil structure and ecosystems function due to its complex physical and chemical properties [61]. Due to its importance, the effect of biochar on carbon emissions and sequestration has attracted much attention in recent years. However, additional research on the implications of biochar from an atmospheric-food perspective is required. Biochar has received widespread attention as a viable solution to solving the problems of food security and climate change in agroecosystems, but many questions surround this strategy's potential influence on crop productivity, soil carbon sequestration, and global warming [62]. The application of biochar offers greater advantages regarding soil properties and crop yields in degraded tropical soils compared to those in temperate regions [63,64]. It also improved crop yields more markedly in nutrient-poor and infertile soils than in healthy and fertile soils [65]. Biochar production can enhance biochar's carbon sequestration potential by up to 45% [66]. Another study described that biochar has the potential to capture and store between 0.6 and 11.9 gigatons of carbon dioxide annually, with the primary determining factor being the accessibility of biomass resources for the production of biochar [67] (Figure 3). Additionally, studies have found that the potential for GHG mitigation from biochar increases over time. Specifically, the studies estimate that by 2030, biochar could mitigate between 1 and 1.8 billion metric tons of CO_2 per year [14,68–70]. By 2050, this potential increases to between 1.8 and 4.8 billion metric tons of CO_2 per year [71–73], and by 2100, the range is estimated to be between 2.6 and 4.8 billion metric tons of CO_2 per year [74]. These estimates suggest that biochar has the potential to be an important tool for mitigating GHG emissions and addressing climate change, especially as its use is scaled up over time. Due to its high stability and large-scale production potential, the application of biochar has a high potential for long-term SOC sequestration and is limited only by the available biomass [75–79].

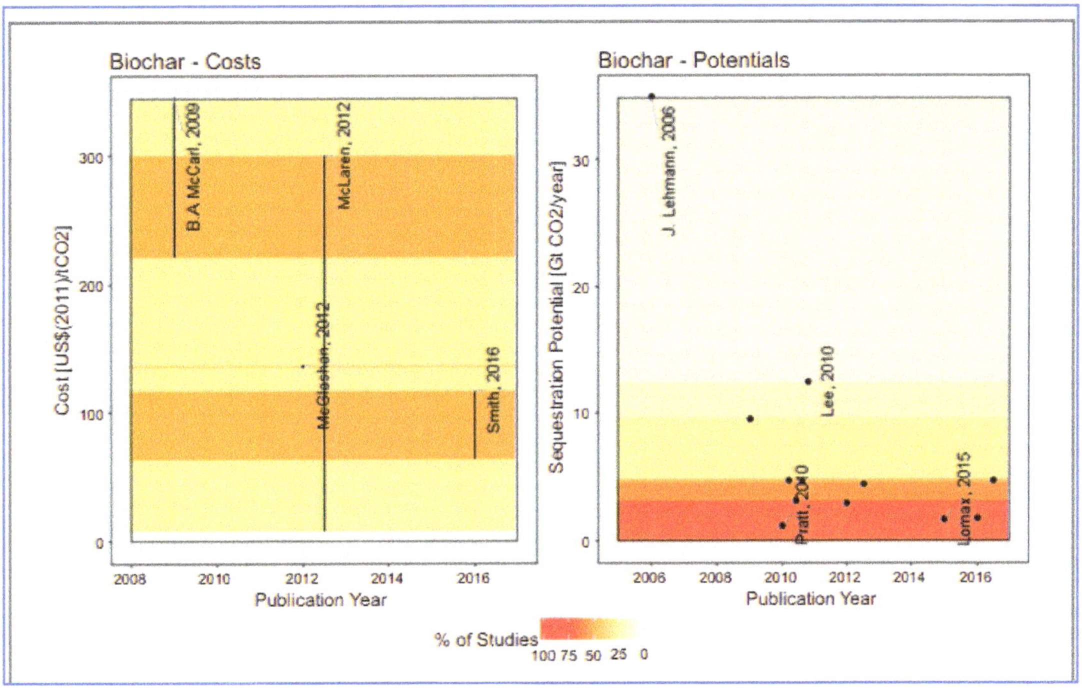

Figure 3. Costs and carbon sequestration potentials for biochar. Estimates and ranges at the top and bottom end of the distribution are labeled; the data can be further explored in our online supporting material available at https://mcc-apsis.github.io/NETs-review/ accessed on 22 May 2018 [67].

4.1. Making, Production, Processing, and Digestion of Biochar

The production, processing, making, and digestion of biochar offer a wide range of opportunities for sustainable agriculture, waste management, and energy production. However, it is important to carefully consider the feedstocks, processing methods, and application rates of biochar to ensure its effectiveness and minimize any negative impacts on the environment (Table 2).

Table 2. Short Summary of Making, Production, Processing, and Digestion of Biochar.

Process	Description	Reference
Making	Biochar can be made from a wide range of feedstocks, including agricultural residues, forestry wastes, and even municipal solid waste. The choice of feedstock and pyrolysis conditions affects the properties of the biochar and its suitability for different applications. For example, biochar made from hardwoods may have a higher carbon content and be more stable than biochar made from softwoods.	[80,81]
Production	Biochar is produced through the process of pyrolysis, which involves heating biomass in the absence of oxygen. This process converts the biomass into a carbon-rich material that is resistant to decomposition. The temperature of pyrolysis affects the properties of the biochar, such as its porosity and surface area.	[13]
Processing	After biochar is produced, it can be processed to improve its properties for specific applications. This may involve crushing, sieving, or adding amendments to the biochar. For example, biochar can be impregnated with nutrients to make it more effective as a soil amendment.	[82]
Digestion	Biochar can be used as a feedstock for anaerobic digestion, a process that converts organic matter into biogas. When biochar is added to an anaerobic digester, it can improve the performance of the system by providing a surface for bacterial growth and removing inhibitory substances from the digester.	[83]

4.2. *Effect of Biochar on Soil Properties*

Biochar is a type of charcoal that is created by heating organic materials, such as wood, agricultural waste, and other types of biomasses in the absence of oxygen. The resulting material is rich in carbon and is used to improve soil fertility, water retention, and carbon sequestration. Biochar has been found to improve soil structure by increasing the amount of pore space and reducing soil compaction. This allows water to penetrate deeper into the soil, reducing runoff and increasing water retention. In one study [84], the addition of biochar to soil increased the water-holding capacity by up to 18% compared to control soils. It has been also used to increase the nutrient use efficiency of plants. For instance, another study [85] revealed that biochar increased the utilization efficiency of nitrogen (NUF), phosphorus (PUE), and potassium (KUE) in quinoa plants and could be utilized as an organic fertilizer for crops [86]. Additionally, according to another study [87], biochar application has a positive effect on soil properties, as evidenced by an increase of 1.5 times Mn in soil and 1.4 times total organic carbon, as well as the neutrality observed in soil pH and the availability of P and cations, total N, and other extractable soil nutrients other than Mn during a two-year trial in Idaho, United States. In connection with this, an optimistic effect was detected when biochar was applied throughout three consecutive years of field trials: a minor increase in soil pH and increased potassium nutrient levels [88]. Moreover, a positive effect was also observed on soil pH, P, available SOC, hydraulic conductivity, water-holding capacity, N content, NH_4 + -N and NO_3-N in soil, soil microbial enhancement, and enzyme activity through the use of biochar during a 5-year field trial in the subtropical region of Jianxi, China [89]. These studies suggest that a biochar amendment is a practical approach to improving soil structure, carbon content, soil availability, the efficiency of nutrient use, and the influence on soil microbes.

The addition of biochar to soil has been shown to increase soil carbon sequestration, which can help mitigate climate change. In a previous study [90], the addition of biochar to soil increased carbon sequestration by up to 0.4 t C/ha/yr. Similarly, biochar significantly improves soil quality, carbon sequestration, and greenhouse gas (CO_2, N_2O, CH_4) emission reduction [91,92]. In addition, other study demonstrated that applying biochar might decrease soil nitrous oxide (N_2O) emissions by 28%, with the total being depleted to 60% [62,93–97]. Moreover, Biochar has been found to enhance soil microbial activity by providing a habitat for beneficial microorganisms and improving soil nutrient availability. According to one study [98], the addition of biochar to soil increased the activity of soil microorganisms by up to 46%. Furthermore, it has unique adsorption and desorption mechanisms that control nutrient leaching [99] and can enhance plant productivity, particularly when coupled with other organic products, such as manure and compost; sewage sludge can be used as a soil amendment [100]. In contrast, there has been a lot of diversity and confusion in the results about the impact of biochar on crop output, soil organic carbon (SOC), and greenhouse gas emissions [101]. This suggests that further research is needed with other agronomic practices, such as soil tillage, to fully understand the effects of biochar on agricultural systems.

The amount of biochar required to increase soil organic carbon stock depends on various factors, including the frequency of application. Studies have shown that repeated applications of biochar increase soil organic carbon stock more than a single application. For example, some studies have found that one application of biochar can increase the soil organic carbon pool by 26%, while consecutive applications can increase it by an average of 55% [79,102]. Moreover, different types of biochar have different effects on soil properties [101]. The structural components of the raw materials used to produce biochar can affect its ability to modify soil properties [103]. Therefore, it is essential to consider the characteristics of the biochar before applying it to the soil.

4.3. Effect of Biochar on Crop Production and Crop Quality

Biochar has been shown to have beneficial effects on crop production and crop quality. It has a high surface area and a porous structure, which allows it to absorb and retain water and nutrients, and to provide a habitat for beneficial microorganisms. This can lead to increased crop yields and improved crop quality. Several studies have reported the positive effects of biochar on crop production. For instance, research conducted by one researcher [94] found that improved crop productivity raised yields by an average of 11.0%. Another study [104] found that the addition of biochar to unfertilized soils at rates of 10 t ha^{-1} and 40 t ha^{-1} increased rice yields by 12% and 14%, respectively; rice yields increased by 8.8% and 12.1% at the same rates of 10 t ha^{-1} ha and 40 t ha^{-1} in soils with N fertilization, respectively. Similarly, it can increase plant productivity by an average of 10%, due to the increased water-holding capacity and nutrient availability of the soil, but the impact on yield can vary depending on factors such as soil type, environment, and management conditions [61,90]. Furthermore, a study conducted in Brazil found that biochar application increased soil pH and improved nutrient availability, resulting in increased sugarcane yield and quality [105]. It can also alter the levels of available nutrients in the soil, which can increase the plant's ability to absorb carbon dioxide from the atmosphere [106]. In addition to its effects on soil and nutrient availability, biochar can also have direct effects on plant physiology and biochemistry, leading to improvements in plant growth and health [107].

Biochar has also been shown to improve crop quality. For example, a study conducted in Malaysia discovered that combining biochar with other soilless media boosted growth and yield without harming the post-harvest quality of lowland cherry tomatoes [108]. In addition to its effects on soil fertility and crop quality, biochar has also improved drought tolerance in maize plants [109]. Overall, the use of biochar has shown promise as a tool for improving crop production, crop quality, and disease suppression, while also potentially contributing to climate change mitigation efforts.

4.4. Effect of Biochar on the Environment

Biochar has been touted as a potential tool for mitigating climate change by sequestering carbon in the soil, improving soil fertility, and reducing greenhouse gas emissions from agricultural practices (Table 3). However, the environmental impact of biochar production and use is not fully understood and requires further research. According to some researchers [110,111], biochar may naturally contain pollutants that were either added by the feedstock (such as heavy metals) or co-produced during (improper) pyrolysis (such as polycyclic aromatic hydrocarbons). This leads to increased soil acidity and the leaching of heavy metals into the environment. Similarly, another study [13] found that biochar produced from feedstocks with high concentrations of heavy metals, such as chicken manure, could potentially lead to increased concentrations of heavy metals in the soil and water. In general, further research is needed to fully understand the environmental implications of biochar production and use.

Table 3. Short Summary of General Effect of Biochar on Soil, Yield, and Environment.

Biochar Effect	Reference
Improves soil fertility	[98,112]
Increases crop yields	[82,90]
Reduces greenhouse gas emissions	[74,81]
Improves water retention	[113,114]
Reduces soil erosion	[113,114]
Enhances nutrient cycling	[81,98]
Suppresses soil-borne pathogens	[90,98]
Improves soil structure	[112,114]
Reduces leaching of nutrients	[105,115]
Reduces leaching of pollutants	[116]
Improves plant growth and health	[117,118]
Increases carbon sequestration	[13,74]

4.5. Biochar Dose Optimization for Crop Yield and Cost

Biochar is a valuable soil amendment that can improve soil fertility, increase its water-holding capacity, and reduce greenhouse gas emissions. However, the optimal dose of biochar required for achieving maximum yield and cost-effectiveness is unclear and depends on various factors, such as soil type, crop type, and biochar properties. Therefore, building an optimization model can help determine the optimal dose of biochar required for maximum yield and cost-effectiveness. Several studies have explored the relationship between biochar dosage, yield, and cost indicators. For example, some researchers [119] have determined the optimal biochar dose for sugar beet yield. They found and recommend that the maximum yield was achieved at a biochar dose of 10 t ha^{-1} $year^{-1}$ with minimum production cost. Additionally, another researcher [120] found the dose of biochar needed for the highest productivity and cost-effectiveness in maize yield. The study discovered that adding 1% biochar resulted in a 22% increase in maize yield, which could generate EUR 620 per hectare in revenue on a maize purchase price of EUR 117 per ton. However, other factors, such as fluctuating costs, government subsidies, environmental restrictions, and market demand, could also affect the economics of biochar production and application. Moreover, a three-year agronomic trial was conducted with maize and mustard farming to study the economic viability of biochar addition. The study found that the optimal dose was determined to be 15 t ha^{-1} based on several considerations, including agronomical factors, such as crop yield; economic factors, such as cost-benefit analysis; and environmental factors, such as carbon sequestration [121].

5. Interactive Effects of Tillage System and Biochar Application on Soil and Crop Productivity

Organic nutrient amendments and tillage systems have been frequently suggested as critical aspects of food security under the changing climate, however, responsible agronomic-based solutions should follow a food system and circular economy approaches. The adaptation of minimum tillage in combination with organic amendments, including biochar, may sustain the soil health indicators i.e., increase nitrogen status, soil organic carbon, and phosphorus [122] to support crop growth and environmental sustainability by reducing soil erosion and carbon emissions [123]. It is also worth mentioning that a combined effect of biochar application and deep tillage has significantly decreased bulk density, increased the air capacity of the soil, and enhanced the soil organic carbon content and available phosphorus, which subsequently enhances the grain yield of crops [124,125]. Higher yield, soil organic matter, soil available nitrogen, phosphorus, and potassium have been also observed under conservation tillage compared with the conventional tillage system [126].

However, the extent and magnitude of the combined effect of tillage and biochar application on crop yield and soil health indicators was found to be inconsistent. For instance, the application of biochar had a more profound effect on crop yield and the productivity of the soil under reduced tillage than under other tillage operations [127]. Higher soil productivity under deep tillage could be due to an invert and a complete mixture of the soil, which further turned the top soil layer into the deeper soil profile, and facilitated the decomposition rate of biochar [128]. These techniques directly influence soil microbes by adding nutrients and indirectly influence them by changing the soil characteristics [129,130]. On the other hand, biochar and conservation farming (CF), which incorporates basin tillage, residue retention, and crop rotation, may assist in reducing the detrimental effects of conventional agriculture [131]. This implies that there is a symbiotic relationship between tillage practices and an exogenous application of biochar, although the magnitude and level of influence of tillage practices rely on the soil and environmental conditions.

6. Conclusions

Agriculture is the backbone of the worldwide economy as it supports different crop yields and animal products. To obtain the required amount and quality of crops, the soil health and the environment should be sustained by undertaking different activities. Among those actions, appropriate agronomic practices, such as proper soil tillage systems and multifunctional use of nutrient application systems, have a significant role in increasing the quality and productivity of the crop yield, improving soil health, and increasing environmental protection as well. Specifically, soil tillage is one of the critical components of agricultural systems and practices, commonly working internationally in croplands to reduce climatic and soil restrictions, even though sustaining several ecosystem services is an issue. It is a crop production factor contributing up to 20% and affecting the sustainable use of soil resources through its influence on the physical and chemical properties of soil. The system of tillage is categorized into conservation tillage or conventional tillage systems. Conservation tillage attains the highest organic matter accumulation, achieves the maximum root mass density (0–15 cm soil depth), improves physical and chemical properties of the soil, sequesters more carbon, and can increase crop yields by 4.6% on average across different crops and regions compared to conventional tillage systems. In addition, biochar is a multifunctional carbon substance, which is used to solve soil fertility and climate change issues. It is considered a novel soil treatment and carbon sequestration pathway that has improved soil structure and ecosystems function due to its complex physical and chemical properties. The use of biochar can enhance soil fertility, reduce soil erosion, decrease soil nitrous oxide (N_2O) emissions by 28% (down to 60%), promote microbial activity, increase carbon sequestration, and increase crop yields by 11% on average.

Therefore, it is clear that the soil tillage system and biochar application have a symbiotic and asymmetric relationship that can promote soil carbon sequestration, improve soil properties, and improve crop yields. The interactive effect of soil tillage and biochar showed a positive effect on crop yield and soil properties in conservation and conventional tillage systems. However, the extent and magnitude of the combined effect of tillage and biochar application on crop yield and soil health indicators was found to be inconsistent. This means it is difficult to detect the most advantageous soil tillage system with the combination of biochar to improve soil health indicators, environmental protection, and crop yield. Thus, to detect the most appropriate soil tillage system with the combination of biochar, further research will be needed in different environments, soil types, seasons, and crop types.

Author Contributions: Conceptualization, A.A.B., A.A.M. and A.P.; Methodology, A.A.B.; Writing Original draft preparation, A.A.B., A.A.M. and A.P.; Writing review and editing, A.A.B., A.A.M. and A.P. All authors have read and agreed to the published version of the manuscript.

Funding: This work was supported by the Tempus Public Foundation (Hungary) within the framework of the Stipendium Hungaricum Scholarship Programme, Hungarian University of Agriculture and life Sciences (MATE), supported by Ministry of Education of Ethiopia and Mekdela Amba University.

Institutional Review Board Statement: Not applicable.

Informed Consent Statement: Not applicable.

Data Availability Statement: Data included in article referenced in article.

Acknowledgments: This review paper is benefited a lot from the work of other scholars working around the subject matter and hence, we give our sincere thanks though we have provided full reference to the information used, we give our thanks also Tempus Public Foundation (Hungary) for providing the opportunity.

Conflicts of Interest: The authors declare no conflict of interest.

References

1. Yadav, G.S.; Lal, R.; Meena, R.S.; Babu, S.; Das, A.; Bhowmik, S.N.; Datta, M.; Layak, J.; Saha, P. Conservation tillage and nutrient management effects on productivity and soil carbon sequestration under double cropping of rice in north eastern region of India. *Ecol. Ind.* **2019**, *105*, 303–315. [CrossRef]
2. Baye, K.N.; Anteneh, A.M.; Amare, A.B. Role of Conservation Tillage as Climate Change Mitigation. *Civ. Environ. Res.* **2019**, *11*, 2224–5790.
3. Skaalsveen, K.; Ingram, J.; Clarke, L.E. The effect of no-till farming on the soil functions of water purification and retention in north-western Europe: A literature review. *Soil Tillage Res.* **2019**, *189*, 98–109. [CrossRef]
4. Bhattacharyya, S.S.; Leite, F.F.G.D.; France, C.L.; Adekoya, A.O.; Ros, G.H.; de Vries, W.; Melchor-Martínez, E.M.; Iqbal, H.M.; Parra-Saldívar, R. Soil carbon sequestration, greenhouse gas emissions, and water pollution under different tillage practices. *Sci. Total Environ.* **2022**, *826*, 154161. [CrossRef]
5. Li, S.; Jiang, X.; Wang, X.; Wright, A.L. Tillage effects on soil nitrification and the dynamic changes in nitrifying microorganisms in a subtropical rice-based ecosystem: A long-term field study. *Soil Tillage Res.* **2015**, *150*, 132–138. [CrossRef]
6. Šimon, T.; Javůrek, M.; Mikanová, O.; Vach, M. The influence of tillage systems on soil organic matter and soil hydrophobicity. *Soil Tillage Res.* **2019**, *105*, 44–48. [CrossRef]
7. Guenet, B.; Gabrielle, B.; Chenu, C.; Arrouays, D.; Balesdent, J.; Bernoux, M.; Bruni, E.; Caliman, J.; Cardinael, R.; Chen, S.; et al. Can N_2O emissions offset the benefits from soil organic carbon storage? *Glob. Chang. Biol.* **2021**, *27*, 237–256. [CrossRef]
8. Hemes, K.S.; Chamberlain, S.D.; Eichelmann, E.; Knox, S.H.; Baldocchi, D.D. A Biogeochemical Compromise: The High Methane Cost of Sequestering Carbon in Restored Wetlands. *Geophys. Res. Lett.* **2018**, *45*, 6081–6091. [CrossRef]
9. Paustian, K.; Six, J.; Elliott, E.T.; Hunt, H.W. Management options for reducing CO2 emissions from agricultural soils. *Biogeochemistry* **2000**, *48*, 147–163. [CrossRef]
10. Prasad, J.V.N.S.; Rao, C.S.; Srinivas, K.; Jyothi, C.N.; Venkateswarlu, B.; Ramachandrappa, B.K.; Dhanapal, G.N.; Ravichandra, K.; Mishra, P.K. Effect of ten years of reduced tillage and recycling of organic matter on crop yields, soil organic carbon and its fractions in Alfisols of semi-arid tropics of southern India. *Soil Tillage Res.* **2016**, *156*, 131–139. [CrossRef]
11. Layek, J.; Narzari, R.; Hazarika, S.; Das, A.; Rangappa, K.; Devi, S.; Balusamy, A.; Saha, S.; Mandal, S.; Idapuganti, R.G.; et al. Prospects of Biochar for Sustainable Agriculture and Carbon Sequestration: An Overview for Eastern Himalayas. *Sustainability* **2022**, *14*, 6684. [CrossRef]
12. Majumder, S.; Neogi, S.; Dutta, T.; Powel, M.A.; Banik, P. The impact of biochar on soil carbon sequestration: Meta-analytical approach to evaluating environmental and economic advantages. *J. Environ. Manag.* **2019**, *250*, 109466. [CrossRef] [PubMed]
13. Lehmann, J.; Joseph, S. (Eds.) *Biochar for Environmental Management: Science, Technology and Implementation*; Routledge: London, UK, 2015.
14. Paustian, K.; Lehmann, J.; Ogle, S.; Reay, D.; Robertson, G.P.; Smith, P. Climate-smart soils. *Nature* **2016**, *532*, 49–57. [CrossRef] [PubMed]
15. Busari, M.A.; Kukal, S.S.; Kaur, A.; Bhatt, R.; Dulazi, A.A. Conservation tillage impacts on soil, crop and the environment. *Int. Soil Water Conserv. Res.* **2015**, *3*, 119–129. [CrossRef]
16. Corsi, S.; Friedrich, T.; Kassam, A.; Pisante, M.; Sà, J.D. *Soil Organic Carbon Accumulation and Greenhouse Gas Emission Reductions from Conservation Agriculture: A Literature Review*; Food and Agriculture Organization of the United Nations (FAO): Rome, Italy, 2012.
17. Abbaspour-Gilandeh, Y.; Sedghi, R. Predicting soil fragmentation during tillage operation using fuzzy logic approach. *J. Terramech.* **2015**, *57*, 61–69. [CrossRef]
18. Romaneckas, K.; Šarauskis, E.; Avižienytė, D.; Buragienė, S.; Arney, D. The main physical properties of planosol in maize (*Zea mays* L.) cultivation under different long-term reduced tillage practices in the Baltic region. *J. Integr. Agric.* **2015**, *14*, 1309–1320. [CrossRef]

19. Hamzei, J.; Seyyedi, M. Energy use and input–output costs for sunflower production in sole and intercropping with soybean under different tillage systems. *Soil Tillage Res.* **2016**, *157*, 73–82. [CrossRef]
20. Vaitauskienė, K.; Šarauskis, E.; Romaneckas, K.; Jasinskas, A. Design, development and field evaluation of row-cleaners for strip tillage in conservation farming. *Soil Tillage Res.* **2017**, *174*, 139–146. [CrossRef]
21. Rasmussen, K.J. Impact of plough less soil tillage on yield and soil quality: A Scandinavian review. *Soil Tillage Res.* **1999**, *53*, 3–14. [CrossRef]
22. Khurshid, K.A.; Iqbal, M.; Arif, M.S.; Nawaz, A. Effect of tillage and mulch on soil physical properties and growth of maize. *Int. J. Agric. Biol.* **2006**, *8*, 593–596.
23. Alam, M.D.; Islam, M.; Salahin, N.; Hasanuzzaman, M. Effect of Tillage Practices on Soil Properties and Crop Productivity in Wheat-Mungbean-Rice Cropping System under Subtropical Climatic Conditions. *Sci. World J.* **2014**, *2014*, 437283. [CrossRef] [PubMed]
24. Lal, R.; Stewart, B.A. (Eds.) *Principles of Sustainable Soil Management in Agroecosystems*; CRC Press: Boca Raton, FL, USA, 2013; Volume 20.
25. Bista, P.; Machado, S.; Ghimire, R.; Yorgey, G.; Wysocki, D. Conservation tillage systems. In *Advances in Dryland Farming in the Inland Pacific Northwest*; Yorgey, G., Kruger, C., Eds.; Washington State University Extension: Pullman, WA, USA, 2017; pp. 99–124.
26. Reicosky, D.C.; Cassel, D.K.; Blevins, R.L.; Gill, W.R.; Naderman, G.C. Conservation tillage in the Southeast. *J. Soil Water Conserv.* **1977**, *32*, 13–19.
27. Hasanuzzaman, M. (Ed.) *Agronomic Crops: Volume 2: Management Practices*, 1st ed.; Springer: Singapore, 2019.
28. CTIC (Conservation Technology Information Center). West Lafayette, Indiana, Conservation in Action Tour 2016, 47906. Available online: https://ctic.org/CIA_Tours/pasttours/2016 (accessed on 24 August 2016).
29. Triplett, G.B.; Dick, W.A. No-tillage crop production: A revolution in agriculture. *Agron. J.* **2008**, *100*, 153–165. [CrossRef]
30. Hansen, N.C.; Allen, B.L.; Baumhardt, R.L.; Lyon, D.J. Research achievements and adoption of no-till, dryland cropping in the semi-arid U.S. Great Plains. *Field Crops Res.* **2012**, *132*, 196–203. [CrossRef]
31. Obour, A.K.; Mikha, M.M.; Holman, J.D.; Stahlman, P.W. Changes in soil surface chemistry after fifty years of tillage and nitrogen fertilization. *Geoderma* **2017**, *308*, 46–53. [CrossRef]
32. Obour, A.K.; Holman, J.D.; Simon, L.M.; Schlegel, A.J. Strategic Tillage Effects on Crop Yields, Soil Properties, and Weeds in Dryland No-Tillage Systems. *Agronomy* **2021**, *11*, 662. [CrossRef]
33. Rusu, T.; Pacurar, I.; Dirja, M.; Pacurar, H.M.; Oroian, I.; Cosma, S.A.; Gheres, M. Effect of tillage systems on soil properties, humus and water conservation. *Agric. Sci.* **2013**, *4*, 35–40. [CrossRef]
34. Abdellaoui, Z.; Teskrat, H.; Belhadj, A.; Zaghouane, O. Comparative study of the effect of conventional tillage, direct seeding and minimum tillage on the behavior of a durum wheat crop in the subhumid zone. In *Third Mediterranean Meetings of Direct Seeding*; Bouzerzour, H., Irekti, H., Vadon, B., Eds.; Ciheam Zaragoza-Spain: Zaragoza, Spain, 2006; pp. 71–87.
35. Dayou, E.D.; Zokpodo, K.L.B.; Kakaï, A.G.; Ganglo, C.J. Impacts of the conventional tillage tools and reduced tillage on the soil fertility preservation: Critical review. *J. Appl. Biosci.* **2017**, *117*, 11684–11695. [CrossRef]
36. Khorami, S.S.; Kazemeini, S.A.; Afzalinia, S.; Gathala, M.K. Changes in Soil Properties and Productivity under Different Tillage Practices and Wheat Genotypes: A Short-Term Study in Iran. *Sustainability* **2018**, *10*, 3273. [CrossRef]
37. Woźniak, A.; Rachoń, L. Effect of Tillage Systems on the Yield and Quality of Winter Wheat Grain and Soil Properties. *Agriculture* **2020**, *10*, 405. [CrossRef]
38. Marongwe, L.S.; Nyagumbo, I.; Kwazira, K.; Kassam, A.; Friedrich, T. *Conservation Agriculture and Sustainable Crop Intensification: A Zimbabwe Case Study*; Food and Agriculture Organization of the United Nations (FAO): Rome, Italy, 2012; Volume 17, pp. xii + 29.
39. 39 Zheng, C.; Jiang, Y.; Chen, C.; Sun, Y.; Feng, J.; Deng, A.; Song, Z.; Zhang, W. The impacts of conservation agriculture on crop yield in China depend on specific practices, crops and cropping regions. *Crop J.* **2014**, *2*, 289–296. [CrossRef]
40. Deines, J.M.; Wang, S.; Lobell, D.B. Satellites reveal a small positive yield effect from conservation tillage across the US Corn Belt. *Environ. Res. Lett.* **2019**, *14*, 124038. [CrossRef]
41. Pittelkow, C.M.; Liang, X.; Linquist, B.A.; Van Groenigen, K.J.; Lee, J.; Lundy, M.E.; Van Gestel, N.; Six, J.; Venterea, R.T.; van Kessel, C. Productivity limits and potentials of the principles of conservation agriculture. *Nature* **2015**, *517*, 365–368. [CrossRef] [PubMed]
42. Fonteyne, S.; Singh, R.G.; Govaerts, B.; Verhulst, N. Rotation, Mulch and Zero Tillage Reduce Weeds in a Long-Term Conservation Agriculture Trial. *Agronomy* **2020**, *10*, 962. [CrossRef]
43. Blevins, R.L. Conservation tillage: An ecological approach to soil management. *Adv. Agron.* **1993**, *51*, 33–78.
44. Womach, J.; Resources, Science, and Industry Division. *Agriculture: A Glossary of Terms, Programs, and Laws*; Library of Congress, Congressional Research Service: Washington, DC, USA, 2005; Volume 97, Issue 905. Available online: https://crsreports.congress.gov/product/pdf/RL/97-905 (accessed on 16 June 2005).
45. Labbaci, T.; Dugué, P.; Kemoun, H.; Rollin, D. Innovation et action collective: Le semis direct des cultures pluviales au Moyen Sébou (Maroc). *Cah. Agric.* **2015**, *24*, 76–83.
46. Tian, K.; Zhao, Y.; Xu, X.; Hai, N.; Huang, B.; Deng, W. Effects of long-term fertilization and residue management on soil organic carbon changes in paddy soils of China: A meta-analysis. *Agric. Ecosyst. Environ.* **2015**, *204*, 40–50. [CrossRef]
47. Mazzilli, S.R.; Kemanian, A.R.; Ernst, O.R.; Jackson, R.B.; Piñeiro, G. Priming of soil organic carbon decomposition induced by corn compared to soybean crops. *Soil Biol. Biochem.* **2014**, *75*, 273–281. [CrossRef]

48. Mazzilli, S.R.; Kemanian, A.R.; Ernst, O.R.; Jackson, R.B.; Piñeiro, G. Greater humification of belowground than aboveground biomass carbon into particulate soil organic matter in no-till corn and soybean crops. *Soil Biol. Biochem.* **2015**, *85*, 22–30. [CrossRef]
49. Dou, X.; Cheng, X.; He, P.; Zhu, P.; Zhou, W.; Wang, L. Dynamics of physically- separated soil organic carbon pools assessed from δ13C changes under 25 years of cropping systems. *Soil Tillage Res.* **2017**, *174*, 6–13. [CrossRef]
50. Dou, X.; He, P.; Cheng, X.; Zhou, W. Long-term fertilization alters chemically-separated soil organic carbon pools: Based on stable C isotope analyses. *Sci. Rep.* **2016**, *6*, 19061. [CrossRef] [PubMed]
51. Meng, F.; Dungait, J.A.J.; Xu, X.; Bol, R.; Zhang, X.; Wu, W. Coupled incorporation of maize (*Zea mays* L.) straw with nitrogen fertilizer increased soil organic carbon in Fluvic Cambisol. *Geoderma* **2017**, *304*, 19–27. [CrossRef]
52. Jastrow, J.D.; Amonette, J.E.; Bailey, V.L. Mechanisms controlling soil carbon turnover and their potential application for enhancing carbon sequestration. *Clim. Chang.* **2007**, *80*, 5–23. [CrossRef]
53. Manns, H.R.; Parkin, G.W.; Martin, R.C. Evidence of a union between organic carbon and water content in soil. *Can. J. Soil Sci.* **2016**, *96*, 305–316. [CrossRef]
54. Lal, R. Carbon sequestration in soil. *CAB Rev. Perspect. Agric. Vet. Sci. Nutr. Nat. Resour.* **2008**, *20*, 1749–8848. [CrossRef]
55. McDaniel, M.D.; Grandy, A.S.; Tiemann, L.K.; Weintraub, M.N. Crop rotation complexity regulates the decomposition of high- and low-quality residues. *Soil Biol. Biochem.* **2014**, *78*, 243–254. [CrossRef]
56. Palombi, L.; Sessa, R. *Climate-Smart Agriculture: Sourcebook*; Food and Agriculture Organization of the United Nations (FAO): Rome, Italy, 2013.
57. Global Soil Partnership. *Global Soil Organic Carbon Map–Leaflet*; Food and Agriculture Organization of the United Nations (FAO): Rome, Italy, 2017.
58. Hazarika, S.; Parkinson, R.; Bol, R.; Dixon, L.; Russell, P.; Donovan, S.; Allen, D. Effect of tillage system and straw management on organic matter dynamics. *Agron. Sustain. Dev.* **2009**, *29*, 525–533. [CrossRef]
59. Ghimire, R.; Lamichhane, S.; Acharya, B.S.; Bista, P.; Sainju, U.M. Tillage, crop residue, and nutrient management effects on soil organic carbon in rice-based cropping systems: A review. *J. Integr. Agric.* **2017**, *16*, 1–15. [CrossRef]
60. Vijay, V.; Shreedhar, S.; Adlak, K.; Payyanad, S.; Sreedharan, V.; Gopi, G.; Sophia, V.T.; Malarvizhi, P.; Yi, S.; Gebert, J.; et al. Review of Large-Scale Biochar Field-Trials for Soil Amendment and the Observed Influences on Crop Yield Variations. *Front. Energy Res.* **2021**, *9*, 710766. [CrossRef]
61. Li, M.; Xiong, Y.; Cai, L. Effects of Biochar on the Soil Carbon Cycle in Agroecosystems: An Promising Way to Increase the Carbon Pool in Dryland. *IOP Conf. Ser. Earth Environ. Sci.* **2021**, *693*, 012082. [CrossRef]
62. Xu, H.; Cai, A.; Wu, D.; Liang, G.; Xiao, J.; Xu, M.; Colinet, G.; Zhang, W. Effects of biochar application on crop productivity, soil carbon sequestration, and global warming potential controlled by biochar C: N ratio and soil pH: A global meta-analysis. *Soil Tillage Res.* **2021**, *213*, 105125. [CrossRef]
63. Zhang, A.; Bian, R.; Pan, G.; Cui, L.; Hussain, Q.; Li, L.; Zheng, J.; Zheng, J.; Zhang, X.; Han, X.; et al. Effects of biochar amendment on soil quality, crop yield and greenhouse gas emission in a Chinese rice paddy: A field study of 2 consecutive rice growing cycles. *Field Crops Res.* **2012**, *127*, 153–160. [CrossRef]
64. Laghari, M.; Mirjat, M.S.; Hu, Z.; Fazal, S.; Xiao, B.; Hu, M.; Chen, Z.; Guo, D. Effects of biochar application rate on sandy desert soil properties and sorghum growth. *Catena* **2015**, *135*, 313–320. [CrossRef]
65. Hussain, M.; Farooq, M.; Nawaz, A.M.; Al-Sadi, A.; Solaiman, Z.M.; Alghamdi, S.S.; Ammara, U.; Ok, Y.S.; Siddique, K.H. Biochar for crop production: Potential benefits and risks. *J. Soils Sediments* **2017**, *17*, 685–716. [CrossRef]
66. Mašek, O.; Buss, W.; Brownsort, P.; Rovere, M.; Tagliaferro, A.; Zhao, L.; Cao, X.; Xu, G. Potassium doping increases biochar carbon sequestration potential by 45%, facilitating decoupling of carbon sequestration from soil improvement. *Sci. Rep.* **2019**, *9*, 5514. [CrossRef] [PubMed]
67. Fuss, S.; Lamb, W.F.; Callaghan, M.W.; Hilaire, J.; Creutzig, F.; Amann, T.; Beringer, T.; de Oliveira Garcia, W.; Hartmann, J.; Khanna, T.; et al. Negative emissions—Part 2: Costs, potentials and side effects. *Environ. Res. Lett.* **2018**, *13*, 063002. [CrossRef]
68. Lomax, G.; Workman, M.; Lenton, T.; Shah, N. Reframing the policy approach to greenhouse gas removal technologies. *Energy Policy* **2015**, *78*, 125–136. [CrossRef]
69. Pratt, K.; Moran, D. Evaluating the cost-effectiveness of global biochar mitigation potential. *Biomass Bioenergy* **2010**, *34*, 1149–1158. [CrossRef]
70. Griscom, B.W.; Adams, J.; Ellis, P.W.; Houghton, R.A.; Lomax, G.; Miteva, D.A.; Schlesinger, W.H.; Shoch, D.; Siikamäki, J.V.; Smith, P.; et al. Natural climate solutions. *Proc. Natl. Acad. Sci. USA* **2017**, *114*, 11645–11650. [CrossRef]
71. Powell, T.W.R.; Lenton, T.M. Future carbon dioxide removal via biomass energy constrained by agricultural efficiency and dietary trends. *Energy Environ. Sci.* **2012**, *5*, 8116. [CrossRef]
72. Smith, P. Soil carbon sequestration and biochar as negative emission technologies. *Glob. Chang. Biol.* **2016**, *22*, 1315–1324. [CrossRef] [PubMed]
73. Moore, J.C.; Jevrejeva, S.; Grinsted, A. Efficacy of geoengineering to limit 21st century sea-level rise. *Proc. Natl. Acad. Sci. USA* **2010**, *107*, 15699–15703. [CrossRef] [PubMed]
74. Woolf, D.; Amonette, J.E.; Street-Perrott, F.A.; Lehmann, J.; Joseph, S. Sustainable biochar to mitigate global climate change. *Nat. Commun.* **2010**, *1*, 56. [CrossRef] [PubMed]
75. Wang, J.; Xiong, Z.; Kuzyakov, Y. Biochar stability in soil: Meta-analysis of decomposition and priming effects. *GCB Bioenergy* **2016**, *8*, 512–523. [CrossRef]

76. Coomes, O.; Miltner, B.C. Indigenous Charcoal and Biochar Production: Potential for Soil Improvement under Shifting Cultivation Systems. *Land Degrad. Dev.* **2017**, *28*, 811–821. [CrossRef]
77. Koçer, A.T.; Mutlu, B.; Özçimen, D. Investigation of biochar production potential and pyrolysis kinetics characteristics of microalgal biomass. *Biomass Convers. Biorefinery* **2020**, *10*, 85–94. [CrossRef]
78. Azzi, E.S.; Karltun, E.; Sundberg, C. Small-scale biochar production on Swedish farms: A model for estimating potential, variability, and environmental performance. *J. Clean. Prod.* **2021**, *280*, 124873. [CrossRef]
79. Gross, A.; Bromm, T.; Glaser, B. Soil Organic Carbon Sequestration after Biochar Application: A Global Meta-Analysis. *Agronomy* **2021**, *11*, 2474. [CrossRef]
80. Elkhlifi, Z.; Iftikhar, J.; Sarraf, M.; Ali, B.; Saleem, M.H.; Ibranshahib, I.; Bispo, M.D.; Meili, L.; Ercisli, S.; Torun Kayabasi, E.; et al. Potential Role of Biochar on Capturing Soil Nutrients, Carbon Sequestration and Managing Environmental Challenges: A Review. *Sustainability* **2023**, *15*, 2527. [CrossRef]
81. Spokas, K.A.; Reicosky, D.C. Impacts of sixteen different biochars on soil greenhouse gas production. *Ann. Environ. Sci.* **2009**, *3*, 179–193.
82. Atkinson, C.J.; Fitzgerald, J.D.; Hipps, N.A. Potential mechanisms for achieving agricultural benefits from biochar application to temperate soils: A review. *Plant Soil* **2010**, *337*, 1–18. [CrossRef]
83. Wang, X.; Yan, R.; Zhao, Y.; Cheng, S.; Han, Y.; Yang, S.; Cai, D.; Mang, H.-P.; Li, Z. Biogas standard system in China. *Renew. Energy* **2020**, *157*, 1265–1273. [CrossRef]
84. Major, J.; Rondon, M.; Molina, D.; Riha, S.J.; Lehmann, J. Maize yield and nutrition during 4 years after biochar application to a Colombian savanna oxisol. *Plant Soil* **2010**, *333*, 117–128. [CrossRef]
85. Rekaby, S.A.; Awad, M.; Majrashi, A.; Ali, E.F.; Eissa, M.A. Corn Cob-Derived Biochar Improves the Growth of Saline-Irrigated Quinoa in Different Orders of Egyptian Soils. *Horticulturae* **2021**, *7*, 221. [CrossRef]
86. Hagemann, N.; Joseph, S.; Schmidt, H.-P.; Kammann, C.I.; Harter, J.; Borch, T.; Young, R.B.; Varga, K.; Taherymoosavi, S.; Elliott, K.W.; et al. Organic coating on biochar explains its nutrient retention and stimulation of soil fertility. *Nat. Commun.* **2017**, *8*, 1089. [CrossRef]
87. Lentz, R.D.; Ippolito, J.A. Biochar and Manure Affect Calcareous Soil and Corn Silage Nutrient Concentrations and Uptake. *J. Environ. Qual.* **2012**, *41*, 1033–1043. [CrossRef]
88. Liang, F.; Li, G.T.; Lin, Q.M.; Zhao, X.R. Crop Yield and Soil Properties in the First 3 Years After Biochar Application to a Calcareous Soil. *J. Integr. Agric.* **2014**, *13*, 525–532. [CrossRef]
89. Jin, Z.; Chen, C.; Chen, X.; Hopkins, I.; Zhang, X.; Han, Z.; Jiang, F.; Billy, G. The crucial factors of soil fertility and rapeseed yield—A five-year field trial with biochar addition in upland red soil, China. *Sci. Total Environ.* **2019**, *649*, 1467–1480. [CrossRef]
90. Jeffery, S.; Verheijen, F.G.A.; van der Velde, M.; Bastos, A.C. A quantitative review of the effects of biochar application to soils on crop productivity using meta-analysis. *Agric. Ecosyst. Environ.* **2011**, *144*, 175–187. [CrossRef]
91. Agegnehu, G.; Srivastava, A.K.; Bird, M.I. The role of biochar and biochar-compost in improving soil quality and crop performance: A review. *Appl. Soil Ecol.* **2017**, *119*, 156–170. [CrossRef]
92. Sahota, S.; Vijay, V.K.; Subbarao, P.M.; Chandra, R.; Ghosh, P.; Shah, G.; Kapoor, R.; Vijay, V.; Koutu, V.; Thakur, I.S. Characterization of leaf waste-based biochar for cost effective hydrogen sulphide removal from biogas. *Bioresour. Technol.* **2018**, *250*, 635–641. [CrossRef] [PubMed]
93. Biederman, L.A.; Harpole, W.S. Biochar and its effects on plant productivity and nutrient cycling: A meta-analysis. *GCB Bioenergy* **2013**, *5*, 202–214. [CrossRef]
94. Liu, X.; Zhang, A.; Ji, C.; Joseph, S.; Bian, R.; Li, L.; Pan, G.; Paz-Ferreiro, J. Biochar's effect on crop productivity and the dependence on experimental conditions—A meta-analysis of literature data. *Plant Soil* **2013**, *373*, 583–594. [CrossRef]
95. Cayuela, M.L.; van Zwieten, L.; Singh, B.P.; Jeffery, S.; Roig, A.; Sánchez-Monedero, M.A. Biochar's role in mitigating soil nitrous oxide emissions: A review and meta-analysis. *Agric. Ecosystem. Environ.* **2014**, *191*, 5–16. [CrossRef]
96. He, Y.; Zhou, X.; Jiang, L.; Li, M.; Du, Z.; Zhou, G.; Shao, J.; Wang, X.; Xu, Z.; Hosseini Bai, S.; et al. Effects of biochar application on soil greenhouse gas fluxes: A meta-analysis. *GCB Bioenergy* **2017**, *9*, 743–755. [CrossRef]
97. Borchard, N.; Schirrmann, M.; Cayuela, M.L.; Kammann, C.; Wrage-Mönnig, N.; Estavillo, J.M.; Fuertes-Mendizábal, T.; Sigua, G.; Spokas, K.; Ippolito, J.A.; et al. Biochar, soil and land-use interactions that reduce nitrate leaching and N2O emissions: A meta-analysis. *Sci. Total Environ.* **2019**, *651*, 2354–2364. [CrossRef]
98. Lehmann, J.; Rillig, M.C.; Thies, J.; Masiello, C.A.; Hockaday, W.C.; Crowley, D. Biochar effects on soil biota—A review. *Soil Biol. Biochem.* **2011**, *43*, 1812–1836. [CrossRef]
99. Xiao, B.; Dai, Q.; Yu, X.; Yu, P.; Zhai, S.; Liu, R.; Guo, X.; Liu, J.; Chen, H. Effects of sludge thermal-alkaline pretreatment on cationic red X-GRL adsorption onto pyrolysis biochar of sewage sludge. *J. Hazard. Mater.* **2018**, *343*, 347–355. [CrossRef]
100. Wang, Z.; Yang, X.; Qin, T.; Liang, G.; Li, Y.; Xie, X. Efficient removal of oxytetracycline from aqueous solution by a novel magnetic clay–biochar composite using natural attapulgite and cauliflower leaves. *Environ. Sci. Pollut. Res.* **2019**, *26*, 7463–7475. [CrossRef]
101. Liu, S.; Zhang, Y.; Zong, Y.; Hu, Z.; Wu, S.; Zhou, J.; Jin, Y.; Zou, J. Response of soil carbon dioxide fluxes, soil organic carbon and microbial biomass carbon to biochar amendment: A meta-analysis. *GCB Bioenergy* **2016**, *8*, 392–406. [CrossRef]
102. Dong, X.; Li, G.; Lin, Q.; Zhao, X. Quantity and quality changes of biochar aged for 5 years in soil under field conditions. *Catena* **2017**, *159*, 136–143. [CrossRef]
103. Aller, M.F. Biochar properties: Transport, fate, and impact. *Crit. Rev. Environ. Sci. Technol.* **2016**, *46*, 1183–1296. [CrossRef]

104. Zhang, A.; Cui, L.; Pan, G.; Li, L.; Hussain, Q.; Zhang, X.; Zheng, J.; Crowley, D. Effect of biochar amendment on yield and methane and nitrous oxide emissions from a rice paddy from Tai Lake plain, China. *Agric. Ecosyst. Environ.* **2010**, *158*, 97–106. [CrossRef]
105. Lehmann, J.; Gaunt, J.; Rondon, M. Bio-char sequestration in terrestrial ecosystems—A review. *Mitig. Adapt. Strateg. Glob. Chang.* **2006**, *8*, 403–427. [CrossRef]
106. Zhang, F.; Liu, M.; Li, Y.; Che, Y.; Xiao, Y. Effects of arbuscular mycorrhizal fungi, biochar and cadmium on the yield and element uptake of Medicago sativa. *Sci. Total Environ.* **2019**, *655*, 1150–1158. [CrossRef] [PubMed]
107. Zhang, N.; Li, J.; Liu, X.H.; Liu, Y.; Wang, Y.P.; Liang, H.Y.; Liao, Y.C. Effects of biochar on growth and yield of summer maize. *J. Agro-Environ. Sci.* **2014**, *33*, 1569–1574. (In Chinese)
108. Ain Najwa, K.; Wan Zaliha, W.; Yusnita, H.; Zuraida, A. Effect of different soilless growing media and biochar on growth, yield and postharvest quality of lowland cherry tomato (*Solanum lycopersicum* var. *cerasiforme*). *Innov. Plant Product. Qual.* **2014**, *22*, 53.
109. Yuan, J.-H.; Xu, R.-K.; Zhang, H. The forms of alkalis in the biochar produced from crop residues at different temperatures. *Bioresour. Technol.* **2011**, *102*, 3488–3497. [CrossRef]
110. Hilber, I.; Bastos, A.C.; Loureiro, S.; Soja, G.; Marsz, A.; Cornelissen, G.; Bucheli, T.D. The different faces of biochar: Contamination risk versus remediation tool. *J. Environ. Eng. Landsc. Manag.* **2017**, *25*, 86–104. [CrossRef]
111. Gupta, D.K.; Gupta, C.K.; Dubey, R.; Fagodiya, R.K.; Sharma, G.; Noor Mohamed, M.B.; Dev, R.; Shukla, A.K. Role of biochar in carbon sequestration and greenhouse gas mitigation. In *Biochar Applications in Agriculture and Environment Management*; Springer: Cham, Switzerland, 2020; pp. 141–165.
112. Glaser, B.; Lehmann, J.; Zech, W. Ameliorating physical and chemical properties of highly weathered soils in the tropics with charcoal—A review. *Biol. Fertil. Soils* **2002**, *35*, 219–230. [CrossRef]
113. Laird, D.A.; Fleming, P.; Davis, D.D.; Horton, R.; Wang, B.; Karlen, D.L. Impact of biochar amendments on the quality of a typical Midwestern agricultural soil. *Geoderma* **2010**, *158*, 443–449. [CrossRef]
114. Novak, J.M.; Lima, I.; Xing, B.; Gaskin, J.W.; Steiner, C.; Das, K.C.; Ahmedna, M.; Rehrah, D.; Watts, D.W.; Busscher, W.J.; et al. Characterization of designer biochar produced at different temperatures and their effects on a loamy sand. *Ann. Environ. Sci.* **2009**, *3*, 195–206.
115. Verheijen, F.G.A.; Jeffery, S.; Bastos, A.C.; van der Velde, M. *Biochar Application to Soils—A Critical Scientific Review of Effects on Soil Properties, Processes and Functions*; EUR 24099 EN; European Commission, Joint Research Centre, Institute for Environment and Sustainability: Ispra, Italy, 2010.
116. Ahmad, M.; Rajapaksha, A.U.; Lim, J.E.; Zhang, M.; Bolan, N.; Mohan, D.; Vithanage, M.; Lee, S.S.; Ok, Y.S. Biochar as a sorbent for contaminant management in soil and water: A review. *Chemosphere* **2014**, *99*, 19–33. [CrossRef] [PubMed]
117. Asai, H.; Samson, B.K.; Stephan, H.M.; Songyikhangsuthor, K.; Homma, K.; Kiyono, Y.; Inoue, Y.; Shiraiwa, T.; Horie, T. Biochar amendment techniques for upland rice production in Northern Laos: 1. Soil physical properties, leaf SPAD and grain yield. *Field Crops Res.* **2009**, *111*, 81–84. [CrossRef]
118. Steiner, C.; Teixeira, W.G.; Lehmann, J.; Nehls, T.; de Macêdo, J.L.V.; Blum, W.E.H.; Zech, W. Long term effects of manure, charcoal and mineral fertilization on crop production and fertility on a highly weathered Central Amazonian upland soil. *Plant Soil* **2007**, *291*, 275–290. [CrossRef]
119. Li, Y.; Yao, N.; Liang, J.; Wang, X.; Jia, Y.; Jiang, F.; Li Liu, D.; Hu, W.; He, H.; Javed, T. Optimum biochar application rate for peak economic benefit of sugar beet in Xinjiang, China. *Agric. Water Manag.* **2022**, *272*, 107880. [CrossRef]
120. Kocsis, T.; Kotroczó, Z.; Kardos, L.; Biró, B. Optimization of increasing biochar doses with soil–plant–microbial functioning and nutrient uptake of maize. *Environ. Technol. Innov.* **2020**, *20*, 101191. [CrossRef]
121. Pandit, N.R.; Mulder, J.; Hale, S.E.; Zimmerman, A.R.; Pandit, B.H.; Cornelissen, G. Multi-year double cropping biochar field trials in Nepal: Finding the optimal biochar dose through agronomic trials and cost-benefit analysis. *Sci. Total Environ.* **2018**, *637–638*, 1333–1341. [CrossRef]
122. Chen, W.; Li, P.; Li, F.; Xi, J.; Han, Y. Effects of tillage and biochar on soil physiochemical and microbial properties and its linkage with crop yield. *Front. Microbiol.* **2022**, *13*, 3349. [CrossRef]
123. Li, Y.; Li, Z.; Chang, S.X.; Cui, S.; Jagadamma, S.; Zhang, Q.; Cai, Y. Residue retention promotes soil carbon accumulation in minimum tillage systems: Implications for conservation agriculture. *Sci. Total Environ.* **2020**, *740*, 140147. [CrossRef]
124. Liu, Z.; Cao, S.; Sun, Z.; Wang, H.; Qu, S.; Lei, N.; He, J.; Dong, Q. Tillage effects on soil properties and crop yield after land reclamation. *Sci. Rep.* **2021**, *11*, 4611. [CrossRef] [PubMed]
125. Wang, Y.K.; Zhang, Z.B.; Jiang, F.H.; Guo, Z.C.; Peng, X.H. Evaluating soil physical quality indicators of a Vertisol as affected by different tillage practices under wheat-maize system in the North China Plain. *Soil Tillage Res.* **2021**, *209*, 104970. [CrossRef]
126. Ahmad, S.; Hussain, I.; Ghaffar, A.; Rahman, M.H.; Saleem, M.Z.; Yonas, M.W.; Hussnain, H.; Ikram, R.M.; Arslan, M. Organic amendments and conservation tillage improve cotton productivity and soil health indices under arid climate. *Sci. Rep.* **2022**, *12*, 14072. [CrossRef] [PubMed]
127. Agbede, T.M. Effect of tillage, biochar, poultry manure and NPK 15-15-15 fertilizer, and their mixture on soil properties, growth and carrot (*Daucus carota* L.) yield under tropical conditions. *Heliyon* **2021**, *7*, e07391. [CrossRef]
128. Qi, J.Y.; Zhao, X.; He, C.; Virk, A.L.; Jing, Z.H.; Liu, Q.Y.; Wang, X.; Kan, Z.R.; Xiao, X.P.; Zhang, H.L. Effects of long-term tillage regimes on the vertical distribution of soil iron/aluminum oxides and carbon decomposition in rice paddies. *Sci. Total Environ.* **2021**, *776*, 145797. [CrossRef]

129. Abujabhah, I.S.; Bound, S.A.; Doyle, R.; Bowman, J.P. Effects of biochar and compost amendments on soil physico-chemical properties and the total community within a temperate agricultural soil. *Appl. Soil Ecol.* **2016**, *98*, 243–253. [CrossRef]
130. Chen, H.; Zheng, C.; Qiao, Y.; Du, S.; Li, W.; Zhang, X.; Zhao, Z.; Cao, C.; Zhang, W. Long-term organic and inorganic fertilization alters the diazotrophic abundance, community structure, and co-occurrence patterns in a vertisol. *Sci. Total Environ.* **2021**, *766*, 142441. [CrossRef]
131. Munera-Echeverri, J.L.; Martinsen, V.; Strand, L.T.; Cornelissen, G.; Mulder, J. Effect of conservation farming and biochar addition on soil organic carbon quality, nitrogen mineralization, and crop productivity in a light textured Acrisol in the sub-humid tropics. *PLoS ONE* **2020**, *5*, 0228717. [CrossRef]

soil systems

Review

Research Evolution on the Impact of Agronomic Practices on Soil Health from 1996 to 2021: A Bibliometric Analysis

Mohamed Houssemeddine Sellami [1,2,3,*] and Fabio Terribile [3,4]

1 Institute for Mediterranean Agricultural and Forestry Systems, National Research Council of Italy, P.le Enrico Fermi 1, Portici, 80055 Naples, Italy
2 ARCA 2010, Contrada Varignano, 81, Acerra, 81100 Naples, Italy
3 Interdepartmental Research Centre on the "Earth Critical Zone", Università degli Studi di Napoli Federico II, Via Università 100, Portici, 80055 Naples, Italy; fabio.terribile@unina.it
4 Department of Agriculture, Università degli Studi di Napoli Federico II, Via Università 100, Portici, 80055 Naples, Italy
* Correspondence: mohamed.sellami@isafom.cnr.it

Abstract: In the last two decades, there has been a significant shift in focus towards soil health by international institutions, organizations, and scholars. Recognizing the vital role of soil in sustaining agriculture, ecosystems, and mitigating climate change, there has been a concerted effort to study and understand soil health more comprehensively. In this study, a bibliometric analysis was performed in order to determine the research trend of the articles published in the Scopus database in the last 26 years on soil health experimental studies and agronomic practices conducted in field conditions on agricultural soils. It has been observed that, after 2013, there has been a significant increase in research articles on soil health, with the USA and India research institutions ranking as the most productive on this topic. There is an asymmetry in international cooperation among research institutions, as well as for scholars. In addition, the research topic is gradually shifting from the effects of soil management strategies, especially nutrient management, on soil organic carbon and yield to the study of the impact of soil management on biochemistry and microbiological soil activities and greenhouse gas emissions. Future research should focus into more integrated approaches to achieve soil indicators enabling to evaluate the impact of sustainable management practices (e.g., cropping practices) on soil health.

Keywords: bibliometric analysis; soil health; agronomic practices

Citation: Sellami, M.H.; Terribile, F. Research Evolution on the Impact of Agronomic Practices on Soil Health from 1996 to 2021: A Bibliometric Analysis. *Soil Syst.* **2023**, *7*, 78. https://doi.org/10.3390/soilsystems7030078

Academic Editor: Adriano Sofo

Received: 26 July 2023
Revised: 31 August 2023
Accepted: 8 September 2023
Published: 11 September 2023

1. Introduction

In the last decade, soil health (from now on abbreviated as SH) has become a key topic in policy and research agenda. This soil health emphasis can be seen as a response to the alarming, degraded state of soils in European Union (EU) and elsewhere and also a push towards innovative sustainable management practices.

EU gave a great contribution to this result by building a series of initiatives. For instance, the EU "Soil Health and Food" Mission (recently renamed as "A Soil Deal for Europe") delivered substantial funding to SH research [1,2]. In addition, the EU established the "new EU Soil Strategy" (COM/2021/699 final), which is already the legal instrument towards soil protection and sustainable soil management in all EU countries; thus, it has a very different formal status from the Soil Mission reporting.

Finally, a new EU soil law proposal has just been proposed to the EU Parliament under the name "Soil Monitoring and Resilience (Soil Monitoring Law)" (COM/2023/416 final). The proposal set the definition of soil health as follows "'soil health' means the physical, chemical, and biological condition of the soil determining its capacity to function as a vital living system and to provide ecosystem services". Thus, a healthy soil is evaluated in view of its provision in delivering a set of ecosystem services which require to be measurable.

Soils health and its delivery of ecosystem services play also a key role in Sustainable Development Goals (SDG) in terms of contributing to food production (SDG2: "zero hunger"), good health and wellbeing (SDG3), water quality (SDG6: "clean water and sanitation"), sustainable production (SDG12: "sustainable consumption and production"), carbon capture and greenhouse gas emission (SDG13: "climate action"), and soil health and biodiversity preservation (SDG15: "life on land").

In last year, there have been many disputes about whether we «should» or «should not» use the term Soil Health [3,4]. Actually, this is not completely new; for instance, similar discussion appeared on the term «Soil & Land Quality» before and after FAO established the definition (1974). For the sake of this specific contribution, we agree with Janzen et al.'s [5] proposal on the use of "soil health metaphor" which states that "as long as soil health helps unearth better ways of knowing and sustaining land, let us use it, honing and redefining it as we learn". Thus here, we claim that the use of "Soil Health" can be a great help in our joint effort to overcome the huge divide between us scientists, experts, and citizens.

One key item in implementing SH is the "how" to measure it. It is well established that SH indicators require to fulfil the following criteria [6,7]: (i) easy to measure; (ii) measurable with practical, rapid, and inexpensive measurement methods; (iii) sensitive to variations in management without being reflective of merely short-term variation; (iv) relevant to soil ecosystem services; and (v) informative for management. In general, soil health indicators are based on physical, chemical, or biological measures [8]. It is not surprising that a multitude of soil-health indicators have been proposed.

Moreover, to these soil indicators are affected by management practices. In fact, agronomic practices have a significant impact on soil health and thus on sustainable agriculture and environmental conservation [9,10]. More specifically it has been demonstrated that practices', such as crop rotation [11], conservation tillage [12], cover cropping [13], and nutrient management [14], impact on soil health.

Attempting to put some order in the entire matter, Jian et al. [15] analysed over 500 studies on soil health and quality on 354 geographic sites (42 countries). They found 42 SH indicators and 46 SH background indicators (e.g., climate, elevation, soil type).

Between them, it is important to highlight those indicators produced after large SH initiatives, between them the Comprehensive Assessment of Soil Health by the Cornell Framework [16], the National Soil Health Institute (www.soilhealthinstitute.org, accessed on 20 November 2022) [17] and the US Department of Agriculture [18]. The EU Soil Mission indeed proposed a list of SH indicators [1] including soil pollutants, % carbon, soil structure, soil biodiversity, soil nutrients, and soil water regimes, but currently no soil moisture depletion, thresholds, and scores are yet established. Most interestingly, the EU Soil Mission, in order to progress further in implementing SH concepts, promoted the joint work of land users and scientists in the so named "Living Labs" where an interdisciplinary approach is foreseen. The aim being to establish by 2030—in EU countries—an effective network of 100 living labs and lighthouses (https://www.soilmissionsupport.eu/ll-lh, on 20 November 2022) to co-create SH knowledge, test solutions, and demonstrate their value in real-life conditions. Being a new initiative, the value of this new end-user engagement approach is yet to be demonstrated.

Considering the complex scenario described above, it is not surprising that there is no unified approach to assess the soil health with indicators, thresholds, scores that can be determined by standard operational methods under practical conditions (e.g., considering costs and time constrains). In fact, different indicator systems are being used by institutions and by scientists with rather separate activities referring to various subdisciplines soil chemistry, soil physics, soil biology, and paedology, while at the same time there is a lack of unified approach. It seems fundamental to put some order to the entire matter which currently the SH research scenario looks rather chaotic with a very large number of different indicators working in different settings. Indeed, outstanding SH reviews have been produced, e.g., [3,19,20], but despite this evidence, the very large development of many and diverse SH papers requires both update of reviews and in-depth analysis of SH indicators from different perspectives. This is crucial if current SH policies must be implemented. Indeed, policies require coherent operational approach to be profitably implemented also to large territories.

Considering this scenario and the key issue of SH indicators, in this specific contribution we aimed to produce some understanding on SH indicators by bibliometric analysis focused on SH research evolution in the period 1996–2021. We produced such analysis giving special emphasis on SH experimental studies and agronomic practices conducted in field condition on agricultural soils because we believe that an in-depth analysis must start from the analysing the large number of SH experimental work already produced in last decades as a preliminary step towards more comprehensive SH assessments. In addition, considering the current development of soil policies in the world (e.g., SDG, USA, EU) we have analysed these results also in view of countries, continents, research institutions, and scientists where SH research was produced.

The remaining sections of this work are organized as follows. Material and Methods are presented in the next main section, which details the search strategy used to gather the relevant literature, criteria for inclusion and exclusion of studies, and data extraction procedure. Then, the study issue is discussed in light of the major findings of the bibliometric analysis, such as publishing trends, authorship patterns, and co-occurrence author's keywords analysis.

2. Materials and Methods

2.1. Methodology

The search was conducted on January 2022, using a bibliometric analysis as a statistical tool to evaluate the scientific literature related to the field of study. A Bibliometric analysis is a quantitative research method that involves the systematic examination of scientific publications to uncover patterns, relationships, and trends within the scientific literature. By applying statistical techniques and data visualization tools, bibliometric analysis provides valuable insights into the structure and dynamics of the academic community, aiding researchers, institutions, and policymakers in decision making and strategic planning [21]. Despite its virtues, bibliometric analysis is still relatively new field in agricultural research. In the last 20 years (from 2003 to 2023), just 110 and 43 research publications on bibliometric analysis, were published in agronomy and soil science, respectively, according to the Web of Science database.

2.2. Data Source and Search Criteria

The bibliometric analysis performed in this study follows the protocol established by the Collaboration for Environmental Evidence (CEE) as defined by Pullin et al. [22]. There-fore, bibliometric analysis was performed using specific inclusion and exclusion criteria for selecting papers. Four selection criteria were evaluated for selecting papers for the bibliometric analysis: (I) studies conducted only under field conditions, but not under green-house conditions, pots, laboratory, and mesocosm; (II) studies that focused on agricultural soils and excluded grassland, forest, mining soils and urban soils; (III) studies included cropland and excluded potted plant, soilless culture, hydroponics, and aquaponic, and (IV) studies that focused on agronomic management except land cover, land use, cropping patterns, and integrated farming system. In this last selection, it was important to exclude both papers not oriented towards farm experiments but more generic towards landscape scenarios and papers involving farm fishing system.

The research question inquiry for this bibliometric analysis (How agricultural practices impact on soil health?) was formulated utilizing the PICOL (Population/Intervention, Comparator/Outcome and Location) model (Table 1), as outlined in the CEE protocol.

Table 1. The eligibility criteria in relation to research question key elements.

PICOL	Description
Population	**Studies:** Study in open-field with experimental design and model and excluded greenhouse, laboratory and mesocosm
	Soil: Agricultural soils and excluded grassland, forest, mining soils and urban soils
	Crops: Study included cropland and excluded potted plant, soilless culture, hydroponics and aquaponic
Intervention	All agronomic management except land cover, land use, cropping patterns and integrated farming system
Comparator	Impacts and/or benefits
Outcome	Soil health indicators
Location	All the world

The PICOL framework was used to incorporate the term of "soil health" into the title, abstract, and keywords of the literature gathered via the Scopus database and to identify articles that were written in English, between 1980 and 2021 in peer-reviewed journals and limited to these Scopus subject area: Agricultural and Biological Sciences, Environmental Science, Biochemistry, Genetics and Molecular Biology, Earth and Planetary Sciences, Immunology and Microbiology, and Multidisciplinary.

2.3. Screening

After performing the initial search, a total of 3327 articles were identified and were subsequently added to Endnote software (Version 20.2.1, Clarivate Analytics). Figure 1 describes the procedure for selecting articles. After applying the aforementioned inclusion criteria to select titles, abstracts, and full-text studies, a total of 2342 articles were excluded. As a consequence, only 985 documents fulfilled the eligibility criteria and were considered for the bibliometric analysis.

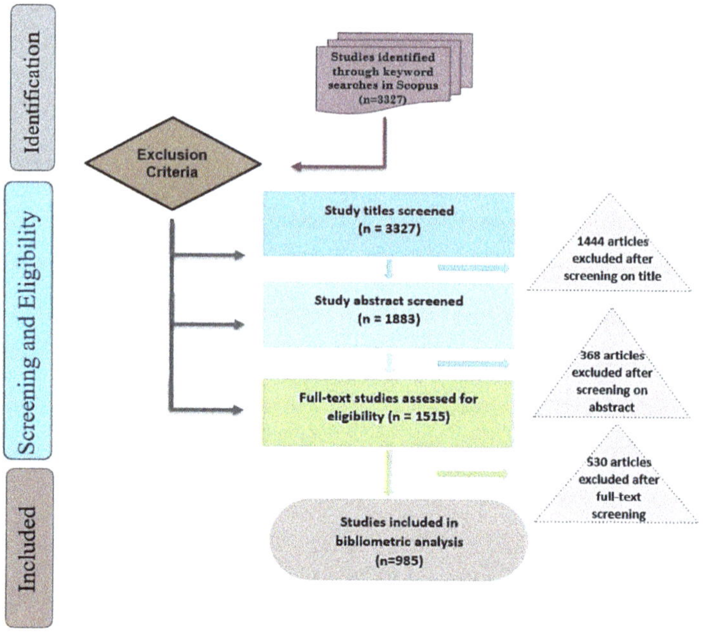

Figure 1. Prisma flowchart depicting the article selection procedure (n represent the number of studies).

2.4. Data Analysis

This study aims to analyse the research evolution of 985 scientific papers focusing on the influence of agronomic practices on soil health over a 26-year period (1996–2021). To achieve this objective, various bibliometric indicators are applied including the number of publications per scholar, number of citations, number of countries, number of institutions and journals, H-index, the journal impact factors as well as keyword co-occurrence analysis across different time periods.

The quantitative analysis was carried out using SciVal data platform, developed by Elsevier, a tool that evaluates the research performance, identify collaboration opportunities, and discover emerging trends within their field based on the Scopus database [23,24]. To create a network of co-occurrence author's keywords, we used VOSviewer [25].

3. Results and Discussion

3.1. Evolution of Scientific Production

A comprehensive selection process was conducted to identify 985 research papers pertaining to soil health in agricultural soils. These papers were chosen based on predetermined eligibility, inclusion, and exclusion criteria for the purpose of conducting a bibliometric analysis. The Table 2 shows the evolution of scientific experimental papers published on soil health under agricultural soils since 1996 until 2021. The analysis of the key attributes of scholarly articles pertaining to soil health in this table is conducted in five distinct periods, each encompassing a duration of five years, except for the final period which spans six years.

Table 2. Characteristics of scientific production from 1996 to 2021.

Period	Articles	Authors	Coutries	TC	TC/A	Journals
1996–2000	7	22	5	410	58.6	5
2001–2005	26	110	13	1386	53.3	22
2006–2010	53	220	16	2993	56.5	38
2011–2015	162	681	29	4145	25.6	83
2016–2021	737	3151	66	12,229	16.6	209

TC: Total citations on 25 June 2023; TC/A: Citations per article.

The data reveal a noticeable increase in research interest regarding soil health in agricultural soils over the analysed period. This trend is particularly evident in the last six years (2016–2021), during which more than 70% and 50% of the total scientific publications and total citations on this topic were published and cited, respectively. It is observed that there has been a more than 100-fold and 30-fold increase in the number of articles and total citations during the recent period, respectively, in comparison to the initial five-year period (1996–2000). The impetus behind this could be after the United Nations General Assembly, at its sixty-eighth session on 20 December 2013, declared 2015 as the International Year of Soils (A/C.2/68/L.21). The scientific production in this research topic experienced a good annual growth rate of 24.26%, with a clear expansional trend of scientific production since 2011 until 2021 (Figure 2). However, in the 2010s, the importance of soil health has been recognized in various international frameworks, such as the Sustainable Development Goals (SDGs), particularly Goal 15: Life on Land. The goal includes a target (15.3) to achieve a land degradation-neutral world by 2030, highlighting the importance of healthy soils for sustainable development [26]. The number of published research papers has fluctuated over the last 26-years, reaching a peak of 246 during 2021, where the total citations peaking at 2662 in 2019 (data not shown).

In spite of the notable expansion in the quantity of articles and citations, there has been a contrasting trend observed in the number of citations per document. Over the past six years, there has been a significant decline of 70% in the citations received per article, in comparison to the initial years (Table 2). The decline in the number of citations received per article over time can be attributed to several factors such as increased competition between scholars. When the number of published articles continues to grow, the pool of potential citations for each article becomes larger. With more research being conducted and published, it becomes increasingly challenging for any single article to stand out and receive a high number of citations. Moreover, over time, researchers have become increasingly specialized in their fields, which may lead to a narrower focus and a decrease in the number of articles they cite. This increased specialization can result in smaller, more focused citation networks. It's important to note that while the number of citations per article may decline, it does not necessarily indicate a decrease in the quality or impact of the research. It may simply be a reflection of the changing landscape of scholarly communication and the growth of the research community.

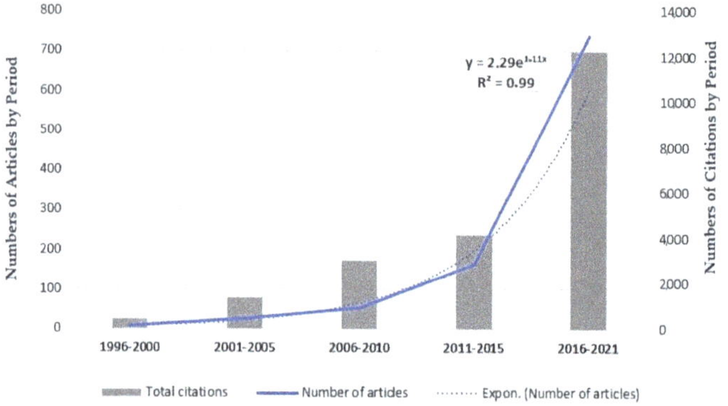

Figure 2. Evolution of the number of articles, total citations, and exponential variation between periods. The dot line refers to the calculated exponential curve fitting the data (number of articles).

The observed increase is noteworthy in relation to various bibliometric indicators, including the number of authors. In the initial period (1996–2000), there were 22 authors, with an average of 3 authors per article. However, in the most recent six-year period, the number of authors has increased 143 times (Table 2). It is worth noting that despite this significant growth, there has been no significant variation in the average number of authors per article (4 authors per article).

It can be noted that in the first five years, articles from only five countries were submitted, but that number has increased more than 13 times in the last six years.

On the other hand, the documents were published in 250 distinct sources. In the initial period, the articles were published in a total of five journals. However, during the subsequent period from 2011 to 2015, there was a substantial increase in the number of journals, reaching a total of 83. Over the course of the last six years, this number further rose to 209 journals.

Most of the bibliometric indicators mentioned above show a very considerable increase in the relevance of this field of study over the previous ten years, further highlighting the strength of the current trend in this line of research.

3.2. *Analysis of Scientific Production*

3.2.1. Subject Area and Journals

Over a span of 26 years (1996–2021), a comprehensive analysis was conducted on 985 research articles that investigated the impact of agronomic practices on soil health in agricultural soils. These articles were classified into 22 distinct subject areas as per the Scopus database. Figure 3 shows that 51.9% of research articles were placed in the subject area of Agricultural and Biological Sciences, while 21.3% were classified in the Environmental Science category. Afterwards, the most important categories are Earth and Planetary Sciences (6.1%), Immunology and Microbiology (4.7%), and Biochemistry, Genetics, and Molecular Biology (3.5%).

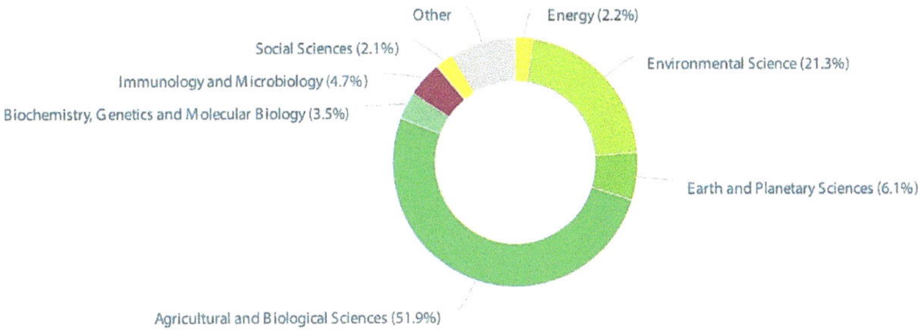

Figure 3. Donut chart of relative publication per Scopus subject Area. Source: SciVal.

Therefore, across the time period studied (1996–2021), Figure 4 illustrates the evolution of the seven key subject areas as Scopus links articles on the research topic. Only four thematic areas (Agricultural and Biological Sciences, Environmental Science, Earth and Planetary Sciences, and Immunology and Microbiology) had articles published during the whole 26-year period under consideration. Agriculture, Biology, and Environmental aspects were believed to be the most pertinent in the analysis of the effect of soil management on soil health, although Microbiology, Biochemistry, Genetic and Molecular biology cannot be overlooked in soil health.

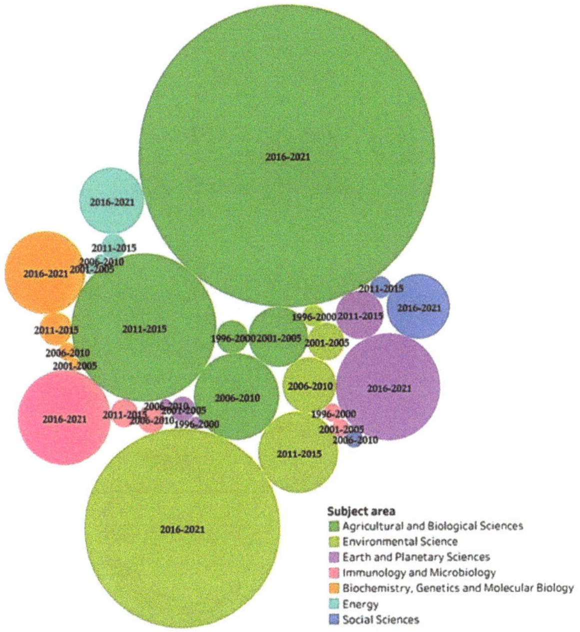

Figure 4. Packed bubbles of the growth trend of the main Scopus subject Area per period. Source: Tableau.

Table 3 shows the most 20 prolific journals for the number of articles on the research topic. A total of 60% of these top 20 journals with the highest scientific production are found in the first quartile (Q1) of CiteScore index in 2022 with impact factor ranging from

2.4 (Archives of Agronomy and Soil science and Journal of Environmental Quality) to 9.7 (Soil Biology and Biochemistry). On the other hand, 55% of this group of journals are located in Europe (mostly Netherlands), 25% in North America (USA), 15% from Asia (India) and the last 5% come from Australia. Since 1996 until 2021, a total of 985 research articles have been published across 250 journals. Notably, the most 20 prolific journals receiving 416 articles, which accounts for 42% of the overall scientific production. During the time frame of our investigation, we identified two journals, Applied Soil Ecology and Soil and Tillage Research, which together accounted for 9.54% and 20.51% of all articles published and overall Total citations, respectively. Moreover, these two journals are considered among journals with high number of citations per article with 41.4 and 51.1 citations per article for Applied Soil Ecology and Soil and Tillage Research, respectively. Moreover, they have the highest H index for articles published on soil health with 24 and 27 for Applied Soil Ecology and Soil and Tillage Research, respectively. Therefore, during the last 6 years analysed (1996–2000), 307 articles on the research topic were published in the 20 most prolific journals, but the five articles on this topic were spread among 5 prolific journals over the first 5 years (1996–2000). In addition, as seen by the increase in research papers and the broad range of journals, impact of soil management on soil health has become an attractive subject for more journals and authors over time.

3.2.2. Most Productive Authors

Using bibliometric metric indicators such as number of publications, total citations, and average citations per article, Table 4 lists the 10 most productive authors on soil health in our research topic. The ten authors in question have published 101 articles and have been cited 2348 times, accounting for 10.25% and 11.09% of total publications and citations, respectively.

It is noteworthy that authors of Asian origin present higher productivity in the field of research, with India (70%) and Pakistan (10%), being particularly prominent. North America, specifically the United States, accounted for 20% of the represented authors.

The most productive author is Das, Anup, from ICAR Research Complex for NEH Region, with a total of 16 articles published during 14 years (2008–2021) and receiving a total of 255 citations, who has the highest H-index of 11, followed by Jat, Mangai Lal with a total of 12 research articles, who is also the author with the highest number of citations and the highest number of citations per articles, with a total of 778 and 64.83, respectively.

Finally, two Asian authors, Farooq, Muhammad A. from University of Agriculture, Faisalabad and Jat, Mangi Lal from International Maize & Wheat Improvement Center, showed a higher percentage of international co-authorship during the period analysed (1996–2021), with 87.5% and 50.0%, respectively (Table 5). The prominence of Asian authors among the top 10 most relevant contributors is evident, a trend that is likely reflected in the numbers and supported by bibliometric analysis. This evidence highlights the importance of soil health research, especially in Indian Research Institutes. However, it is crucial to maintain perspective regarding their publications, primarily confined to local journals and limited international impact. This serves as a reminder that while numerical metrics highlight certain patterns, considering the qualitative aspects of influence is equally vital.

3.2.3. Most Productive Countries and Affiliations

Only 37.44% of the nations across the globe (73 countries) have contributed to this particular field of research. Figure 5 shows that 48.34% of research articles are found in Asia (642 publications), followed by North America (338 publications, 25.45%), and Europe (174 publications, 13.10%). These three continents together represent 87% of papers published in the soil health research field. In contrast, our bibliometric analysis reveals that Oceania, Africa and South America exhibit the lowest number of publications with 75, 51, and 48 articles, respectively.

Table 3. Top 20 prolific journals (1996–2021).

Journal	A	TC	TC/A	Hi (A)	Hi (J)	IF (J)	CiteScore 2022	C	FA	LA	R (A)				
											1996–2000	2001–2005	2006–2010	2011–2015	2016–2021
APPLIED SOIL ECOLOGY	48	1988	41.4	24	136	4.8	8.7 (Q1)	Netherlands	1999	2021	1 (2)	1 (3)	1 (8)	4 (7)	3 (28)
SOIL AND TILLAGE RESEARCH	46	2352	51.1	27	162	6.5	12.7 (Q1)	Netherlands	1998	2021	5 (1)	3 (2)	36 (1)	3 (10)	1 (32)
COMMUNICATIONS IN SOIL SCIENCE AND PLANT ANALYSIS	35	463	13.2	11	75	1.8	3.0 (Q2)	USA	2005	2021	-	10 (1)	-	1 (14)	7 (20)
INDIAN JOURNAL OF AGRICULTURAL SCIENCES	32	168	5.3	8	30	0.4	0.9 (Q4)	India	2004	2021	-	2 (2)	6 (2)	5 (7)	6 (21)
AGRONOMY	30	270	9.0	11	67	3.7	5.2 (Q1)	Switzerland	2019	2021	-	-	-	-	2 (30)
SOIL SCIENCE SOCIETY OF AMERICA JOURNAL	26	845	32.5	13	184	2.9	4.9 (Q2)	USA	2005	2021	-	21 (1)	35 (1)	77 (1)	5 (23)
AGRONOMY JOURNAL	23	186	8.1	7	145	2.1	4.3 (Q2)	USA	2017	2021	-	-	-	-	4 (23)
INDIAN JOURNAL OF AGRONOMY	18	81	4.5	5	25	0.036	0.5 (Q4)	India	2008	2019	-	-	7 (2)	2 (10)	31 (6)
GEODERMA	17	766	45.1	14	190	6.1	12.9 (Q1)	Netherlands	2004	2021	-	13 (1)	-	-	8 (16)
AGRICULTURE, ECOSYSTEMS AND ENVIRONMENT	16	788	49.3	13	200	6.6	10.2 (Q1)	Netherlands	2000	2020	3 (1)	-	10 (1)	8 (3)	12 (11)
ARCHIVES OF AGRONOMY AND SOIL SCIENCE	16	245	15.3	9	49	2.4	5.5 (Q1)	United Kingdom	1998	2021	4 (1)	5 (1)	12 (1)	9 (3)	14 (10)
SUSTAINABILITY (SWITZERLAND)	15	129	8.6	7	136	3.9	5.8 (Q1)	Switzerland	2018	2021	-	-	-	-	9 (15)
FRONTIERS IN MICROBIOLOGY	13	437	33.6	9	201	5.2	7.8 (Q1)	Switzerland	2016	2021	-	-	-	-	10 (13)
SCIENTIFIC REPORTS	13	428	32,9	12	282	4.6	7.5 (Q1)	United Kingdom	2016	2021	-	-	-	-	11 (13)
JOURNAL OF ENVIRONMENTAL QUALITY	12	116	9.7	6	183	2.4	6.6 (Q1)	USA	2012	2021	-	-	-	63 (1)	13 (11)
SOIL BIOLOGY AND BIOCHEMISTRY	12	588	49.0	11	250	9.7	14.3 (Q1)	United Kingdom	2004	2021	-	20 (1)	34 (1)	31 (2)	21 (8)
SOIL RESEARCH	11	142	12.9	5	92	1.6	3.6 (Q2)	Australia	2014	2021	-	-	-	32 (2)	17 (9)
EUROPEAN JOURNAL OF SOIL BIOLOGY	11	422	38.4	8	84	4.2	5.9 (Q1)	France	2008	2021	-	-	4 (2)	18 (2)	24 (7)
HORTSCIENCE	11	167	15.2	7	100	1.9	3.2 (Q2)	USA	2007	2020	-	-	5 (2)	6 (5)	48 (4)
JOURNAL OF ENVIRONMENTAL BIOLOGY	11	61	5.5	5	57	0.7	1.4 (Q3)	India	2013	2021	-	-	-	7 (4)	25 (7)

(A): number of articles; (TC): number of citations; (TC/A): number of citations per article; Hi (A): h-index of articles; Hi (J): h-index of journal; (C): Country; (FA): First article; (LA): Last article; (Q): Quartile; (R): Rank; IF (J): Impact factor (IF) produced by InCites Journal of Citation Reports (www.clarivate.com) on 6 July 2023.

Table 4. Top 10 most relevant authors on soil health from 1996 to 2021.

Authors	A	TC	TC/A	Institution	C	FA	LA	H Index
Das, Anup	16	255	15.94	ICAR Research Complex for NEH Region	India	2008	2021	11
Jat, Mangi Lal	12	778	64.83	International Maize & Wheat Improvement Center (CIMMYT)	India	2004	2020	10
Babu, Subhash	11	136	12.36	ICAR Research Complex for NEH Region	India	2013	2021	8
Sainju, Upendra M.	11	117	10.64	United States Department of Agriculture (USDA)	USA	2018	2021	5
Ghimire, Rajan P.	10	131	13.10	New Mexico State University	USA	2019	2021	5
Yadav, Gulab Singh	9	168	18.67	ICAR Research Complex for NEH Region	India	2013	2021	7
Dwivedi, Brahma S.	8	252	31.50	ICAR—Indian Agricultural Research Institute	India	2003	2020	7
Farooq, Muhammad A.	8	135	16.88	University of Agriculture, Faisalabad	Pakistan	2017	2021	5
Singh, Vijendra K.	8	242	30.25	Central Research Institute for Dryland Agriculture India	India	2003	2021	8
Kumar, Sandeep	8	134	16.75	Indian Institute of Technology Guwahati	India	2018	2021	6

(A): number of articles; (TC): number of citations; (TC/A): number of citations per article; (C): Country; (FA): First article; (LA): Last article.

Table 5. International co-authorship of top 10 most relevant authors on soil health from 1996 to 2021.

Authors	C	IC	IC (%)	Cited Publications
Das, Anup	India	3	18.8	16
Jat, Mangi Lal	India	6	50.0	12
Babu, Subhash	India	1	9.1	11
Sainju, Upendra M.	USA	3	27.3	8
Ghimire, Rajan P.	USA	1	10.0	8
Yadav, Gulab Singh	India	2	22.2	9
Dwivedi, Brahma S.	India	1	12.5	8
Farooq, Muhammad A.	Pakistan	7	87.5	6
Singh, Vijendra K.	India	2	25.0	8
Kumar, Sandeep	India	1	12.5	8

(C): Country; (IC): International collaboration.

There is a growing recognition of the importance of soil health in policy and research arenas, and many countries and organizations have adopted soil health targets and strategies. For example, in Asia, soil degradation and nutrient depletion are major concerns due to intensive agriculture, population pressure, and climate change. The Food and Agriculture Organization (FAO) has identified several hotspots of soil degradation in Asia, including the Indus–Ganges basin, The Mekong Delta, and the Yangtze River Basin [27,28]. To address these issues, governments and research institutions in the region are investing in soil conservation and management practices, such as no-till farming, crop rotation, and agroforestry [29–31]. On the other hand, in North America, soil health has become a major focus for sustainable agriculture and conservation efforts in recent years. The USDA's Natural Resources Conservation Service (NRCS) has launched a Soil Health Partnership to promote the adoption of soil health practices, such as cover cropping, reduced tillage, and nutrient management [32]. There is also a growing body of research on the impacts of soil health on crop yields, carbon sequestration, and ecosystem services.

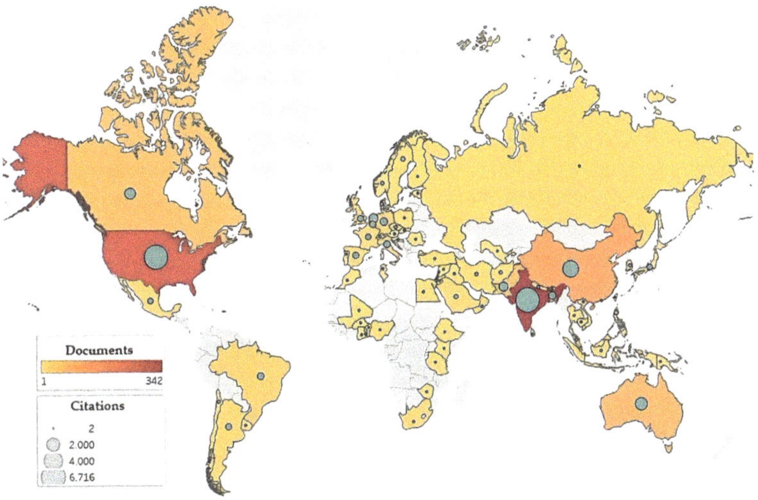

Figure 5. World map displaying the distribution of publications and total citations by country. Source: Tableau.

In Europe, soil health is a key priority under the EU's Common Agricultural Policy (CAP) and the European Green Deal. The EU has set targets for improving soil organic matter content, reducing erosion, and promoting biodiversity in agricultural landscapes. There are also several research networks and initiatives focused on soil health, such as the Soil Care project (https://www.soilcare-project.eu/, accessed on 20 July 2023) and the European Soil Partnership (https://www.europeansoilpartnership.org/, accessed on 20 July 2023).

Table 6 shows the most 10 prolific countries in terms of number of publications, total citations, and number of citations per article. India has the highest number of articles published and cited on the research topic, with a total of 342 articles and 6716 citations. The United States follows closely with 279 articles and 6506 citations, while China ranks third with 114 articles and 3050 citations. Additionally, it is noteworthy that India, the United States, and China possess the highest H-index values in relation to articles published on soil health, with respective scores of 41, 40, and 31. This observation suggests that these three countries play a prominent role in driving scientific research related to soil health. These three countries collectively accounted for 55% of the total published articles.

On the other hand, it is worth noting that Canada and Bangladesh are ranked first in terms of the number of citations per article, with a rate of 27.3. However, during each of the time periods analysed, India, the USA, and Australia consistently published scientific articles. Since the third 5 year period (2006–2010) under consideration, both China and the United Kingdom have made contributions to this research field, but Bangladesh did not do so until after 2011.

Based on the data presented in Table 7, it can be observed that Germany exhibits the highest level of international collaboration, accounting for 92.59% of its articles (25 out of 27) being written in collaboration. The United Kingdom follows closely with 81.82% (18 articles), while Bangladesh demonstrates a collaboration rate of 80% (20 articles). Brazil and Australia also engage in international collaboration, with rates of 72% (18 articles) and 60.81% (45 articles), respectively.

Table 6. Top 10 most productive countries on soil health from 1996 to 2021.

Country	A	TC	TC/A	H Index	R (A)				
					1996–2000	2001–2005	2006–2010	2011–2015	2016–2021
India	342	6716	19.6	41	5 (1)	1 (11)	1 (26)	1 (92)	2 (212)
United States	279	6506	23.3	40	1 (3)	2 (6)	2 (10)	2 (21)	1 (239)
China	114	3050	26.8	31	-	-	3 (5)	7 (5)	3 (104)
Australia	74	1747	23.6	24	2 (2)	3 (2)	5 (3)	3 (13)	4 (54)
Pakistan	60	955	15.9	17	-	11 (1)	14 (1)	5 (8)	5 (50)
Canada	52	1422	27.3	19	-	4 (2)	6 (3)	4 (8)	6 (39)
Germany	27	647	24.0	13	-	5 (2)	-	8 (4)	8 (21)
Bangladesh	25	682	27.3	15	-	-	-	6 (5)	9 (20)
Brazil	25	488	19.5	13	-	6 (1)	10 (1)	13 (2)	7 (21)
United Kingdom	22	450	20.5	13	-	-	16 (1)	12 (3)	11 (18)

(A): number of articles; (TC): number of citations; (TC/A): number of citations per article; R: Rank.

Table 7. Top 10 most productive countries and international collaboration from 1996 to 2021.

Country	NC	Main Collaborators	IC(%)	TC/A	
				IC	NIC
India	24	United States, Australia, China, Canada, Germany	14.91	34.5	17.1
United States	34	India, Pakistan, China, Australia, Canada	26.88	28.8	21.3
China	32	India, United States, Pakistan, Australia, Canada	53.51	29.0	24.1
Australia	26	United States, India, China, Pakistan, Canada	60.81	25.6	20.6
Pakistan	19	United States, China, Australia, Canda, Germany	53.33	22.0	8.9
Canada	10	United States, India, China, Australia, Pakistan	40.38	23.3	30.1
Germany	28	United States, India, China, Australia, Pakistan	92.59	22.7	39.5
Brazil	15	United States, India, Australia, Canda, Germany	72.00	19.8	18.7
Bangladesh	15	India, China, Australia, United Kingdom, Netherlands	80.00	25.9	33.0
United Kingdom	19	United States, India, China, Australia, Pakistan	81.82	23.6	6.5

NC = number of collaborations; IC(%) = percentage of articles made with international collaboration; TC/A: number of citations per article; IC: International collaboration; NIC: no international collaboration.

At only 14.91%, India has the lowest percentage of international cooperation. It should be noted that, with the exception of Canada, Germany, and Bangladesh, all countries included in the list of the most 10 productive countries exhibit a higher number of citations for articles produced through international collaboration compared to those produced without collaboration. Scientific collaboration is a reaction to the growing professionalization of science as noted by Beaver and Rosen [33]. As a result, international co-authored articles receive more citations than domestically co-authored articles because they receive more citations overall [34,35].

Figure 6 illustrates the international cooperation network among major countries, which is established through co-authorship analysis. According to Figure 6, the number of countries engaged in international collaboration related to the research topic is limited to 70. The countries have been grouped into seven clusters based on their specific fields of collaboration. These nations are categorized into seven clusters based on their areas of cooperation. Eighteen nations made up the first cluster (red), which was led by India. This cluster, which also included two of the top ten prolific countries namely Pakistan and Bangladesh. There was a total of 38% published articles coming from this cluster. The second cluster (green) is led by South Africa and includes 15 countries with 80 articles, which represents 6% of the total number of published articles, and included countries such as France, Belgium, and Mexico. The third cluster in blue, with a total of nine countries and it is led by Germany, with 86 articles (6.5% of the total articles). This group included Spain, Argentina, and Denmark. The fourth cluster (yellow), is led by Brazil and represent 4.97% of the total. This cluster included United Kingdom, in addition to Switzerland, Indonesia, Poland and Iraq. The fifth cluster (purple), is led by China with 250 articles (18.83% of the total articles). The fifth cluster included Australia and Canada. The sixth cluster in Sky blue, is led by Italy and include Netherlands, Russian Federation and Mali. This cluster represents 3.46% of published articles. Finally, the seventh cluster (Orange) is led by the United States and represents the 21.76% of published articles on research topic during the last 26 years (1996–2021). This cluster included Nigeria, Jordan, and Sri Lanka.

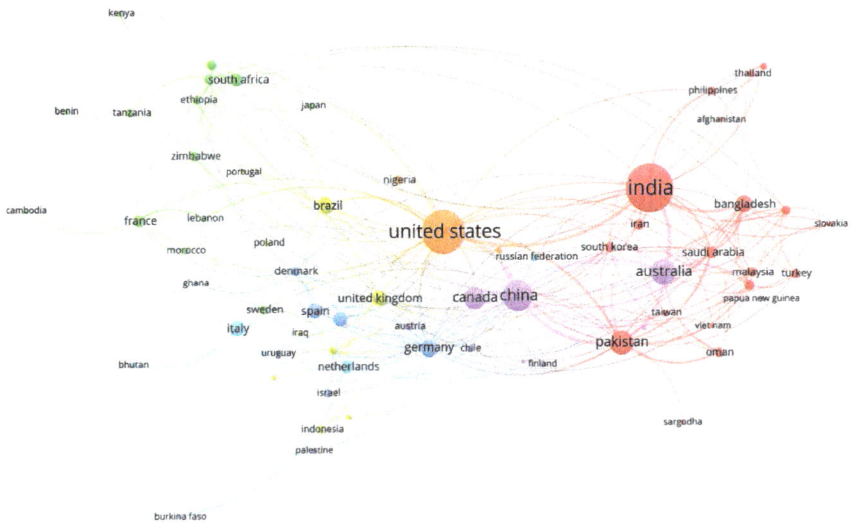

Figure 6. Country collaboration network visualization map. Colour: represents a cluster of country collaboration in the soil health research field; Nodes: represent countries (node size based on number of publications); Links: represent collaboration between two countries. Source: VOSviewer (clusters resolution 0.5; minimum cluster size 1 and no merge small clusters).

Our research topic on the impact of soil management on soil health under agricultural soils during the last 26 years (1996–2021) produced by 250 different affiliations. According the Table 8, the United States Department of Agriculture (USDA) form the United States is the most prolific institutions with 88 published articles and 2025 citations in this research topic. This institution shares the highest H-index of 24 with ICAR—Indian Agricultural Research Institute from New Delhi, India. In addition, 12.5% of articles from USDA were published with international co-authorship.

The nation with the greatest presence in this ranking is India, with five affiliations. Among the institutions considered, ICAR -Indian Agricultural Research Institute from New Delhi is the institution with the second highest number of articles of 71, which 19.7% from international co-authorship. In addition, this institution has the second most citations with 1612 total citations. The other four Indian institutions exhibit varying proportions of international co-authorship. For instance, Indian Council of Agricultural Research and ICAR—Indian Institute of Soil Science; Bhopal have published more than a third of their articles with international co-authorship. Conversely, ICAR—Research Complex for North Eastern Hill Region in Umiam and Punjab Agricultural University have the lowest percentage of international co-authorship, with 9.5 and 5.6%, respectively. The majority of articles having international co-authorship (more than 60%) are published by the Chinese Academy of Sciences and the University of Agriculture in Faisalabad, Pakistan.

Table 8. Top 10 most productive affiliations on soil health from 1996 to 2021.

Institution	C	A	TC	TC/A	H Index	IC(%)	TC/A	
							IC	NIC
United States Department of Agriculture (USDA)	USA	88	2025	23.01	24	12.5	21.1	23.3
ICAR—Indian Agricultural Research Institute, New Delhi	India	71	1612	22.70	24	19.7	41.0	18.2
Indian Council of Agricultural Research	India	60	1278	21.30	21	30.0	38.6	13.9
ICAR—Research Complex for North Eastern Hill Region; Umiam	India	31	379	12.23	12	9.7	33.0	10.0
Agriculture and Agri-Food Canada	Canada	29	558	19.24	14	27.6	24.3	17.3
University of Agriculture; Faisalabad	Pakistan	28	483	17.25	13	64.3	23.8	5.4
Chinese Academy of Sciences	China	22	593	26.95	11	68.2	31.1	18.0
Cornell University	USA	21	534	25.43	12	28.6	11.5	31.0
Punjab Agricultural University	India	18	261	14.50	7	5.6	2.0	15.2
ICAR—Indian Institute of Soil Science; Bhopal	India	18	579	32.17	12	33.3	30.3	33.1

(A): number of articles; (TC): number of citations; (C): Country; IC (%) = percentage of articles made with international collaboration; TC/A: number of citations per article; IC: International collaboration; NIC: no international collaboration.

It is worth mentioning that, apart from two institutions based in the United States and two institutions based in India, all affiliations listed among the top 10 most productive institutions demonstrate a greater number of citations per articles resulting from international collaboration relative to those produced without collaboration.

3.3. Keywords Co-Occurrence Network Analysis

A total of 985 articles containing 2460 author keywords were taken into account for this study. The 20 most frequently used author keywords in the 985 research articles related to soil health during the period 1996–2021 are shown in Table 8. We expected soil health to be one of the most prominent keywords on the list (240 articles, 24.4% of total articles), as it was one of the most frequently searched terms. In the first five years of the analysis (1996–2000), this keyword occurred in five articles. It peaked in the last six years (2016–2021), when 169 documents included this term. On the other hand, it is interesting that we additionally identified yield, soil organic carbon and soil enzyme activity as three soil health indicators among the top five keywords. The term "yield" ranked second in the list with 100 documents (10.2% of all articles). Therefore, the yield is considered as part of the soil ecosystem service and it is an important indicator of soil health, as it directly relates to the capacity of the soil to sustain agricultural production and provide food for human populations [36,37]. It occurred for the first time with one document in the second year period (2001–2005). However, it has increased to 44 papers since 2016. Moreover, both soil organic carbon and soil enzyme activity provide information about nutrient cycling, soil structure, water retention, carbon sequestration, and the overall biological activity in the soil [38–40].

Monitoring these indicators helps assess soil health, make informed management decisions, and promote sustainable agricultural practices. Soil organic carbon first appeared in 1996–2000 with one document, while soil enzyme activity first appeared in 2001–2005 with one document, but by the fifth analysed period (2016–2021), they had increased to 60 and 55 documents, respectively.

According to our analysis, the half of the top 20 author keywords is associated with soil management (Table 9). This significant contribution of these 10 soil managements highlights the scientific community's interest in analysing the impact of these soil management strategies on soil health. Our findings indicate that the term "manure" as cropping practice is ranked fourth in terms of relevance, with a total of 75 documents (7.6%) involving this keyword, and it appears for the first time with 1 document in the first 5-year period (1996–2000), until it reaches 43 papers since 2016.

Table 9. Main keywords on Soil health (1996–2021).

Keyword	1996–2021		1996–2000		2001–2005		2006–2010		2011–2015		2016–2021	
	A	(%)	A	(%)	A	(%)	A	(%)	A	(%)	A	(%)
soil health	240	24.4%	5	71.4%	5	19.2%	15	28.3%	46	28.4%	169	22.9%
yield	100	10.2%	0	0.0%	1	3.8%	7	13.2%	48	29.6%	44	6.0%
soil organic carbon	85	8.6%	1	14.3%	5	19.2%	3	5.7%	16	9.9%	60	8.1%
manure	75	7.6%	1	14.3%	1	3.8%	3	5.7%	27	16.7%	43	5.8%
soil enzyme activity	68	6.9%	0	0.0%	1	3.8%	6	11.3%	6	3.7%	55	7.5%
soil microbial biomass	63	6.4%	0	0.0%	4	15.4%	8	15.1%	11	6.8%	40	5.4%
conservation agriculture	60	6.1%	0	0.0%	1	3.8%	2	3.8%	9	5.6%	48	6.5%
sustainability	59	6.0%	2	28.6%	1	3.8%	5	9.4%	10	6.2%	41	5.6%
compost	56	5.7%	1	14.3%	2	7.7%	4	7.5%	11	6.8%	38	5.2%
cover crop	56	5.7%	1	14.3%	0	0,0%	1	1.9%	5	3.1%	49	6.6%
soil quality	56	5.7%	0	0.0%	2	7.7%	8	15.1%	12	7.4%	34	4.6%
no tillage	55	5.6%	0	0.0%	2	7.7%	1	1.9%	8	4.9%	44	6.0%
soil microbial community	54	5.5%	0	0.0%	0	0.0%	3	5.7%	2	1.2%	49	6.6%
fertilization	53	5.4%	2	28.6%	0	0.0%	4	7.5%	16	9.9%	31	4.2%
soil properties	52	5.3%	0	0.0%	1	3.8%	5	9.4%	8	4.9%	38	5.2%
crop residue management	51	5.2%	1	14.3%	2	7.7%	4	7.5%	9	5.6%	35	4.7%
tillage	50	5.1%	1	14.3%	0	0.0%	8	15.1%	12	7.4%	29	3.9%
integrated nutrient management	40	4.1%	0	0.0%	0	0.0%	1	1.9%	11	6.8%	28	3.8%
wheat	37	3.8%	0	0.0%	3	11.5%	4	7.5%	10	6.2%	20	2.7%
crop rotation	35	3.6%	1	14.3%	2	7.7%	6	11.3%	7	4.3%	19	2.6%

A: number of articles; %: Percentages of articles in which it appears.

Table 9 shows another interesting item. The time evolution of SH papers moved from an early emphasis (e.g., years 1996–2000) on farm management (e.g., fertilization, tillage, rotation) towards a more recent (years 2016–2021) biochemistry and microbiological approaches. This trend depicts the progress in soil biology towards estimating soil health.

Figure 7a shows the keywords co-occurrence network analysis of 100 most relevant keywords which appeared at least two times in 985 articles, which includes 98 nodes, 1055 links, and 2283 total link strength. Each node in the network represents a keyword, and the size of the node reflects the number of times the keyword appeared. According to Yang and Zhuang [41], the presence of keywords with higher occurrences within specific time periods suggests that the corresponding topics are of significant interest and focus during those periods. The network is organized into four clusters of keywords that share similar topics, with the red cluster relating to management strategies and soil health indicators. Because it is a well-developed and important theme with 57 keywords that focuses on various aspects of soil management and soil indicators, this cluster is known as the "motor themes". On the other hand, the green cluster is in second spot in terms of keyword density (26 keywords), and it includes topics linked to the crop yield and nutrient management. This cluster focuses spatially to enhancing crop productivity and soil health using different sustainable crop nutrition strategies such as green manure, compost, bio-fertilizer, and integrated nutrient management. This particular cluster is categorized as fundamental and pertinent across various research fields. The third cluster, represented by the colour blue and containing eleven keywords, deals with soil health and biological activity. This cluster examines the effects of wastewater irrigation and heavy metal contamination on soil biological activity. This group reflects an isolated theme with limited relevance to our research topic. Finally, the yellow cluster can be identified by means of four specific keywords. This particular cluster exhibits a correlation between soil health and greenhouse gas emissions. This cluster was viewed as a marginal and underdeveloped subject.

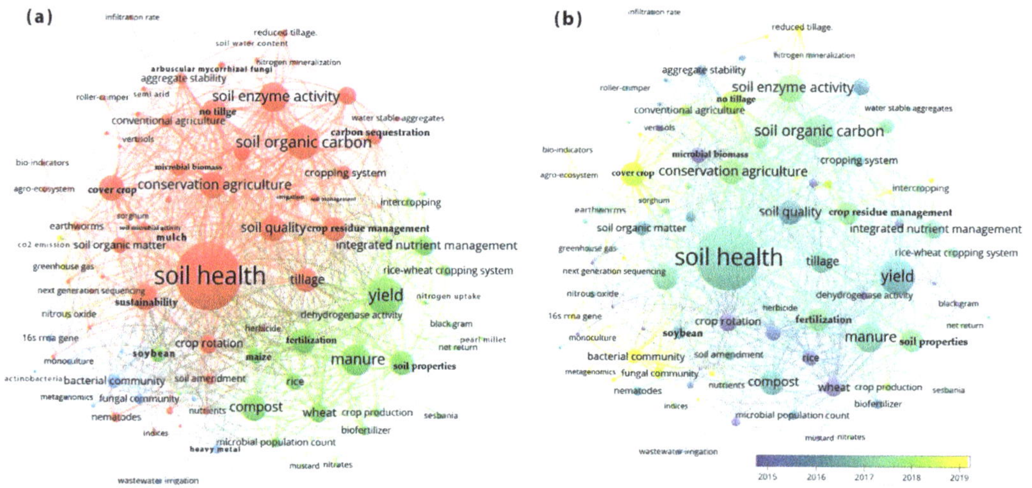

Figure 7. (**a**) Co-occurrence network visualization of 100 most relevant keywords which appeared at least 2 times in 985 articles. Colour: represents a cluster of keywords (please see the text for more details); Nodes: represent keywords (node size based on number of occurrences); (**b**) Evolution of the authors' keywords on research topic from 2015 to 2019. Source: VOSviewer (clusters resolution 0.8; minimum cluster size 1 and no merge small clusters).

From the time frame 1996–2021 here in Figure 7b, we report the focus on the evolution of the authors' keywords in the co-occurrence network between 2015 to 2019. Our analysis revealed that keywords related to the impact of soil management such as crop rotation, irrigation, and heavy metal contamination on agricultural soils appeared earlier (before 2015), while keywords related to the impact of cropping practices such as reduced tillage on biochemistry and microbiological soil activities appeared later (in 2019).

The keywords co-occurrence analysis (1996–2021), examining the interplay between agronomic practices and soil health, reveals a landscape rich with insights yet dotted with significant research gaps. The overall scenario (e.g., Figure 7) depicts a strong fragmentation of approaches with many separated analyses about the interaction between soil health and management practices. Basically, a plethora of diverse approaches with limited to almost none attempt of synthesis. In more detail, while the literature showcases strides in understanding short-term impacts of individual practices, it falls short in assessing their long-term implications. A notable opportunity lies in unravelling the effects of holistic approaches that combine various practices, as well as in exploring understudied crops and regions. Amid the intricate tapestry of soil health, microbial interactions emerge as a focal point deserving deeper exploration, promising novel pathways to enhance agricultural sustainability. There is an evident lack on the evaluation of ecosystem services which are becoming crucial in soil health policy across the globe (e.g., EU, USA).

Moreover, the analysis hints at the need to broaden the scope beyond scientific aspects, venturing into economic feasibility, social acceptance, and adaptation challenges. As climate change looms large, the resilience of soil health warrants a spotlight, urging researchers to investigate practices that fortify soil against evolving climatic pressures. The synthesis of the existing literature through meta-analyses is a missing piece that could provide a comprehensive overview, complementing field-scale studies that bridge the gap between controlled experiments and real-world complexities. Collaborative efforts across disciplines hold the potential to yield more holistic insights. Embracing technological and scientific advancements, such as precision agriculture, remote sensing and spatially explicit decision support systems, could further propel the understanding of agronomic impacts on soil health. In conclusion, these research gaps present an invigorating pathway forward, beckoning scientists, practitioners, and policymakers to collectively foster a more resilient and sustainable agricultural future.

4. Conclusions

This bibliometric analysis aims to provide a comprehensive review of the research topic concerning the effects of agronomic practices on soil health. This topic has gained significant importance in both political and scientific areas since the beginning of the second millennium. Based on our data, scholarly production on this research topic has increased significantly over the previous 26 years, with a spike occurring after 2013. These findings supported those of Liu et al. [42], who discovered that since 2013, researchers have become increasingly conscious of the significance of soil health research, and the number of publications published has significantly grown. According to our bibliometric analysis, it can be determined that the journal Applied Soil Ecology and Soil and Tillage Research exhibited the highest level of productivity. The majority of research institutions that investigate our chosen research topic are situated in the United States and India. These nations are consistently ranked among of the most international cooperative in their respective fields. However, when it comes to research institutions, the United States has not achieved the desired level of bilateral cooperation, unlike certain Indian research institutions where international institutional cooperation reached 41 percent. According to the keywords co-occurrence analysis, our research topic is gradually shifting from the effects of soil management strategies, such as nutrient management, on soil indicators, particularly soil organic carbon and yield to the study of the effects of cropping practices on soil biology and biochemistry and greenhouse gas emissions. The last two themes only emerged in the last decade and were seen to be marginal and underdeveloped themes; however, they might provide a promising topic for further study. Overall, our analysis depicts a very large number of soil health research work lacking more integrated and holistic approaches especially in view of analysing the connection between soil health and soil-based ecosystem services.

This study has one limitation, which is that our analysis included only articles from Scopus, and therefore our research cannot cover the entire literature on our research topic. However, the data presented in this study still hold significant potential for understanding the evolving patterns before and after the increase in research on the topic. Finally, this study emphasizes the importance of incorporating additional novel approaches, such as systematic reviews, in order to acquire a more comprehensive understanding of the effects of soil management practices on soil health, and to establish a framework for future research.

Author Contributions: Conceptualization, M.H.S. and F.T.; methodology, M.H.S. and F.T; software, M.H.S.; investigation, M.H.S.; data curation, M.H.S.; writing—original draft preparation, M.H.S. and F.T.; writing—review and editing, M.H.S. and F.T.; visualization, M.H.S.; supervision, F.T. All authors have read and agreed to the published version of the manuscript.

Funding: This research was funded by European Union's Horizon Europe research and innovation program, under Grant Agreement: 101091010, Project BENCHMARKS.

Institutional Review Board Statement: Not applicable.

Informed Consent Statement: Not applicable.

Data Availability Statement: Not applicable.

Conflicts of Interest: The authors declare no conflict of interest.

References

1. Veerman, C.; Pinto, C.; Bastioli, C.; Biro, B.; Bouma, J.; Cienciala, E.; Emmett, B.; Frison, E.A.; Grand, A.; Wittkowski, R.; et al. *Caring for Soil is Caring for Life—Ensure 75% of Soils Are Healthy by 2030 for Food, People, Nature and Climate, Independent Expert Report*; European Commission Publications Office of the European Union: Luxembourg, 2020.
2. Dro, C.; Kapfinger, K.; Rakic, R. *European Missions: Delivering on Europe's Strategic Priorities*; R&I Paper Series; Policy Brief EU-DG Science and Innovation: Brussels, Belgium, 2022.
3. Harris, J.A.; Evans, D.L.; Mooney, S.J. A new theory for soil health. *Eur. J. Soil Sci.* **2022**, *73*, e13292. [CrossRef]
4. Bouma, J. Feeding existing expertise into a possibly 'new' theory for soil health? *Eur. J. Soil Sci.* **2022**, *73*, e13312. [CrossRef]
5. Janzen, H.H.; David, W.J.; Edward, G.G. The 'Soil Health' Metaphor: Illuminating or Illusory? *Soil Biol. Biochem.* **2021**, *159*, 108167. [CrossRef]
6. Doran, J.W.; Zeiss, M.R. Soil health and sustainability: Managing the biotic component of soil quality. *Appl. Soil Ecol.* **2000**, *15*, 3–11. [CrossRef]
7. Rinot, O.; Levy, G.J.; Steinberger, Y.; Svoray, T.; Eshel, G. Soil health assessment: A critical review of current methodologies and a proposed new approach. *Sci. Total Environ.* **2019**, *648*, 1484–1491. [CrossRef]
8. Bünemann, E.K.; Bongiorno, G.; Bai, Z.; Creamer, R.E.; De Deyn, G.; de Goede, R.; Fleskens, L.; Geissen, V.; Kuyper, T.W.; Mäder, P.; et al. Soil Quality—A Critical Review. *Soil Biol. Biochem.* **2018**, *120*, 105–125. [CrossRef]
9. Young, M.D.; Ros, G.H.; de Vries, W. Impacts of Agronomic Measures on Crop, Soil, and Environmental Indicators: A Review and Synthesis of Meta-Analysis. *Agric. Ecosyst. Environ.* **2021**, *319*, 107551. [CrossRef]
10. Yang, T.; Siddique, K.H.; Liu, K. Cropping systems in agriculture and their impact on soil health—A review. *Glob. Ecol. Conserv.* **2020**, *23*, e01118. [CrossRef]
11. Chahal, I.; Hooker, D.C.; Deen, B.; Janovicek, K.; Van Eerd, L.L. Long-term effects of crop rotation, tillage, and fertilizer nitrogen on soil health indicators and crop productivity in a temperate climate. *Soil Tillage Res.* **2021**, *213*, 105121. [CrossRef]
12. Cooper, R.J.; Hama-Aziz, Z.Q.; Hiscock, K.M.; Lovett, A.A.; Vrain, E.; Dugdale, S.J.; Sünnenberg, G.; Dockerty, T.; Hovesen, P.; Noble, L. Conservation tillage and soil health: Lessons from a 5-year UK farm trial (2013–2018). *Soil Tillage Res.* **2020**, *202*, 104648. [CrossRef]
13. Mitchell, J.P.; Shrestha, A.; Mathesius, K.; Scow, K.M.; Southard, R.J.; Haney, R.L.; Schmidt, R.; Munk, D.S.; Horwath, W.R. Cover cropping and no-tillage improve soil health in an arid irrigated cropping system in California's San Joaquin Valley, USA. *Soil Tillage Res.* **2017**, *165*, 325–335. [CrossRef]
14. Singh, A.; Pandey, A.K.; Singh, U. Effect of Integrated Nutrient Management Practices on Soil Chemical Properties: A Review. *Int. J. Chem. Stud.* **2020**, *8*, 3438–3441. [CrossRef]
15. Jian, J.; Du, X.; Stewart, R.D. A database for global soil health assessment. *Sci. Data* **2020**, *7*, 13. [CrossRef] [PubMed]
16. Moebius-Clune, B.N.; Moebius-Clune, D.J.; Gugino, B.K.; Idowu, O.J.; Schindelbeck, R.R.; Ristow, A.J.; van Es, H.M.; Thies, J.E.; Shayler, H.A.; McBride, M.B.; et al. *Comprehensive Assessment of Soil Health—The Cornell Framework*, 3rd ed.; Cornell University: Geneva, NY, USA, 2016; p. 134.
17. Norris, C.E.; MacBean, G.; Cappellazi, S.B.; Cope, M.; Greub, K.L.H.; Liptzin, D.; Rieke, E.L.; Tracy, P.W.; Morgan, C.L.S.; Honeycutt, C. Introducing the North American project to evaluate soil health measurements. *Agron. J.* **2020**, *112*, 3195–3215. [CrossRef]
18. NRCS-USDA. *Soil Health*; National Resources Conservation Services of the US Department of Agriculture: Washington, DC, USA, 2019. Available online: https://www.nrcs.usda.gov/wps/portal/nrcs/main/soils/health (accessed on 26 July 2023).
19. Creamer, R.E.; Barel, J.M.; Bongiorno, G.; Zwetsloot, M.J. The life of soils-integrating the who and how of multifunctionality. *Soil Biol. Biochem.* **2022**, *166*, 108561. [CrossRef]
20. Lehmann, J.; Bossio, D.A.; Kögel-Knabner, I.; Rillig, M.C. The concept and future prospects of soil health. *Nat. Rev. Earth Environ.* **2020**, *1*, 544–553. [CrossRef] [PubMed]
21. Donthu, N.; Satish, K.; Debmalya, M.; Nitesh, P.; Weng, M.L. How to Conduct a Bibliometric Analysis: An Overview and Guidelines. *J. Bus. Res.* **2021**, *133*, 285–296. [CrossRef]
22. Pullin, A.S.; Frampton, G.; Livoreil, B.; Petrokofsky, G. Guidelines and Standards for Evidence Synthesis in Environmental Management: Version 5.1. 2022. Available online: https://environmentalevidence.org/information-for-authors/ (accessed on 7 July 2023).
23. Vardell, E.; Feddern-Bekcan, T.; Moore, M. SciVal experts: A collaborative tool. *Med. Ref. Serv. Q.* **2011**, *30*, 283–294. [CrossRef]
24. Dresbeck, R. SciVal. *J. Med. Libr. Assoc.* **2015**, *103*, 164–166. [CrossRef]
25. Van Eck, N.J.; Waltman, L. How to normalize co-occurrence data? An analysis of some well-known similarity measures. *J. Am. Soc. Inf. Sci.* **2009**, *60*, 1635–1651.
26. United Nations. Millennium Development Goals Report 2010. 2010. Available online: https://www.un.org/en/development/desa/publications/millennium-development-goals-report-2010.html (accessed on 7 July 2023).
27. FAO; ITPS. *Status of the World's Soil Resources (SWSR)—Main Report*; Food and Agriculture Organization of the United Nations and Intergovernmental Technical Panel on Soils: Rome, Italy, 2015.
28. FAO. *The State of the World's Land and Water Resources for Food and Agriculture 2021—Systems at Breaking Point*; FAO: Rome, Italy, 2022. [CrossRef]

29. Naresh, R.K.; Singh, S.P.; Dwivedi, A.; Kumar, P.; Kumar, L.; Singh, V.; Kumar, V.; Gupta, R.K. Soil Conservation Practices for Sustainability of Rice-Wheat Systemin Subtropical Climatic Conditions: A Review. *Int. J. Pure App. Biosci.* **2016**, *4*, 133–165. [CrossRef]
30. Bera, T.; Sharma, S.; Thind, H.S.; Sidhu, H.S.; Jat, M.L. Soil biochemical changes at different wheat growth stages in response to conservation agriculture practices in a rice-wheat system of north-western India. *Soil Res.* **2018**, *56*, 91–104. [CrossRef]
31. Jat, H.S.; Datta, A.; Sharma, P.C.; Kumar, V.; Yadav, A.K.; Choudhary, M.; Choudhary, V.; Gathala, M.K.; Sharma, D.K.; Jat, M.L.; et al. Assessing soil properties and nutrient availability under conservation agriculture practices in a reclaimed sodic soil in cereal-based systems of North-West India. *Arch. Agron. Soil Sci.* **2017**, *64*, 531–545. [CrossRef] [PubMed]
32. Karlen, D.L.; Nicholas, J.G.; Kristen, S.V.; Matt, A.Y. On-Farm Soil Health Evaluations: Challenges and Opportunities. *J. Soil Water Conserv.* **2017**, *72*, 26A–31A. [CrossRef]
33. Beaver, D.; Rosen, R. Studies in scientific collaboration Part III. Professionalization and the natural history of modern scientific co-authorship. *Scientometrics* **1979**, *1*, 231–245. [CrossRef]
34. Narin, F. Globalization of research, scholarly information, and patents–10-year trends. *Ser. Libr.* **1991**, *21*, 33–44. [CrossRef]
35. Persson, O.; Glänzel, W.; Danell, R. Inflationary bibliometric values: The role of scientific collaboration and the need for relative indicators in evaluative studies. *Scientometrics* **2004**, *60*, 421–432. [CrossRef]
36. Bommarco, R.; David, K.; Simon, G.P. Ecological Intensification: Harnessing Ecosystem Services for Food Security. *Trends Ecol. Evol.* **2013**, *28*, 230–238. [CrossRef]
37. Kremen, C. Ecological Intensification and Diversification Approaches to Maintain Biodiversity, Ecosystem Services and Food Production in a Changing World. *Emerg. Top Life Sci.* **2020**, *4*, 229–240. [CrossRef]
38. Adetunji, A.T.; Lewu, F.B.; Mulidzi, R.; Ncube, B. The Biological Activities of β-Glucosidase, Phosphatase and Urease as Soil Quality Indicators: A Review. *J. Soil Sci. Plant Nutr.* **2017**, *17*, 794–807. [CrossRef]
39. Dick, R.P. *Soil Enzyme Activities as Indicators of Soil Quality*; SSSA Special Publications: Madison, WI, USA, 2015; pp. 107–124. [CrossRef]
40. Ngatia, L.W.; Moriasi, D.; Grace, J.M., III; Fu, R.; Gardner, C.S.; Taylor, R.W. Land Use Change Affects Soil Organic Carbon: An Indicator of Soil Health. In *Environmental Health*; IntechOpen: London, UK, 2021. [CrossRef]
41. Yang, Q.; Zhuang, R. Mapping intellectual structures of administrative division research in recent 20 years. *World Reg. Stud.* **2018**, *27*, 141–150.
42. Liu, Y.; Wu, K.; Zhao, R. Bibliometric analysis of research on soil health from 1999 to 2018. *J. Soils Sediments* **2019**, *20*, 1513–1525. [CrossRef]

soil systems

MDPI

Article

Soil Carbon Stock and Indices in Sandy Soil Affected by Eucalyptus Harvest Residue Management in the South of Brazil

Jackson Freitas Brilhante de São José [1], Luciano Kayser Vargas [1,*], Bruno Britto Lisboa [1], Frederico Costa Beber Vieira [2], Josiléia Acordi Zanatta [3], Elias Frank Araujo [4] and Cimelio Bayer [5]

1 Department of Agricultural Research and Diagnosis, Secretariat of Agriculture, Livestock, Sustainable Production and Irrigation of Rio Grande do Sul. 570, Porto Alegre 90130-060, Brazil; jackson-jose@agricultura.rs.gov.br (J.F.B.d.S.J.); bruno-lisboa@agricultura.rs.gov.br (B.B.L.)
2 Soil and Forest Ecology Laboratory, Universidade Federal Do Pampa, São Gabriel 97307-020, Brazil; fredericovieira@unipampa.edu.br
3 Embrapa Florestas, Estrada da Ribeira, Colombo 83411-000, Brazil; josileia.zanatta@embrapa.br
4 CMPC, Celulose Riograndense, Rua São Geraldo, Guaíba 92703-470, Brazil; elias.araujo@cmpcrs.com.br
5 Department of Soil Science, Faculty of Agronomy, Universidade Federal do Rio Grande do Sul, Porto Alegre 91540-000, Brazil; cimelio.bayer@ufrgs.br
* Correspondence: luciano-kayser@agricultura.rs.gov.br

Abstract: There has been limited research on the effect of eucalyptus harvest residue management on soil organic carbon (SOC) in subtropical environments. This research evaluated the effect on soil C indices of the following eucalyptus harvest residue managements: AR, with all forest remnants left on the soil; NB, where bark was removed; NBr, in which branches were removed; NR, which removed all residues; and NRs, which is same as NR but also used a shade net to prevent the litter from the new plantation from reaching the soil surface. C stocks within the soil depths of 0–20 cm and 0–100 cm increased linearly with the C input from eucalyptus harvest residues. In the layer of 0–20 cm, the lowest soil C retention rate was 0.23 Mg ha^{-1} year^{-1}, in the NR treatment, while in the AR treatment, the retention rate was 0.68 Mg ha^{-1} year^{-1}. In the 0–100 cm layer, the highest C retention rate was obtained in the AR (1.47 Mg ha^{-1} year^{-1}). The residues showed a high humification coefficient (k_1 = 0.23) and a high soil organic matter decomposition rate (k_2 = 0.10). The carbon management index showed a close relationship with the C input and tree diameter at breast height.

Keywords: carbon management index; forest residues; soil health; plantation forestry; carbon sequestration

Citation: de São José, J.F.B.; Vargas, L.K.; Lisboa, B.B.; Vieira, F.C.B.; Zanatta, J.A.; Araujo, E.F.; Bayer, C. Soil Carbon Stock and Indices in Sandy Soil Affected by Eucalyptus Harvest Residue Management in the South of Brazil. *Soil Syst.* **2023**, *7*, 93. https://doi.org/10.3390/soilsystems7040093

Academic Editors: Antonella Lavini, Mohamed Houssemeddine Sellami and Nick B. Comerford

Received: 24 August 2023
Revised: 16 October 2023
Accepted: 19 October 2023
Published: 21 October 2023

1. Introduction

Forest ecosystems have a high potential for soil carbon sequestration, around 0.4 Pg C year^{-1} [1]. Among these ecosystems, eucalyptus plantations occupy roughly 25 Mha worldwide [2], of which 7.53 Mha are in Brazil [3]. In the next decade, this forest base should be expanded given Brazil's low carbon emission agriculture plan, which foresees an expansion of ~4 Mha by 2030 with planted forests, with the potential to mitigate 510 million Mg CO_2 eq of greenhouse gas emissions [4]. Much of this potential results from long-term carbon sequestration through increased soil organic matter (SOM) [5], which ensures the sustainability of forest production under subtropical conditions [6], especially in fragile soils such as sandy ones.

Forest companies have considered the removal of eucalyptus harvesting residues, given their possible use as an energy source [7]. Nevertheless, it is a practice that reduces nutrient availability [8,9], soil microbiota activity [10], soil quality [11], and forest productivity [11–14]. The impact of this practice on soil carbon stocks is still inconclusive. While certain research has indicated that the removal of eucalyptus residues can reduce soil carbon stocks [14–17], others have indicated that the impact is almost zero [18,19]. These differences may be related to climate [20], the number of rotations in which the residue is removed from the area [14], soil texture and mineralogy [10,21], and the quality of the residue contributing to the soil [22,23].

Traditionally, it has been believed that recalcitrant residues were more efficient in accumulating C in soil, and recent evidence has revealed that higher-quality residues (lower C/N and lignin/N ratio) can show higher efficiency in accumulating C in soil than low-quality residues (higher C/N and lignin/N ratio) [24]. However, this effect depends on the available mineral surface for carbon stabilization, so that in sandy soils, much of the OM will be stabilized in the form of particulate organic C (POC) originating from more recalcitrant plant inputs [25]. Most of the eucalyptus harvest residues consist of branches and bark and are considered more recalcitrant because of the high content of phenolic compounds, notably tannin, lignin, and polyphenols [26]. These components are important in soil C retention in eucalyptus areas [10,17], especially in sandy soils characterized by a limited capacity of stabilizing C in soil organic matter [27].

Despite the substantial volume of information from long-term experiments concerning the effect of eucalyptus harvest residue management on the stocks of soil organic carbon [15,18,19], few initiatives have sought to take advantage of these data to estimate, for management, the rates of humification and soil organic matter losses. In this context, using a monocompartmental model based on first-order kinetics allows the prediction of the forthcoming development of SOC stocks until they stabilize and the amount of carbon that should be added to maintain the initial stocks [28].

In short or medium periods, the total organic C content is not a sensitive indicator to assess the effect of agricultural or forestry managements in the soil. Therefore, researchers have sought to combine the lability of organic matter in this evaluation by estimating the carbon management index (CMI). This index aggregates a carbon stock index (CSI) and a carbon lability index (LI), making it a sensitive tool to evaluate the impact of management practices, which also have a strong relationship with biological, chemical, and physical soil attributes [29,30]. The CMI has been efficient in evaluating crops and forest systems [31–33].

Our initial hypothesis is that the maintenance of eucalyptus harvest residues increases the retention of atmospheric C in soil organic matter after the rotation cycle compared to their removal. Hence, the objectives of this research were to assess the impact of managing eucalyptus harvest residues on the soil organic C content and stock, to evaluate eucalypt harvest residue management using the carbon management index and to determine the relationship between CMI and eucalyptus growth in sandy soil in subtropical Brazil.

2. Materials and Methods

2.1. Field Experiment and Treatments

The experiment was carried out in Rio Grande do Sul, the southernmost state of Brazil, in the city of Barra do Ribeiro. The experimental area was located around the coordinates 30°23′ S and 51°07′ W, with an altitude of about 30 m above the sea level. The climate in the local area is defined as humid subtropical (Cfa) according to the Köppen classification, and the region has an annual precipitation average of approximately 1400 mm, with no dry season. The highest average monthly temperature never exceeds 25 °C, and the lowest average monthly temperature is about 14 °C with slight frosts. The soil is a Quartzipsament, with sandy texture, weak structure, low capacity for water storage [34], and low cation exchange capacity (Table S1). The experiment was implemented in 2010 using *Eucalyptus saligna* (clone 2864). Each plot measuring 30 × 30 m was planted with 100 trees arranged in a grid of 10 rows and 10 plants per row. For the measurements, we considered an inner subplot of 18 × 18 m, containing 6 rows × 6 plants each. The experimental setup followed a complete randomized block design with four replicates and five treatments. The treatments were five eucalyptus residue managements, described as follows:

1. AR: all forest residues remained on the soil (i.e., bark, branches, leaves, and the litter layer from the previous rotation), and only trunk wood was removed.
2. NB: where bark was also removed.

3. NBr: in which branches were also removed.
4. NR: in which all eucalyptus residues were removed.
5. NRs: same as NR, but a shade net was also used to prevent litter from the new plantation from reaching the soil surface.

2.2. Estimation of Carbon Input and Biochemical Composition Analysis of Eucalyptus Harvesting Residue Components

When the experiment was implemented, the branches, bark, and leaves from the previous rotation were collected and quantified in terms of their mass and ground. Based on information about accumulated litter production in experiments conducted by Celulose Riograndense with *Eucalyptus saligna* near the study region, the addition of litter accumulated until the sixth year of the current cultivation, was estimated using Equation (1), as adjusted by Witschoreck [35]:

$$\text{ABP (Mg ha}^{-1}) = 8.875950 + 0.100160 \text{ x (DBH x age)}, \quad (1)$$

where ABP is the accumulated litter production (Mg ha^{-1}), DBH is the average diameter at 1.30 m height obtained from trees, and age is the number of years since the eucalyptus plantation started. For C input by the eucalyptus roots, we considered the average data obtained by Londero et al. [36] at six years in *E. saligna* plantations in the nearby city of Guaíba, and an average C content of 37.84% from the data of Ribeiro et al. [37].

All components of the eucalyptus harvest residues were submitted to biochemical characterization. In the case of the litter from the current rotation, samples were collected from the NR treatment plots by placing a screen approximately one meter above the soil surface, and the litter was gathered about every two months. The C and N concentrations were determined by dry combustion in a Flash 2000 analyzer (Fisher Scientific Inc., Waltham, Massachusetts, EUA). Additionally, the lignin, cellulose, and hemicellulose contents of each component were determined following the conventional procedure described by Van Soest [38], in which the constituents of the plant tissue are separated via sequential filtration after heating with neutral and acid detergents. Hemicellulose is solubilized after the second filtration and is obtained by the difference between the two filtrates. The cellulose is calcined in a muffle, leaving the lignin.

2.3. Soil Organic C Stock and Fractionation

In July 2016, during the sixth year of the eucalyptus plantation, a 1 m deep and 1 m wide trench was opened per plot, and soil was collected on two opposite sides of each trench, totalizing eight samples per treatment in the following layers: 0–2.5, 2.5–5, 5–10, 10–30, 20–30, 30–50, 50–75, and 75–100 cm. As the spacing between plants was 3 × 3 m, and the trenches were opened at 0.5 m from a eucalyptus plant, one side of the trench configured the collection in the planting line, and the opposite side configured the collection between the planting lines. Equal amounts of soil from each sub-sample were pooled to form a composite sample representing the overall average of the spaces between rows and the planting row. The soil samples were oven-dried at 60 °C, grinded in a ball mill, and samples of ~50 g were removed. These subsamples were grinded again in a porcelain mortar and submitted to C analysis in a Shimadzu TOC-VCSH analyzer.

The bulk density in the same soil layer was evaluated by collecting metal rings [39]. Metal rings of 93.48 cm^3 were used in the 0–2.5 cm and 2.5–5 cm layers and 102.73 cm^3 for the other layers. The method of equivalent soil masses to the soil mass of the AR treatment was used to calculate soil organic C stocks [40]. Since the soil organic C contents of the experiment period in 2010 had been obtained using the wet digestion method (Walkley-Black), it was necessary to convert these contents to equivalent contents using the dry combustion method, which was the method used for the organic C analyses in this study. Thus, a set of 10 samples of this soil with a wide variation in organic C content was analyzed using the two methods, and a conversion coefficient of 0.66 was determined, which corresponds to the angular coefficient of the relationship between the C contents determined via the wet combustion method and the total C contents via the dry combustion method (Figure S1).

The soil samples from the layer of 0–20 cm were also physically fractionated. For that, 20 g of air-dried and sieved to <2 mm soil and 70 mL of sodium hexametaphosphate solution were arranged in 100-mL capacity vials and were shaken in horizontal position for 15 h (60 cycles min^{-1}). The supernatant was sieved on a 53-µm sieve, and the fraction <53 µm, consisting of mineral-associated organic C (MAC), was dried at 50 °C, quantified in terms of its mass, and the organic C content was analyzed in a Shimadzu TOC VCSH analyzer. Particulate organic C (POC) was estimated as the difference between total organic C (TOC) and MAC (<53 µm) [41].

2.4. Carbon Management Index

The CMI was calculated in accordance with Blair et al. [29], assuming that the POC was labile C in the soil and the MAC was the non-labile fraction [42]. From this, the CMI is obtained by multiplying the C stock index (CSI) and the C lability index (LI) as follows:

$$\mathrm{CMI} = \mathrm{CSI} \times \mathrm{LI} \times 100, \quad (2)$$

where CMI is the product of the CSI and LI. The CSI is obtained by the quotient between the stock of C in the treatment under evaluation and the stock of C in the soil of the reference treatment in the layer of 0–20 cm (Equation (3)). The LI (Equation (4)) is the quotient between the C lability (L) (Equation (5)) in the treatment under evaluation and the L in the reference treatment. NRs was considered to be the reference treatment for the estimation of CMI (CMI = 100).

$$\mathrm{CSI} = \text{C stock in treatment}/\text{C stock in reference treatment} \quad (3)$$

$$\mathrm{LI} = \text{L in treatment}/\text{L in reference treatment} \quad (4)$$

$$\mathrm{L} = \mathrm{POC}/\mathrm{CAM} \quad (5)$$

2.5. Annual Soil C Retention Rates

Annual soil C retention rates were calculated by subtracting the C stock in the treatment from C stock in the NRs treatment (Equation (6)), considering a period of six years.

$$\Delta\,\text{C soil (Mg ha}^{-1}\ \text{yr}^{-1}) = (\text{C soil treatment} - \text{C soil NRs})/(6\ \text{years}) \quad (6)$$

2.6. Estimate of the Organic Matter Humification (k_1) and Decomposition (k_2) Coefficients

The fraction of added C retained in soil organic matter (k_1), called the humification coefficient, was roughly calculated from the angular coefficient of the linear regression relating the amounts of C added annually to the annual rate of variation (dC/dt) in the soil organic C stock in the 0–20 cm layer.

From the values of the effective addition of C to the soil by the management of eucalyptus harvest residues (k_1A) and the stocks of organic C in the 0–20 cm soil layer, we estimated the annual rate of soil organic matter loss (k_2) using the equation $dC/dt = k_1A\text{-}k_2C$ in the condition dedC/dt = zero according to the process reported by Bayer et al. [43] and Vieira et al. [30]. In this condition, $k_1A = k_2C$ and $k_2 = k_1A/C$, where C represents the soil organic C stock in the initial condition (13.02 Mg ha^{-1}) and A represents the annual C addition rate required to maintain the initial soil organic C stock unchanged over time (i.e., dC/dt = zero).

2.7. Statistical Analysis

The effects of the treatments on soil organic C levels were evaluated using analysis of variance and the difference between the means of the treatments using the Tukey test at 5%.

Linear regression analyses were utilized to verify the relationship between C addition in the different managements of eucalyptus harvest residues and the stocks of total organic C and mineral-associated organic C (POC and MAC), as well as concerning CSI, LI, and CMI. Furthermore, linear regression analyses were employed to gain deeper insights into the relationship between CMI and tree diameter at breast height (DBH). On the soil sampling date, we assessed 36 trees from each plot to estimate these parameters.

3. Results

3.1. C Contribution and Composition of Eucalyptus Harvest Residues, Litter, and Roots

The C contribution to the soil varied according to the eucalyptus harvest residue management; it varied from 2.04 Mg ha^{-1} year^{-1} in the NRs treatment, in which all residues were removed, to 5.04 Mg ha^{-1} year^{-1} in the AR treatment, in which all residues but the trunk wood were kept (Table 1). Of these C contributions, leaves totaled 0.03 Mg ha^{-1} year^{-1}, branches accounted for 0.96 Mg ha^{-1} year^{-1}, bark 0.45 Mg ha^{-1} year^{-1}, the litter from the current rotation totaled 1.56 Mg ha^{-1} year^{-1}, and roots totaled 2.04 Mg ha^{-1} year^{-1}.

Table 1. Total biomass and C contribution to soil and biochemical composition of harvest residues from the previous eucalyptus rotation and litter from the current eucalyptus rotation.

Eucalyptus Residue Management	Litter [a]	Composition of Harvest Residues				
		Leaves	Branches	Bark	Root	Total
			Residue biomass (Mg ha^{-1} year^{-1})			
NRs	0.00	0.00	0.00	0.00	5.41 [b]	5.41
NR	3.05	0.00	0.00	0.00	5.41	8.46
NBr	3.05	0.07	0.00	1.10	5.41	9.62
NB	3.05	0.07	2.17	0.00	5.41	10.70
AR	3.05	0.07	2.17	1.10	5.41	11.80
			Carbon input (Mg ha^{-1} year^{-1})			
NRs	0.00	0.00	0.00	0.00	2.04 [c]	2.04
NR	1.56	0.00	0.00	0.00	2.04	3.60
NBr	1.56	0.03	0.00	0.45	2.04	4.08
NB	1.56	0.03	0.96	0.00	2.04	4.59
AR	1.56	0.03	0.96	0.45	2.04	5.04
			Biochemical composition			
C (g kg^{-1})	501.07 ± 25.97	474.77 ± 7.1	442.53 ± 2.7	413.35 ± 9.2	-	-
N (g kg^{-1})	11 ± 0.1	15.74 ± 0.7	1.99 ± 1.1	3.74 ± 0.2	2.8 [d]	-
C/N	47.1 ± 2.0	30.2 ± 0.9	316.2 ± 255.8	110.6 ± 9.0	162.7 [e]	-
Lignin (g kg^{-1})	274.4 ± 0.3	227.1 ± 2.8	142.5 ± 2.7	87.1 ± 0.3	-	-
Lignin/N	25.8 ± 2.0	14.4 ± 1.9	112.3 ± 111.0	23.2 ± 0.8	82.9 [e]	
Hemicellulose (g kg^{-1})	104.4 ± 0.3	122.2 ± 0.2	192.4 ± 0.4	158.6 ± 1.0	-	-
Cellulose (g kg^{-1})	204.4 ± 0.9	143.3 ± 0.9	580.2 ± 2.4	450.1 ± 1.5	-	-

[a] Addition estimated according to Witschoreck [35]; [b] root addition according to Londero et al. [36]. [c] C content in roots 37.84%, obtained from Ribeiro et al. [37], [d] obtained by Guimarães et al. [44], and [e] obtained by Demolinari et al. [17].

The values of C/N and lignin/N ratio in the eucalyptus harvest residue management are shown in Table 2. In general, we observed that the C/N ratio values in the eucalyptus residue managements ranged from 111.7 in the NBr treatment to 162.7 in the NRs treatment. As for the lignin/N values, the values ranged from 58.1 in the NR management to 69.1 in the NB treatment.

3.2. Effect of Eucalyptus Harvest Residue Management on Soil Organic C Content and Stocks

Eucalyptus harvest residue managements had no statistically significant effect on soil organic C levels (Figure S2). Nevertheless, there was a trend for C contents to increase with the maintenance of the eucalyptus harvest residues. When the layers up to 10 cm depth were averaged, the soil in the NRs treatment presented an organic C content that was 39% lower than that observed in the AR treatment.

The soil organic C stocks of the 0–20 cm (r^2 = 0.80, p = 0.03) and 0–100 cm (r^2 = 0.82, p = 0.03) layers increased linearly with the increase in C input by the maintenance of eucalyptus harvest residues (Figure 1a,b). Each 1 Mg ha^{-1} year^{-1} of residue resulted in a differential accumulation of 1.41 and 2.95 Mg ha^{-1} in the layers of 0–20 cm and 0–100 cm, respectively, at the end of 6 years of cultivation. The partial and total maintenance of the eucalyptus harvest residues in the soil (NBr, NB, and AR) determined higher stocks of C than in the other management methods, presenting, in the layer from 0 to 20 cm depth, organic C contents that were 30, 31, and 27% higher than in the treatment in which the harvest residues from the current harvest were completely removed (NR) (Figure 1a). These findings demonstrate that these three types of management, despite having low quality components, such as bark and branches, resulted in higher soil organic carbon stocks.

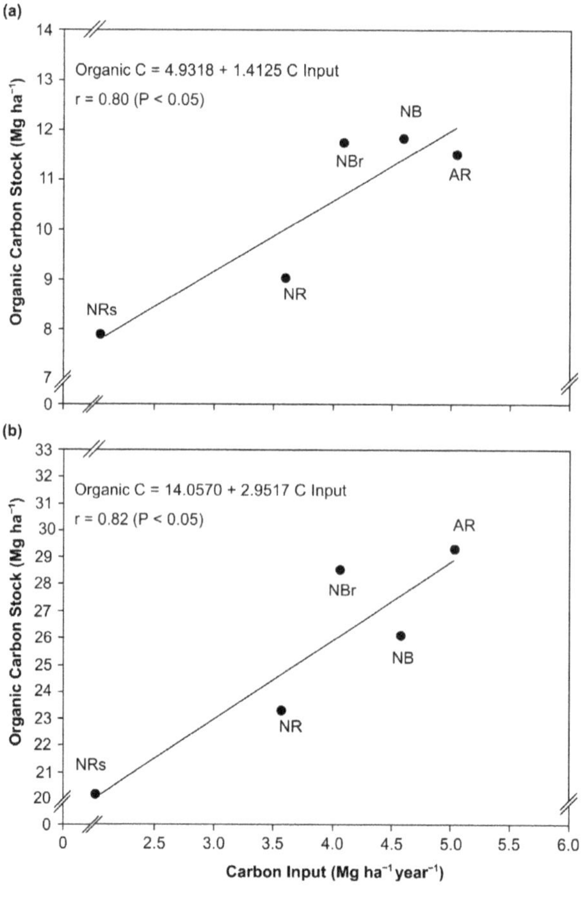

Figure 1. Relationship between C input by eucalyptus harvesting residue management and organic C stock in the 0–20 cm depth layer (**a**) and in the 0–100 cm depth layer (**b**).

Table 2. C/N and lignin/N ratio of eucalyptus harvesting residue management.

Residue Management	C/N [1]	Lignin/N [1]
NRs	162.7	82.9
NR	112.6	58.1
NBr	111.7	53.9
NB	154.6	69.1
AR	150.7	65.0

[1] Ratio weighted by the amount of waste added in each treatment.

3.3. Annual Soil C Retention Rates

Annual soil C retention rates varied from 0.23 to 0.68 Mg ha^{-1} $year^{-1}$ in the 0–20 cm layer (Figure 2a). In this soil layer, the NR treatment exhibited the lowest soil C retention rate (0.23 Mg ha^{-1} $year^{-1}$), while with the AR treatment, the retention rate was 0.68 Mg ha^{-1} $year^{-1}$ (i.e., about threefold greater). In turn, the maintenance of the bark (NBr) and branches (NB) determined accumulation rates that were quite similar to treatment AR. Similar trends were observed in the 0–100 cm layer, with the annual soil C retention rates reaching 1.47 Mg ha^{-1} in the AR treatment. In the intermediate treatments with the maintenance of bark (NBr) or branches (NB), annual soil C retention rates were 0.8 and 1.0 Mg ha^{-1} $year^{-1}$, respectively (Figure 2b). Although few studies have related soil C retention rates to forest residue management, maintaining residues on the soil surface is one of the main approaches to increase soil C retention rates in various rotations. The annual soil C retention rate went from 0.5 to 1.6 Mg ha^{-1} in the 0–100 cm layer, reinforcing that more than half of the C stabilization occurs in the 20–100 cm layer (Figure 2b).

3.4. Estimate of the Humification (k_1) and Decomposition (k_2)

Figure 3 shows the relationship between the amount of C inputted (A) by the management of eucalyptus harvest residues and the annual rate of change (dC/dt) of the organic C stocks in the 0 to 20 cm layer in relation to the initial stock of organic C at the beginning of the field experiment (13.02 Mg ha^{-1}). The angular coefficient of the equations represents the k_1, that is, the fraction (or percentage) of added C that effectively remains in the soil. The value of k_1 is 0.23, meaning that approximately 23% of the added C was integrated into soil organic matter after one year. Knowing the value of k_1, the decomposition rate (k_2) was estimated to be 0.10 $year^{-1}$ (i.e., ~10% of the soil C is released to the atmosphere as CO_2 by microbial decomposition).

3.5. Carbon Lability, Carbon Management Index, and Relationship with Eucalyptus Growth

The highest C stock index (CSI) values were obtained with full (AR) and partial (NBr and NB) maintenance of eucalyptus harvest residues, which promoted 46, 50, and 49% increases, respectively, regarding the reference (NRs) treatment. On the other hand, the NR treatment increased CSI by only 14% compared to NRs management (Table 3 and Figure 4). There was a linear relationship between POC and the addition of C by eucalyptus harvest residue management. The mineral-associated organic C (MAC) showed reduced values compared to POC and with a slight tendency to increase values with C input by the management of eucalyptus harvest residues. The higher proportions of organic C in POC form in the soil in NBr and AR determined higher lability (L) and, consequently, LI compared to NRs (Table 3). The LI in the 0–20 cm depth layer was associated with the C input from the eucalyptus harvest residue management and ranged from 1.00 to 1.35 (Table 3). Higher values of CMI resulted in a higher diameter at breast height (DBH) of eucalyptus, as shown by the linear regression between the two variables (r^2 = 0.98, $p < 0.001$) (Figure 5).

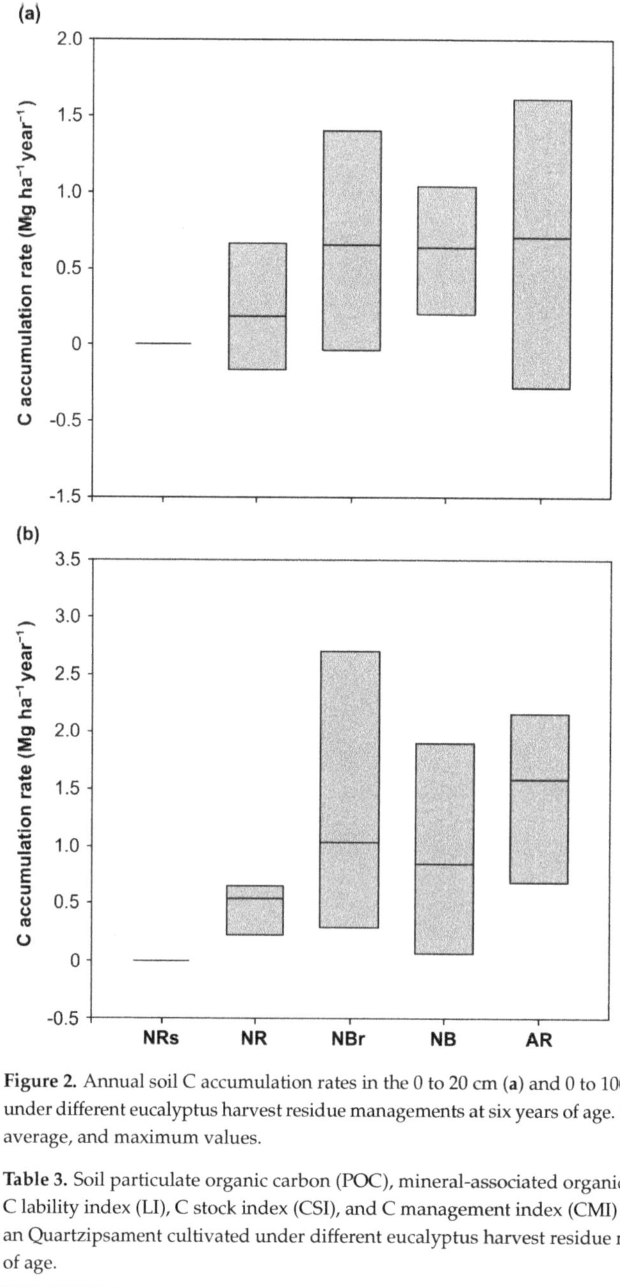

Figure 2. Annual soil C accumulation rates in the 0 to 20 cm (**a**) and 0 to 100 cm (**b**) layers cultivated under different eucalyptus harvest residue managements at six years of age. Boxplot shows minimum, average, and maximum values.

Table 3. Soil particulate organic carbon (POC), mineral-associated organic C (MAC), C lability (L), C lability index (LI), C stock index (CSI), and C management index (CMI) in the 0 to 20 cm layer of an Quartzipsament cultivated under different eucalyptus harvest residue management, at six years of age.

Management	POC	MAC	L	LI	CSI	CMI
	g kg^{-1}					
NRs	2.92	0.033	88.4	1.00	1.00	100
NR	3.33	0.032	104.0	1.17	1.14	133
NBr	4.32	0.036	120.0	1.35	1.50	202
NB	4.47	0.038	117.6	1.33	1.49	203
AR	4.42	0.039	113.3	1.28	1.46	186

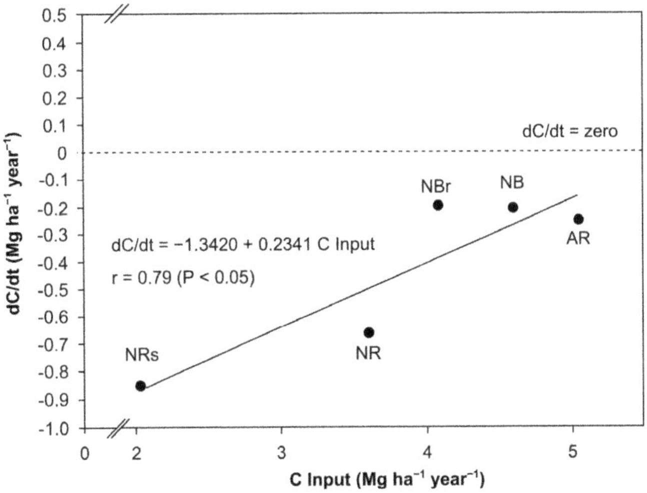

Figure 3. Relationship between the amounts of C added and the variation (dC/dt) of the soil organic C stocks in the 0–20 cm layer cultivated under different managements of eucalyptus harvesting residues at six years of age.

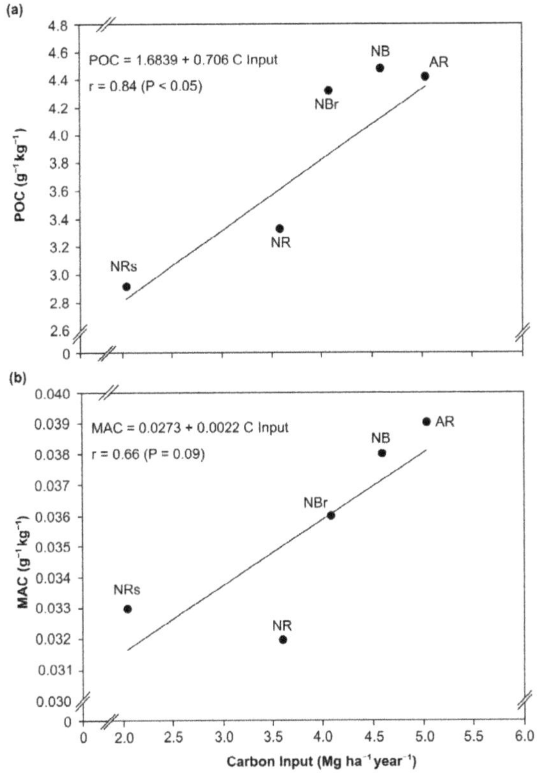

Figure 4. Relationship between the amounts of carbon added and the particulate organic C (POC) (**a**) and the mineral-associated organic C (MAC) (**b**) of the SOC stocks in the layer of 0–20 cm of a soil cultivated under different managements of eucalyptus harvesting residues at six years of age.

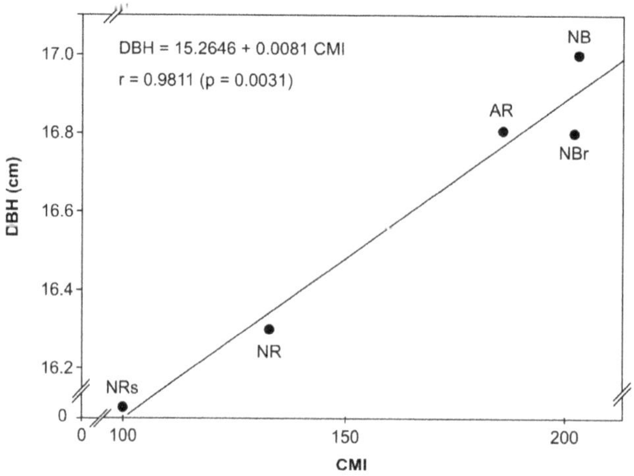

Figure 5. Relationship between the C management index of the soil (CMI) and the diameter at breast height (DBH) under different managements of eucalyptus harvesting residues at six years of age.

4. Discussion

The eucalyptus harvest residue managements resulted in different quantities of C inputted annually into the soil. As hypothesized, the input of C was inversely proportional to the removal of harvest residues, resulting in the AR exceeding the NRs by 3.0 Mg ha^{-1} $year^{-1}$. These results closely resembled the ones obtained by Demolinari et al. [17]. The authors emphasized the significance of retaining the eucalyptus harvest residues in the field for the maintenance of soil C, particularly the bark, as it constitutes the most abundant component of harvest residues.

Our results also corroborate a study conducted on sandy soils under eucalyptus in Congo in which eucalyptus harvest residue removal also altered soil C contents by ±40% [15]. However, in soils with elevated clay content, this effect could not be observed in the first production cycle [19,44], requiring more than one cycle with successive removal for this effect to be observed [18]. This occurs due to the higher efficiency of the physical C protection mechanism in clayey soils, which reduces the access of the microbial community and, thus, the mineralization rate of SOM [28,45]. Compared with the NR treatment, maintaining the bark (NBr) and branches (NB) promoted increments of 18 and 44% in organic C contents in the 0–2.5 cm soil layer, respectively. Nonetheless, comparisons with the NRs and NR treatments revealed that the removal of the forest litter promoted a 49% reduction in the SOC contents in the 0–2.5 cm soil layer. These results possibly have to do with the amount and the composition of the C contributed by the different components of the eucalyptus harvest residues.

The litter is a constituent that presents in its chemical composition a low C/N ratio in comparison to the bark and branches, although it has higher lignin levels (Table 1). The impacts of the quantity and composition of the C contributed by forest residues on soil C retention can be variable. On the one hand, the removal of eucalyptus litter promotes a reduction in soil C contents [46]. On the other hand, a study by Wang et al. [23] demonstrated that, irrespective of the amount and the composition of the litter contributed by eucalyptus species on the soil surface, there is no impact on the SOC contents. This is because the contribution in quantity and quality of C in soils under forests may not always consistently lead to an increase in SOC due to differences in the biochemical composition of residues, in soil mineralogy, and in physical protection mechanisms of SOC [47].

Typically, litter removal in forested areas reduces the organic C contents in surface and subsurface soil layers [48]. This practice causes more significant changes in organic C content in tropical and subtropical soils [49]. In fact, this was verified by Sena et al. [50], who observed, in sandy soil, a decline in the carbon content and stock in the 20 to 40 cm layers after 90 days of complete or partial eucalyptus litter removal, evincing that removing the litter that would accumulate during six years could rapidly reduce the soil carbon stock. Likewise, our results demonstrate that the top layers of sandy soils are extremely sensitive to removing residues and litter. This practice may directly affect soil C storage in this subtropical ecosystem.

Our results indicate that roughly half of the contributed C was accumulated in the 20–100 cm layer. This may be linked to the contribution of recalcitrant residues and the presence of the deep root system of eucalyptus [1,51]. Roots are more efficient than leaf litter in increasing organic carbon stocks in the soil profile. This effect was demonstrated by Pegoraro et al. [52], who observed that carbon contribution from eucalypt roots equals 50% of the total harvest residue. Furthermore, dissolved organic C can alleviate the soil profile and stabilize subsurface layers more efficiently than surface soil layers due to the higher soil saturation deficit [53]. However, in this extremely sandy soil (33 g kg^{-1} clay), it is unlikely that the organic–mineral interaction is an effective mechanism for stabilizing organic C in soil, with organic matter stabilization being more related to its biochemical recalcitrance [27].

In the treatments NBr, NB, and AR, the maintenance of the eucalyptus harvest residues resulted in higher soil organic carbon stocks, despite the low quality components, such as bark and branches. Therefore, in the first moment, these results lead us to believe that there is a divergence from the theory that higher quality residues, characterized by fast mineralization rates, narrow C/N ratio, and reduced phenol concentrations, lead to greater microbial efficiency in carbon accumulation in the soil compared to low-quality residues, characterized by low mineralization rates and higher C/N and elevated phenol concentrations [24].

When analyzing the annual soil C retention rates, we observed that more than half of the C stabilization occurred in the 20–100 cm layer. This result reinforces the need for sampling of deep soil layers (up to 100 cm), since sampling restricted to superficial layers (20 cm) may underestimate the environmental impact of forest management systems. Similar findings have been obtained in agricultural areas, where some researchers have demonstrated that C accumulation in the 20–100 cm soil layer is equivalent to 30–50% of the C accumulated in the 0–30 cm layer [32]. However, studies related to soil C retention in areas with *Eucalyptus* have concentrated on the superficial soil layers and have no thickness standardization, making it challenging to establish comparisons between different studies. In areas of *Eucalyptus*, Lima et al. [54] found an average annual retention rate of 0.22 Mg ha^{-1} $year^{-1}$ of C over thirty years in the 0–10 cm layer, with the highest rate obtained between the second and third rotation in areas of altitude reaching 0.57 Mg ha^{-1} $year^{-1}$ of C. In contrast, Cook et al. [55] and Hernández et al. [56] found a retention rate of 0.20 Mg ha^{-1} $year^{-1}$ of C in the 0–15 cm layer in eucalyptus plantations. Possibly, the results of these studies are underestimated due to the sampling being restricted to the superficial soil layers. Thus, subsurface layer sampling is recommended for determining C retention rates in soils under *Eucalyptus* forests.

The obtained value of k_1 of 0.23 $year^{-1}$ is well above the average values of k_1 in agricultural soils [30,43]. The intrinsic characteristics of the residue, such as its origin (aerial part and root) and composition (lignin content and C/N ratio), will directly affect the soil's residence time. In this sense, possibly, the quality of the eucalyptus harvesting residues contributing to the soil in the present study favors a higher humification coefficient due to the lower lability of the material, given the presence of recalcitrant components, like bark and branch, with an elevated C/N ratio (Table 2). Maintaining these residues on the soil surface has resulted in a higher C transfer from these components to the soil [17]. In addition, studies have shown that roots have a humification coefficient 2.3 times higher than surface-contributed residues due to the higher C/N ratio [57].

Regarding the values of k_2, the value of 0.10 $year^{-1}$ was much higher than the average values of k_2 of 0.019 and 0.040 $year^{-1}$ obtained in agricultural soils under no-till and conventional tillage, respectively [43]. The sandy soil texture is possibly one of the main factors related to this high annual decomposition rate, because of the diminished aggregation and physical protection of organic matter and the low stabilization capacity by organic–mineral interactions [45].

Full (AR) and partial (NBr and NB) maintenance of eucalyptus harvest residues resulted in the highest C stock index (CSI) values. These results reinforce the significant contribution of eucalyptus harvest residue maintenance to soil C accumulation [14,58]. Almost all the organic carbon in eucalyptus harvest residue management is particulate organic carbon (POC) (Table 3). This proportion of POC is higher than that obtained in a study carried out by Oliveira Filho et al. [59] in forest and sugarcane areas under Quartzarenic Neosol, who observed that approximately 60% of total organic C is in the form of POC. This difference between the results may be related to the fact that their study presented clay contents six times greater than in our study, which could favor a higher stabilization of C in the mineral fraction of the soil.

The carbon lability index (LI) expresses the ability of management systems to preserve labile SOM compared to the reference management [60], and it was related to the C input from the eucalyptus harvest residue management in our research. This effect was also evidenced by Rocha et al. [14] in eucalypt areas, who observed that removing eucalypt harvest residues for two consecutive rotations reduced the soil's labile C fraction. Thus, these results demonstrate the significance of maintaining the eucalyptus harvest residues to preserve the SOM.

The LI of NBr and AR were 32% higher than NRs. These results show that although the bark and branch components present in the NBr and AR managements are of low quality, they favored the LI in the soil. The increase in LI values observed in the current investigation may be related to biochemical recalcitrance, which contributes to carbon accumulation in the labile fraction [61], since the soil in the present study has a low mineral surface area available for soil C stabilization [62]. These effects were also found by Puttaso et al. [63], who observed a more significant increase in soil carbon in the labile fraction when the frequent application of residues with moderate contents of nitrogen, lignin, and polyphenols and low cellulose contents was performed in sandy soil. This fraction has been utilized as an indicator to evaluate alterations in the SOM arising from the change in use or management practices in forest systems [33].

The higher CSI and LI were reflected in the CMI. The CMI is an index that compares the alterations in CSI and LI, reflecting the C sequestration and nutrient cycling potential. LI was less sensitive than CSI for these two indices, and this can be seen in the smaller range of values for LI (35%) compared to CSI (50%). On average, the three managements with the highest CMI (NBr, NB, and AR) together had a CMI that was 97% higher compared to the reference management NRs, demonstrating that an improvement in the soil quality has occurred, and indicating that forest systems in which eucalyptus harvest residues are maintained on the soil surface are more sustainable in the long term. Therefore, this index is an excellent tool for assessing the soil quality and production systems [30].

The highly significant linear regression between the DBH and CMI demonstrates that the CMI is a reliable quality index for soil management evaluation and show that both lability and carbon stocks are important in soil sustainability and maintenance of eucalyptus growth. The CMI efficiently estimates the influence of conservation management practices [31]. Management practices that allow the return of residues to the soil surface contribute to increasing CMI values. In fact, this was observed in an investigation by Chatterjee et al. [64], who observed that maintaining 10 Mg ha^{-1} of wheat residues in maize-grown areas resulted in a higher CMI when compared to sites without residues. An improvement in the CMI has resulted in increased yields of maize and rice [32]. Thus, maintaining eucalypt harvest residues is a forest management strategy that improves the SOM and forest productivity [11,14,46].

Given our results, we can classify the AR, NB, and NBr as superior for soil quality and NRs and NR as inferior. Thus, adopting management strategies that promote the removal of bark and branches to favor silvicultural operations or the use of these components for bioenergy production is a forest management strategy that should be avoided by forest companies, especially in fragile soils such as sandy ones.

5. Conclusions

The maintenance of eucalyptus harvest residues is a strategy that promotes increases in soil carbon stocks in sandy soils over the removal of harvest residues. The bark and branches are components that help to increase the rates of carbon retention of the soil, and it is crucial to maintain these components in the field. Therefore, extremely sandy soils depend more on the contribution of recalcitrant residues (bark and branches) that provide great resistance to decomposition by chemical recalcitrance.

The soil humification coefficient under eucalyptus harvest residue management was high, as was the annual rate of SOM loss. The CMI is a sensitive indicator to evaluate the quality of eucalyptus harvest residue management, showing a close relationship with the addition of C and tree diameter at breast height, demonstrating that management that reduced the soil quality also reduced eucalyptus growth.

Supplementary Materials: The following supporting information can be downloaded at: https://www.mdpi.com/article/10.3390/soilsystems7040093/s1, Table S1. Soil chemical and physical attributes in the experimental area; Figure S1: Relationship between C evaluated by Walkley-Black (wet digestion) method and dry combustion method; Figure S2. Total organic carbon content of a Quartzarenic Neosol in the 0–100 cm layer, cultivated under different management of eucalyptus harvest residues at six years of age.

Author Contributions: Conceptualization, J.F.B.d.S.J., L.K.V., B.B.L., F.C.B.V., J.A.Z., E.F.A. and C.B.; methodology, J.F.B.d.S.J., E.F.A., and C.B.; formal analysis, J.F.B.d.S.J., F.C.B.V., J.A.Z. and C.B.; investigation, J.F.B.d.S.J., F.C.B.V., J.A.Z., E.F.A. and C.B.; resources, E.F.A.; data curation, J.F.B.d.S.J.; writing—original draft preparation, J.F.B.d.S.J. and L.K.V.; writing—review and editing, J.F.B.d.S.J., L.K.V. and B.B.L.; supervision, C.B.; project administration, J.F.B.d.S.J. and C.B.; funding acquisition, E.F.A. All authors have read and agreed to the published version of the manuscript.

Funding: This research was funded by Celulose Riograndense—CMPC, Foundation for Research Support of Rio Grande do Sul State (Fapergs, Innovation and Technology Network of Low Carbon Agriculture and adapted to Climate Change in Rio Grande do Sul State), the National Council for Scientific and Technological Development (CNPq), Research Centre for Greenhouse Gas Innovation (RCGI), hosted by the University of São Paulo (USP), and sponsored by FAPESP—São Paulo Research Foundation (2020/15230-5) and Shell Brasil.

Institutional Review Board Statement: Not applicable.

Informed Consent Statement: Not applicable.

Data Availability Statement: The data presented in this study are available on request from the corresponding author. The data are not publicly available due to being part of a broader study in progress.

Acknowledgments: The authors are grateful to CMPC for enabling assessment of the field experiment and for funding.

Conflicts of Interest: The authors declare no conflict of interest.

References

1. Lal, R. Forest soils and carbon sequestration. *For. Ecol. Manag.* **2005**, *220*, 242–258. [CrossRef]
2. Zhou, X.; Zhu, H.; Wen, Y.; Goodale, U.M.; Zhu, Y.; Yu, S.; Li, C.; Li, X. Intensive management and declines in soil nutrients lead to serious exotic plant invasion in *Eucalyptus* plantations under successive short-rotation regime. *Land Degrad. Dev.* **2020**, *31*, 297–310. [CrossRef]
3. Brazilian Tree Industry: Annual Report. 2022. Available online: https://www.iba.org/datafiles/publicacoes/relatorios/relatorio-anual-iba2022-compactado.pdf (accessed on 7 December 2022).
4. Ordinance No. 323, of October 21, 2021, of the Ministry of Agriculture, Livestock and Supply of Brazil. Available online: https://in.gov.br/en/web/dou/-/portaria-mapa-n-323-de-21-de-outubro-de-2021-353829462 (accessed on 20 June 2023).

5. Mayer, S.; Wiesmeier, M.; Sakamoto, E.; Hübner, R.; Cardinael, R.; Kühnel, A.; Kögel-Knabner, I. Soil organic carbon sequestration in temperate agroforestry systems—A meta-analysis. *Agric. Ecosyst. Environ.* **2022**, *323*, 107689. [CrossRef]
6. Gonçalves, J.L.M.; Alvares, C.A.; Higa, A.R.; Silva, L.D.; Alfenas, A.C.; Stahl, J.; Ferraz, S.F.B.; Lima, W.P.; Brancalion, P.H.S.; Hubner, A.; et al. Integrating genetic and silvicultural strategies to minimize abiotic and biotic constraints in Brazilian eucalypt plantations. *For. Ecol. Manag.* **2013**, *301*, 6–27. [CrossRef]
7. U.S. Energy Information Admnistration. Internaltional Energy Outlook 2021 with Projections to 2050. 2021. Available online: https://www.eia.gov/outlooks/ieo/ (accessed on 8 November 2022).
8. Hernández, J.; del Pino, A.; Hitta, M.; Lorenzo, M. Management of forest harvest residues affects soil nutrient availability during reforestation of *Eucalyptus grandis*. *Nutr. Cycl. Agroecosyst.* **2016**, *105*, 141–155. [CrossRef]
9. Rocha, J.H.T.; Marques, E.R.G.; Gonçalves, J.L.M.; Hübner, A.; Brandani, C.B.; Ferraz, A.V.; Moreira, R.M. Decomposition rates of forest residues and soil fertility after clear-cutting of *Eucalyptus grandis* stands in response to site management and fertilizer application. *Soil Use Manag.* **2016**, *32*, 289–302. [CrossRef]
10. Oliveira, F.C.; Ferreira, G.W.; Dungait, J.A.; Araújo, E.F.; Soares, E.M.; Silva, I. R Eucalypt harvest residue management influences microbial community structure and soil organic matter fractions in an afforested grassland. *Soil Tillage Res.* **2021**, *205*, 104787. [CrossRef]
11. São José, J.F.B.; Cherubin, M.R.; Vargas, L.K.; Araújo, E.F.; Bayer, C. A soil quality index for subtropical sandy soils under different *Eucalyptus* harvest residue managements. *J. For. Res.* **2022**, *34*, 243–255. [CrossRef]
12. São José, J.F.B.; Vargas, L.K.; Bayer, C.; Lisboa, B.B.; Araújo, E.F.D. Initial growth and nutrition of eucalyptus under different management of harvest residues. *Floresta Ambiente* **2020**, *27*, e20180161. [CrossRef]
13. Rocha, J.H.T.; Gonçalves, J.L.M.; Gava, J.L.; Godinho, T.O.; Melo, E.A.S.C.; Bazani, J.H.; Hubner, A.; Arthur, J.C.; Wichert, M.P. Forest residue maintenance increased the wood productivity of a *Eucalyptus* plantation over two short rotations. *For. Ecol. Manag.* **2016**, *379*, 1–10. [CrossRef]
14. Rocha, J.H.T.; Gonçalves, J.L.M.; Gava, J.L.; Godinho, T.O.; Melo, E.A.S.C.; Bazani, J.H.; Hubner, A.; Arthur Junior, J.C.; Wichert, M.P. Forest residue removal decreases soil quality and affects wood productivity even with high rates of fertilizer application. *For. Eco. Manag.* **2018**, *430*, 188–195. [CrossRef]
15. Epron, D.; Mouanda, C.; Mareschal, L.; Koutika, L. Impacts of organic residue management on the soil C dynamics in a tropical eucalypt plantation on a nutrient-poor sandy soil after three rotations. *Soil Biol. Biochem.* **2015**, *85*, 183–189. [CrossRef]
16. Menegale, M.L.C.; Rocha, J.H.T.; Harrison, R.; Goncalves, J.L.D.M.; Almeida, R.F.; Piccolo, M.D.C.; Hubner, A.; Arthur Junior, J.C.; De Vicente Ferraz, A.; James, J.N.; et al. Effect of timber harvest intensities and fertilizer application on stocks of soil C, N, P, and S. *Forests* **2016**, *7*, 319–333. [CrossRef]
17. Demolinari, M.S.M.; Sousa, R.N.; Silva, I.R.; Teixeira, R.S.; Neves, J.C.L.; Mendes, G.O. Effect of mineral nitrogen on transfer of ^{13}C-carbon from *Eucalyptus* harvest residue components to soil organic matter fractions. *Rev. Bras. Cienc. Solo* **2017**, *41*, e0160203. [CrossRef]
18. Mendham, D.S.; Ogden, G.N.; Short, T.; O'Connell, T.M.; Grove, T.S.; Rance, S.J. Repeated harvest residue removal reduces *E. globulus* productivity in the 3rd rotation in south-western Australia. *For. Ecol. Manag.* **2014**, *329*, 279–286. [CrossRef]
19. Kumaraswamy, S.; Mendham, D.S.; Grove, T.S.; O'Connell, A.M.; Sankaran, K.V.; Rance, S.J. Harvest residue effects on soil organic matter, nutrients and microbial biomass in eucalypt plantations in Kerala, India. *For. Ecol. Manag.* **2014**, *328*, 140–149. [CrossRef]
20. Kramer, M.G.; Chadwick, O.A. Climate-driven thresholds in reactive mineral retention of soil carbon at the global scale. *Nat. Clim. Chang.* **2018**, *8*, 1104–1108. [CrossRef]
21. Wan, X.; Xiao, L.; Vadeboncoeur, M.A.; Johnson, C.E.; Huang, Z. Response of mineral soil carbon storage to harvest residue retention depends on soil texture: A meta-analysis. *For. Ecol. Manag.* **2018**, *408*, 9–15. [CrossRef]
22. Almeida, L.F.J.; Hurtarte, L.C.C.; Souza, I.F.; Soares, E.M.B.; Vergütz, L.; Silva, I.R. Soil organic matter formation as affected by eucalypt litter biochemistry—Evidence from an incubation study. *Geoderma* **2018**, *312*, 121–129. [CrossRef]
23. Wang, Y.; Zheng, J.; Xu, Z.; Abdullah, K.M.; Zhou, Q. Effects of changed litter inputs on soil labile carbon and nitrogen pools in a eucalyptus-dominated forest of southeast Queensland, Australia. *J. Soils Sediments* **2019**, *19*, 1661–1671. [CrossRef]
24. Cotrufo, M.F.; Wallenstein, M.D.; Boot, C.M.; Denef, K.; Paul, E. The Microbial Efficiency-Matrix Stabilization (MEMS) framework integrates plant litter decomposition with soil organic matter stabilization: Do labile plant inputs form stable soil organic matter? *Glob. Chang. Biol.* **2013**, *19*, 988–995. [CrossRef]
25. Angst, G.; Mueller, K.E.; Castellano, M.J.; Vogel, C.; Wiesmeier, M.; Mueller, C.W. Unlocking complex soil systems as carbon sinks: Multi-pool management as the key. *Nat. Commun.* **2023**, *14*, 2967. [CrossRef]
26. García, A.; Alriols, M.G.; Labidi, J. Evaluation of different lignocellulosic raw materials as potential alternative feedstocks in biorefinery processes. *Ind. Crops Prod.* **2014**, *53*, 102–110. [CrossRef]
27. Cotrufo, M.F.; Soong, J.L.; Horton, A.J.; Campbell, E.E.; Haddix, M.L.; Wall, D.H.; Parton, W.J. Formation of soil organic matter via biochemical and physical pathways of litter mass loss. *Nat. Geosci.* **2015**, *8*, 776–779. [CrossRef]
28. Dalal, R.C.; Mayer, R.J. Long-term trends in fertility of soils under continuous cultivation and cereal cropping in southern Queensland: II Total organic carbon and its rate of loss from the soil profile. *Aust. J. Soil Res.* **1986**, *24*, 265–279. [CrossRef]
29. Blair, G.J.; Lefroy, R.D.B.; Lisle, L. Soil carbon fractions based on their degree of oxidation, and the development of a carbon management index for agricultural systems. *Aust. J. Agric. Res.* **1995**, *46*, 1459–1466. [CrossRef]

30. Vieira, F.C.B.; Bayer, C.; Zanatta, J.A.; Dieckow, J.; Mielniczuk, J.; He, Z.L. Carbon management index based on physical fractionation of soil organic matter in an Acrisol under long-term no-till cropping systems. *Soil Tillage Res.* **2007**, *96*, 195–204. [CrossRef]
31. Ghosh, B.N.; Meena, V.S.; Alam, N.M.; Dogra, P.; Bhattacharyya, R.; Sharma, N.K.; Mishra, P.K. Impact of conservation practices on soil aggregation and the carbon management index after seven years of maize–wheat cropping system in the Indian Himalayas. *Agric. Ecosyst. Environ.* **2016**, *216*, 247–257. [CrossRef]
32. Ghosh, A.; Bhattacharyya, R.; Meena, M.C.; Dwivedi, B.S.; Singh, G.; Agnihotri, R.; Sharma, C. Long-term fertilization effects on soil organic carbon sequestration in an Inceptisol. *Soil Tillage Res.* **2018**, *177*, 134–144. [CrossRef]
33. Sá, J.C.M.; Gonçalves, D.R.P.; Ferreira, L.A.; Mishra, U.; Inagaki, T.M.; Furlan, F.J.F.; Morog, R.S.; Florianih, N.; Briedisf, C.; Ferreira, A.O. Soil carbon fractions and biological activity based indices can be used to study the impact of land management and ecological successions. *Ecol. Indic.* **2018**, *84*, 96–105. [CrossRef]
34. Reichert, J.M.; Amado, T.J.C.; Reinert, D.J.; Rodrigues, M.F.; Suzuki, L.E.A.S. Land use effects on subtropical, sandy soil under sandization/desertification processes. *Agric. Ecosyst. Environ.* **2016**, *233*, 370–380. [CrossRef]
35. Witschoreck, R. Recomendação de Fertilizantes Para *Eucalyptus saligna* Sm. Com Base no Balanço Nutricional na Região de GUAÍBA-RS. Ph.D. Thesis, Universidade Federal de Santa Maria, Santa Maria, Brazil, 2014.
36. Londero, E.K.; Schumacher, M.V.; Szymczak, D.A.; Araújo, E.F. Calibração do modelo 3-PG para *Eucalyptus saligna* Smith na região de Guaíba-RS. *Cienc. Florest* **2015**, *25*, 293–305. [CrossRef]
37. Ribeiro, S.C.; Soares, C.P.B.; Fehrmann, L.; Jacovine, L.A.G.; von Gadow, K. Aboveground and belowground biomass and carbon estimates for clonal eucalyptus trees in southeast Brazil. *Rev. Arvore* **2015**, *39*, 353–363. [CrossRef]
38. Van Soest, P.J. Development of a comprehensive system of feed analysis and its application to forages. *J. Anim. Sci.* **1967**, *26*, 119–128. [CrossRef]
39. EMBRAPA. Manual de métodos de análise de solo. In *Empresa Brasileira de Pesquisa Agropecuária*, 2nd ed.; EMBRAPA: Rio de Janeiro, Brazil, 1997.
40. Ellert, B.H.; Bettany, J.R. Calculation of organic matter and nutrients stored in soils under contrasting management regimes. *Can. J. Soil Sci.* **1995**, *75*, 529–538. [CrossRef]
41. Cambardella, C.A.; Elliott, E.T. Particulate soil organic matter changes across a grassland cultivation sequence. *Soil Sci. Soc. Am. J.* **1992**, *56*, 777–783. [CrossRef]
42. Dieckow, J.; Mielniczuk, J.; Knicker, H.; Bayer, C.; Dick, D.P.; Kögel-Knabner, I. Carbon and nitrogen stocks in physical fractions of a subtropical Acrisol as influenced by long-term no-till cropping systems and N fertilisation. *Plant Soil* **2005**, *268*, 319–328. [CrossRef]
43. Bayer, C.; Lovato, T.; Dieckow, J.; Zanatta, J.A.; Mielniczuk, J. A method for estimating coefficients of soil organic matter dynamics based on long-term experiments. *Soil Tillage Res.* **2006**, *91*, 217–226. [CrossRef]
44. Guimarães, C.C.; Schumacher, M.V.; Witshoreck, R.; Souza, H.P.; Santo, J.C. BIomass and nutrients of *Eucalyptus dunnii* Maiden stand in pampa gaúcho. *Rev. Árvore* **2015**, *39*, 873–882. [CrossRef]
45. Dieckow, J.; Bayer, C.; Conceição, P.C.; Zanatta, J.A.; Martin-Neto, L.; Milori, D.B.M.; Salton, J.C.; Macedo, M.M.; Mielniczuk, J.; Hernani, L.C. Land use, tillage, texture and organic matter stock and composition in tropical and subtropical Brazilian soils. *Eur. J. Soil Sci.* **2009**, *60*, 240–249. [CrossRef]
46. Cao, J.B.; He, X.X.; Chen, Y.Q.; Chen, Y.P.; Zhang, Y.J.; Yu, S.Q.; Zhou, L.X.; Liu, Z.F.; Zhang, C.L.; Fu, S.L. Leaf litter contributes more to soil organic carbon than fine roots in two 10-year-old subtropical plantations. *Sci. Total Environ.* **2020**, *704*, 135–341. [CrossRef] [PubMed]
47. Lajtha, K.; Jones, J. Forest harvest legacies control dissolved organic carbon export in small watersheds, western Oregon. *Biogeochemistry* **2018**, *140*, 299–315. [CrossRef]
48. Tanner, E.V.J.; Sheldrake, M.W.A.; Turner, B.L. Changes in soil carbon and nutrients following 6 years of litter removal and addition in a tropical semi-evergreen rain forest. *Biogeosciences* **2016**, *13*, 6183–6190. [CrossRef]
49. Xu, S.; Liu, L.L.; Sayer, E.J. Variability of above-ground litter inputs alters soil physicochemical and biological processes: A meta-analysis of litterfall manipulation experiments. *Biogeosciences* **2013**, *10*, 7423–7433. [CrossRef]
50. Sena, K.N.; Maltoni, K.L.; Troleis, M.J.B.; Faria, G.A. Forest harvest management systems and residual phytomass affecting physical properties of a sandy soil. *Rev. Bras. Cienc.* **2021**, *45*, e0200190. [CrossRef]
51. Laganiére, J.; Angers, D.A.; Paré, D. Carbon accumulation in agricultural soils after afforestation: A meta-analysis. *Glob. Chang. Biol.* **2010**, *16*, 439–453. [CrossRef]
52. Pegoraro, R.F.; Silva, I.R.; Souza, I.F.; Novais, R.F.; Barros, N.F.; Fonseca, S. Carbon accumulation and partitioning above and belowground under coppiced and replanted eucalypt plantations. *For. Sci.* **2022**, *68*, 162–171. [CrossRef]
53. Castellano, M.J.; Mueller, K.E.; Olk, D.C.; Sawyer, J.E.; Six, J. Integrating plant litter quality, soil organic matter stabilization, and the carbon saturation concept. *Glob. Chang. Biol.* **2015**, *21*, 3200–3209. [CrossRef] [PubMed]
54. Lima, A.M.; Silva, I.R.; Neves, J.C.; Novais, R.F.; Barros, N.F.; Mendonça, E.S.; Leite, F.P. Soil organic carbon dynamics following afforestation of degraded pastures with eucalyptus in southeastern Brazil. *For. Ecol. Manag.* **2006**, *235*, 219–231. [CrossRef]
55. Cook, R.L.; Binkley, D.; Mendes, J.C.T.; Stape, J.L. Soil carbon stocks and forest biomass following conversionof pasture to broadleaf and conifer plantations in southeastern Brazil. *For. Ecol. Manag.* **2014**, *324*, 37–45. [CrossRef]

56. Hernandez, J.; del Pino, A.; Vance, E.D.; Califra, Á.; Del Giorgio, F.; Martínez, L.; González-Barrios, P. Eucalyptus and Pinus stand density effects on soil carbon sequestration. *For. Ecol. Manag.* **2016**, *368*, 28–38. [CrossRef]
57. Kätterer, T.; Bolinder, M.A.; Andrén, O.; Kirchmann, H.; Menichetti, L. Roots contribute more to refractory soil organic matter than above-ground crop residues, as revealed by a long-term field experiment. *Agric. Ecosyst. Environ.* **2011**, *141*, 184–192. [CrossRef]
58. Oliveira, F.C.C.; Silva, I.R.; Ferreira, G.W.D.; Soares, E.M.B.; Silva, S.R.; Silva, E.F. Contribution of eucalyptus harvest residues and nitrogen fertilization to carbon stabilization in Ultisols of southern Bahia. *Rev. Bras. Cienc.* **2018**, *42*, e0160340. [CrossRef]
59. Oliveira Filho, J.D.S.; Pereira, M.G.; Aquino, B.F. Organic matter labile fractions and carbon stocks in a typic quartzipsamment cultivated with sugarcane harvested without burning. *Rev. Caatinga* **2017**, *30*, 24–31. [CrossRef]
60. De Bona, F.D.; Bayer, C.; Dieckow, J.; Bergamaschi, H. Soil quality assessed by carbon management index in a subtropical Acrisol subjected to tillage systems and irrigation. *Aust. J. Soil Res.* **2008**, *46*, 469–475. [CrossRef]
61. Chivenge, P.; Vanlauwe, B.; Gentile, R.; Six, J. Comparison of organic versus mineral resource effects on short-term aggregate carbon and nitrogen dynamics in a sandy soil versus a fine textured soil. *Agric. Ecosyst. Environ.* **2011**, *140*, 361–371. [CrossRef]
62. Zinn, Y.L.; Lal, R.; Resck, D.V. Changes in soil organic carbon stocks under agriculture in Brazil. *Soil Tillage Res.* **2005**, *84*, 28–40. [CrossRef]
63. Puttaso, A.; Vityakon, P.; Rasche, F.; Saenjan, P.; Treloges, V.; Cadisch, G. Does organic residue quality influence carbon retention in a tropical sandy soil? *Soil Sci. Soc. Am. J.* **2013**, *77*, 1001–1011. [CrossRef]
64. Chatterjee, S.; Bandyopadhyay, K.K.; Pradhan, S.; Singh, R.; Datta, S.P. Effects of irrigation, crop residue mulch and nitrogen management in maize (*Zea mays* L.) on soil carbon pools in a sandy loam soil of Indo-gangetic plain region. *Catena* **2018**, *165*, 207–216. [CrossRef]

MDPI
St. Alban-Anlage 66
4052 Basel
Switzerland
www.mdpi.com

Soil Systems Editorial Office
E-mail: soilsystems@mdpi.com
www.mdpi.com/journal/soilsystems

www.ingramcontent.com/pod-product-compliance
Lightning Source LLC
Chambersburg PA
CBHW061520210926
51850CB00008B/1981
9783036597065